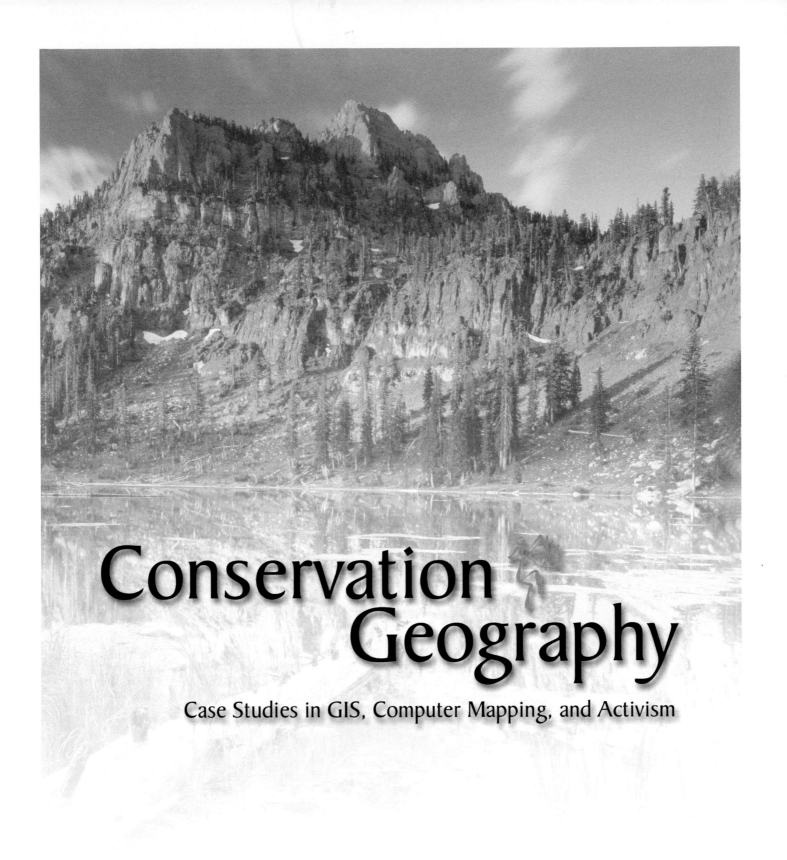

Conservation Geography

Case Studies in GIS, Computer Mapping, and Activism

Edited by Charles L. Convis, Jr.

ESRI PRESS
REDLANDS, CALIFORNIA

ESRI

Conservation Geography: Case Studies in GIS, Computer Mapping, and Activism

ISBN 1-58948-024-4

First printing June 2001

Printed in the United States of America

Published by ESRI, 380 New York Street, Redlands, California 92373-8100.

Books from ESRI Press are available to resellers worldwide through Independent Publishers Group (IPG). For information on volume discounts, or to place an order, call IPG at 1-800-888-4741 in the United States, or at 312-337-0747 outside the United States.

Table of Contents

You can help!
Articles, submissions, photos, maps, and news for a planned volume II are welcome any time. Mail your hardcopy materials to: ESRI Conservation Program, 380 New York Street, Redlands, CA, 92373-8100; phone: (909) 793-2853, x2488, or FTP to ftp.esri.com/pub/incoming. Send e-mail queries to ecp@esri.com.

ECP donates ESRI products to nonprofits. Get ECP grant information by sending a blank e-mail to ecpgrant@esri.com.

Charles L. Convis, Jr., Editor
www.esri.com/conservation

Acknowledgments

Special thanks go to the following donors and sponsors, without whom this book would not exist:

- Janice Thomson of The Wilderness Society Center for Landscape Analysis.

The Wilderness Society, founded in 1935, works to protect America's wilderness and develop a nationwide network of wildlands through public education, scientific analysis, and advocacy. The goal is to ensure that future generations enjoy the clean air and water, beauty, wildlife, and opportunities for recreation and spiritual renewal provided by the nation's pristine forests, rivers, deserts, and mountains. Integral to the Society's land protection work is its Center for Landscape Analysis, which houses the Society's conservation GIS program.

- Nina Jablonski and the California Academy of Sciences, who cosponsored the wildlife and taxonomy sections of the book.

The California Academy of Sciences, founded in 1853, is the oldest scientific institution in the West. In its early days, spurred by its concern over the natural environment during the California Gold Rush, the academy consisted of a group of naturalists who met weekly in an office on Clay Street in San Francisco. Through this forum, scientific papers were presented on topics of interest to a growing membership of local citizens. As the collection of specimens from the field grew in number and scope, the important scientific work of identifying, classifying, and naming species (known as "systematics") began. Today, as one of the ten largest natural history museums in the world, the academy brings the message of research to nearly one and one-half million visitors each year. Like its sister institutions, the Smithsonian Institution, the American Museum of Natural History in New York, and the Field Museum in Chicago, the California Academy of Sciences is devoted to the study, display, and interpretation of scientific collections that inspire people of all ages to explore the rich variety of life on Earth.

- Tony Lavoi and the National Oceanic and Atmospheric Administration Coastal Services Center, who cosponsored the marine geography section.

The mission of the NOAA Coastal Services Center is to foster and sustain the environmental and economic well being of the nation's coast by linking people, information, and technology. The NOAA Coastal Services Center works with various branches of NOAA and other federal agencies to bring information, services, and technology to the nation's coastal resource managers. The Center is a partner in over a hundred ongoing projects geared to resolve site specific coastal issues.

Thanks also to Scott T. Smith, for his enthusiastic donation of the cover photography and other photographs.

And thanks to Frans Lanting, Inc., for a generous donation of photographs.

Thanks to the following staff, contributors, artists, and organizers:

Steve Pablo, for creating and developing the initial design; Sheila Huskinson, Wendy Brown, and the entire staff at ESRI word processing.

Joe Breman, marine archaeologist, SCGIS board member and ESRI technical support staff, for serving as guest editor and fundraiser for the *Marine Geography* section, with assistance from Barbara Shields of ESRI's marketing department.

Gillian Woolmer and the Wildlife Conservation Society, for organizing the section on *Mapping Nature's Diversity: The Wildlife Conservation Society.*

The ESRI Press staff: Jennifer Galloway, who worked long hours adapting the design and producing the book; and Christian Harder, Dave Boyles, and R. W. Greene for helping to shepherd the project into book form.

Special thanks to Jack Dangermond, president of ESRI, for his steadfast support, encouragement, contributions, and inspiration. Without it, none of the nonprofit successes you'll read about in these pages would be possible.

This book is dedicated to my dad Chuck, who taught me to love nature and justice, and my mom Mary Anne, who taught me to recognize beauty.

Charles L. Convis, Jr.
Redlands, California

A Special Acknowledgment

Dear Conservation Geographer:

In 1989 we started the ESRI Conservation Program to try to help change the way nonprofit activists carried out their missions of nature conservation and social change. This kind of thing is normally the province of a charity and it's often been an odd fit for a commercial firm to be conducting this sort of nonprofit work. At ESRI, we do it and we'll keep doing it because we respect what you do and we believe that it is very important to the future of our shared earth. As a group who bridge a gap between technology, science and activism, you and your work are important to us in several ways:

- You are great individuals: In most cases you have chosen to set aside more rewarding private careers to follow a passionate commitment to service within difficult and often tenuous employment circumstances. You have chosen to place a mission of service to nature and to others ahead of many personal needs others take for granted. For this commitment and self-sacrifice, we honor each of you.

- You are uniquely skilled: Having a strong background in the field sciences usually doesn't lend itself to deep computer skills. Field skills revolve around observation, living in and studying natural phenomena from up close. Computer skills require great amounts of time indoors. Finally, a dedication to activism requires a passionate partisanship about nature, which can itself conflict with the desire for careful, methodical fact-finding and hypothesis-testing of science. To possess interests and skills in each of these areas makes you unique and very important in the world.

- Nature needs you: As I travel the globe more, it becomes more obvious to me how broken many of our global and local natural systems are. Erosion, flooding, fires, famine, and extinctions seem to be so much blah blah blah on the evening news. When you see it for yourself and realize how severely this affects the majority of our rural populations and how there are no places on earth left that are undamaged, you begin to lose patience with the short-sightedness and outright greed that seems to govern too many resources decisions. You want to be able to do more, to work effectively, but from the best science. You want to change so many decisions before the chance to decide is lost. It's an impossible task for one person, but as a group you are doing this, all over the world. I've seen it first-hand, and I can assure you that the changes you are beginning are real and global. You are making this happen, slowly but surely. You need to persevere, to do more, and we need to make sure you have the help and the tools to do it.

I am constantly reminded of how much I am indebted to you, and I hope that some day many more will appreciate and honor the sacrifices you make for the earth and for the underprivileged. I know the work you do is often difficult, unappreciated and wrought with political conflict. Take heart, and remember that there is at least one person out there who appreciates you, who respects you, who honors you, and who will remember what you did.

Regards,

Jack Dangermond, President, ESRI

Preface

Geography is the science of the earth's surface and its varied occupants. Conservation is the practice of protecting resources of value so that future generations can enjoy them. Taken together, conservation and geography offer a concise and powerful framework for human survival.

While not a traditional discipline, the community of people engaged in conservation action using the tools of geographic science is large, dedicated, and growing. While conservationists and ecologists have always relied on theories of biogeography and evolution, the spark that ignited a distinct global community of practitioners was the growth of computerized geography tools in the late 1970s following the first Earth Day. Collectively called geographic information systems, or GIS for short, these computer tools attracted a great deal of attention among nonprofit organizations throughout the 1980s, but the typical $30,000 price tag left the tools out of reach for many. In 1989, ESRI decided to initiate a grants program to make software and training available at no charge to those groups devoted to conservation and public service. This program now encompasses over four thousand nonprofit organizations worldwide, spread among many sectors of conservation action, sustainable development, social justice, and human rights.

This book presents the stories of these groups, in their own words, showing their own maps, their own research, and their own community work. It is organized around the different types of groups and the different ways they work on conservation and social justice:

Environmental Justice shows how GIS is used as a tool for social equity within urban settings and communities of color, tracking toxic releases and helping collect the evidence needed to bring offenders to account for their actions. The chapters also show how GIS is helping to build parks and homes and monitor street conditions. *Citizen Science* shows how GIS is being used in volunteer water quality monitoring work, where young people and schools are involved in monitoring and mapping environmental quality in the communities and watersheds where they live.

Watershed Geography presents the work of groups devoted to protecting rivers, bays, and watersheds.

Forest GIS shows how GIS is used in forest conservation, including sustainable forest practices and community-based forestry, whereby mapping, planning, and sound husbandry techniques are used to ensure local economies in forested regions can grow in a sensible and sustainable manner.

Conservation Design digs more into the science behind conservation geography, with presentations by several groups involved in GIS applications for biogeography and reserve design.

Landscape Conservation presents a broader look at GIS applications in biogeography, as they are applied across larger landscapes and a wider variety of conservation challenges, including protection of whole mountain ranges and entire forest regions.

Species Geography narrows in on the taxonomic viewpoint in conservation, examining how GIS tools are used at zoos and botanical gardens to study and protect biological species worldwide.

Bird GIS continues the taxonomic viewpoint, presenting the work of groups devoted to the protection of birds and showing how they apply the tools of GIS to their mission.

Marine Geography is devoted to the applications of GIS in the marine environment, covering over three-quarters of the earth's surface. With help from the National Oceanic and Atmospheric Administration's Coastal Services Center, this is a first effort of what we plan to launch as an independent publication in the coming years.

Mapping Nature's Diversity is devoted to the work of the Wildlife Conservation Society (WCS), which began introducing GIS to its worldwide projects and partners only a year ago and which organized a large effort to collect all the WCS mapping stories for this volume.

Preface *CONTINUED*

Conservation Worldwide reviews the international work of the Society for Conservation GIS, the independent, all-volunteer organization devoted to helping people learn to do better conservation with the best science and mapping tools. Following an introduction and a world map of members and their projects, you'll see a collection of stories from SCGIS members, scholarship recipients, and grantees from partner GIS programs all over the world.

Other ESRI Conservation Program publications focus on GIS in land trusts and GIS for indigenous peoples. Future titles will be devoted to specific techniques, showing by step-by-step tutorials and worked examples how GIS tools are used in specific conservation and social analyses, and how specific maps can be produced and included in effective community organizing and activism work.

Conservation is what we do to survive on Earth, and to ensure that other species, our fellow travelers, survive as well. Geography and GIS bring the power of the computer revolution directly into the hands of conservationists, offering for the first time tools able to encompass the incredible diversity of information, data, analyses, and publications required for modern conservation to succeed. Together, we hope they will play a role in stemming the global extinction crisis.

Introduction

Galileo had the telescope and Newton had the calculus. Physics, biology, astronomy, and many other disciplines became rigorous sciences thanks to similar breakthroughs that revolutionized them and formalized them. Ecologists and conservationists have never been so lucky. Darwin's ideas of natural selection provide a conceptual framework, but a fundamental tool, like Newton's calculus, that would integrate thousands of ecosystem observations into a single predictive model, has yet to be found.

Photograph donated courtesy of Scott T. Smith, ©2001 Scott T. Smith

Besides the lack of a central tool, ecologists have another burden that no other science has had to confront to the same dangerous degree: extinction. In this case, extinction isn't just an academic concern, it threatens our very survival. As human populations exceed the limits of biological resources everywhere, and greed for quick profits pushes exploitation without limit, the fabric of our natural systems is threatened and human civilization itself lies in peril. If humans are to survive, we must become stewards of our home rather than demolishers. We must realize that biodiversity—the wide variety of living species—is a critical factor in the health of our planet. We live in a time of incredibly rapid change; even so, there are still opportunities for concerted action to protect and manage what remains of our natural heritage so that we may ensure the survival and enlightenment of future generations. To be effective, however, action must be based upon accurate information and the best possible ecological and biological science.

Geography has struggled for centuries as a discipline that borrowed from many other natural and social sciences but had little to offer back. It wasn't until the 1930s that it began to develop its own mathematics. Beginning with Von Neumann's first use of U.S. Army Signal Corps computers for weather mapping in 1950, geography began a long revolution that would eventually see the emergence of the first-ever computer tools capable of integrating information from all of the natural and social sciences based on location.

Those tools, collectively called geographic information systems (GIS), have in a decade or two become the predominant platform for analysis, modeling, and management in thousands of ecology labs, planning departments, parks, agencies, and nonprofit organizations worldwide. This book presents dozens of case studies and accounts of the use of GIS in nonprofit conservation and science organizations. It is a snapshot of GIS advancement, twelve years after the founding of the world's first GIS support program for nonprofits in 1989. Starting with one organization and many months of collaborative work, this grant program has grown to encompass more than four thousand conservation-oriented organizations globally and has led to the founding of two new nonprofit organizations, the Society for Conservation GIS (SCGIS) and the Conservation Technology Support Program (CTSP). Many of the stories you will read here come from participants in these programs.

The point of all of these stories is to show what is possible when the best technological tools and the most current scientific methods are made available to concerned and passionate conservationists around the world. Before they had access to these expensive GIS tools, nonprofit organizations were at a disadvantage when trying to bring a conservation viewpoint or even a local community viewpoint to the table at increasingly technological resources discussions. Important development decisions typically went to the side that had the best tools and the slickest presentation, regardless of whether their science was any good or whether they had even made the attempt to answer environmental issues. Since 1989, conservation groups have increasingly been able to level the playing field by attending hearings, meetings, and court cases armed with the same powerful analytical tools as the "big boys" have. In such cases, however, conservationists are able to use GIS to present the best science illustrating the real extinction threats posed by even small developments that are poorly planned and located. Giving decision makers and the public a chance to examine equally and fairly all the sides of important and complex resources decisions is what a democracy is all about. One definition of the "Information Age" is that control over the sources and means of information will replace control over means of production as the definition of national power. It's important, therefore, for citizens and voters to be able to have access to and an understanding of the important work of the thousands of nonprofit organizations struggling for social justice and environmental sanity.

In communicating decisions, nothing beats a map. Photos are visually powerful but can only communicate general ideas or passions. Charts and text can communicate detailed quantitative data but can make it difficult to obtain an overview of a complex argument. Only maps combine quantitative and statistical explicitness with the visceral punch of photojournalism. When properly done, a map can leave as vivid and unforgettable a concept as the most impassioned speech. Because a GIS allows maps to be created that integrate any amount of complex and incompatible information on the simple basis of location, they permit scientists and activists to present the findings of years of study in simple, powerful form.

As organizations grow in their ability to use the power of GIS for good, other problems arise unique to the realm of conservation. We hope that these stories will help present some of these issues and help point the way toward solutions. One of the most pressing of these problems currently is the sharing and dissemination of data. As an issue, sharing and distribution of information is a complex one. There are many reasons why groups choose not to make their mapping work available to the public. Reasons include concerns over the security of endangered species or rare archaeological artifacts, political sensitivity over land conservation proposals, negotiation issues in land acquisition, and plain old competition for limited grant funds between groups doing similar work. There are many reasons why groups do not publish even if they want to, including shortages of time and funds, demands for emergency action that never let up, prioritizing basic research instead of publishing, lack of time to properly document a dataset for release, and lack of knowledge of cartography.

The World Wide Web has fostered the removal of many traditional publishing barriers. It is increasingly coming to represent a true global peer-to-peer network where any desktop can be immediately linked to any others—sharing data, work, and ideas regardless of separation in time or space. Lots of us irreverently call this the "Mapster" idea, being able to swap maps desktop-to-desktop. As it rapidly becomes real, it represents a fundamental advance in democracy, permitting every citizen access to knowledge, opinion, research, and data far outstripping the best resources of the best library on Earth. The Web removes the traditional barriers to knowledge from race, class, and neighborhood, and permits a participatory society where every citizen can have access to the discussions, decisions, and findings formerly hidden behind bureaucratic and economic obstacles.

Dr. Randy Webb, senior ecologist at Net Work Associates–Ecological Consulting (randy-webb@worldnet.att.net) has considered this problem as well, and offers the following:

"I'm an environmental lawyer, so I have thought a lot about the sensitive data problem. As to the risks of sharing data in litigation, judges aren't trained in this area and a GIS analysis can easily be reduced to a 'battle of the experts'—something environmental lawyers always try to avoid. Because of the deference doctrine and other reasons, too, the environmental group is likely to lose those battles. Another risk is that many of the groups with GIS capability are larger and more moderate, so if they share data or output, other, more hardcore groups would then have access and might sue when the more moderate group thinks it's better to wait, or better to work out a cooperative management plan. Finally, my concern is to not alert an agency as to how advanced our analysis is. I don't want them to have time to work up a bunch of phony conservation plans. I want real, on-the-ground results. A listing under the Endangered Species Act is a way to get that. Of course, if agencies and industries have acted proactively to conserve species, or prevent pollution, or whatever, then I'm very happy that there is no need to do a lawsuit or publicity campaign, etc. But this is still pretty rare.

"There are lots of related issues also. One is that when an environmental group does even a minor GIS effort, this can 'scare' government agencies into doing a larger, more comprehensive, more fine-scaled effort. Ultimately, though, litigants always run into a legal doctrine called 'factual deference'—the courts will always defer to agency conclusions in favor of independent scientists. There are a number of cases on this. In one case, about twenty of the world's top population biologists said one thing and a few agency folks said another; the court went with the agency, allowing a very destructive logging plan to proceed. But, as Huck Finn once said, 'Truth will out' eventually. That's the power of citizen GIS, allowing citizen groups to work with government agencies via GIS."

David M. Theobald at Colorado State University (davet@nrel.colostate.edu) is managing a survey on the issue of sensitive data (http://ndis.nrel.colostate.edu/davet/datareport.html), and offers the following:

"Increasingly conservationists and GIS practitioners must deal with the issue of 'sensitive' data. Though a precise definition of sensitive data is still wanting, I offer the following working definition. Sensitive data is data that, for some reason or for some context, should be used or shown differently depending on the audience. Therefore, sensitive data requires special techniques or instructions when analyzing, producing cartographic output, or distributing to others. It is not the data that creates a sensitive situation, it is the context in which it is used. For example, avian biologists typically collect GPS data on bird nest locations and want maps that show the locations precisely. However, biologists and policy makers alike are often concerned when these same data sets are used to produce maps for public consumption or disseminated publicly that also portray the precise locations of the nests. As data continues to be collected with a high degree of spatial precision (e.g., using GPS), displayed on zoom-capable interactive maps, and accessible to anonymous users across the World Wide Web, we will be challenged to deal constructively with situations that involve sensitive spatial data.

"At a recent conference of the Society for Conservation GIS, there was widespread agreement that sensitive data was an important issue. Generally speaking, there was a wide range of perspectives on the issue, bounded on one side by a cautious perspective (hide or remove data deemed sensitive) and a brash perspective (show it, share it, and damn the consequences). To be sure, most of the discussion focused on identifying some, more comfortable, middle ground.

"A strong theme of discussion was the 'purposeful take vs. accidental harm' dilemma. A primary reason for removing or clouding the precision of sensitive locations is to reduce the likelihood of loss due to purposeful take. A classic example of this line of reason is that providing precise locational information for the nest sites of rare birds (e.g., falcons, eagles) increases the likelihood that eggs will be removed, fledglings taken, or adults hassled. Another example is that accurate maps of the location of archaeological artifacts likely increases the chance that sensitive sites will be looted. Conversely, sensitive locations may be accidentally disturbed (accidental harm) if data about them is *not* shared or displayed. Losses due to accidental harm are particularly tragic, because information about the location of a rare species existed but was not shared appropriately. The crux of this dilemma frequently falls to a judgment of whether society and biodiversity will gain or lose more if sensitive data is displayed or hidden. An important question here, but probably difficult to answer, is how frequently have sensitive species been disturbed either because they were easier to locate because of the map, or, they were not mapped and therefore not known about? How many purposeful takings do we know of? How many accidental 'harms' do we know of?"

Another central issue in conservation GIS is the role of taxonomic and collections data in conservation action and mapping. Lack of data is typically the most critical factor in conducting authoritative ecological analyses and trying to develop sound conservation management plans. At the same time, the world's largest, most continuous and most authoritative source of biodiversity data lies largely unrecognized in the thousands of botanic gardens, herbarium collections, museums, and academies of sciences found in every country and most cities. A central challenge of both conservation geography and the taxonomic sciences is to find ways to make this incredible treasure trove of biogeographic data available for conservation projects and mapping analysis applications.

Computer progress in genetic research and statistical analysis have greatly advanced the practice of taxonomy and ecology. As analytical tools, however, most of these advances led toward a finer, deeper understanding of smaller and smaller bits of nature, when what was needed was a tool that had the same analytical power but which was primarily synthetic in nature, that acted to combine and integrate data from these other disciplines and tools to create inclusive visualizations of whole populations and ecosystems. GIS fills that bill nicely. A GIS built upon the relational data model is founded on the same mathematical principles that underlie modern statistical theory. Even the most complex statistical analyses can be modeled and implemented in a GIS in a straightforward manner. More importantly, a GIS allows the direct manipulation of space in a way other tools cannot, and that is the secret to its great integrative power. In nature, location is everything. From the movements of fugitive species to the way plants and soils act to integrate over time all of the environmental factors affecting them, location is one of the most fundamental causes and effects in nature. Nearly every aspect of ecological data has a spatial component that is important if not dominant.

There are many chapters in this book devoted to the use of GIS in taxonomic and species-based research and action, especially in *Species Geography, Bird GIS,* and *Mapping Nature's Diversity.*

Besides the issues of sensitive data and museum collections, you will find other discussions of nonprofit GIS issues, problems and challenges in the pages of this book. We hope that you will be inspired to see how the tools of geography can help you in your own public service work, taking advantage of the grant programs organized by the ESRI Conservation Program and our partners. We also hope that you will feel emboldened to contact any of the various groups described here in this book to offer your help, your questions, and your membership.

Charles L. Convis, Jr.

Environmental Justice and Citizen Science

The Community Builders, Inc.

Ronald Wong, Planning Project Manager   RonW@tcbinc.org

Building communities and building futures

Mission and History

The Community Builders, Inc. (TCB), is a 35-year-old nonprofit organization with a mission of rebuilding urban neighborhoods to serve families and individuals of all incomes. We began as a neighborhood-based organization (then known as South End Community Development, Inc.) determined to prove that abandoned, dilapidated brownstones in Boston's South End neighborhood could be reclaimed as excellent housing for low-income families. Since then, TCB has completed more than fourteen thousand units of affordable and mixed-income housing in cities throughout the northeast and mid-Atlantic states. Today, our revitalization projects integrate a range of elements found in successful urban neighborhoods including housing, offices, retail space, and community service functions such as day care, health clinics, and youth programs.

Reducing Sprawl by Rebuilding America's Urban Neighborhoods

The negative effects of urban sprawl on native wildlife are numerous and well documented, but little has been done to slow the rapid encroachment of development on forests, coastal areas, and fragile natural habitats. Policy initiatives have curbed sprawl in a few metropolitan areas, but growth continues unchecked and unregulated in many more. The unfortunate reality is that Americans will continue to make longer commutes to work and to develop more natural land until we can resurrect our core urban areas as desirable places to live.

The Community Builders, Inc., believes that our work in urban core areas directly contributes to responsible patterns of development. Today, we are developing compact, pedestrian-oriented neighborhoods that make use of the infrastructure and public transportation found in older urban areas. In our current neighborhood-scale projects, subsidized and market-rate homes are interspersed and incorporate the same level of design and construction quality. We market these developments to households who might otherwise leave cities for newly built suburban areas. Our vision for the future is one in which vibrant urban neighborhoods attract families of all incomes.

Applying ArcView® GIS 3.1 to Neighborhood Planning and Redevelopment

The projects we undertake continue to grow in scale and complexity, and as a result, it is increasingly important for us to organize and manipulate information on the demographics, cultural resources, land ownership, zoning, buildings, streets, and utilities of particular neighborhoods. In October 1998, the ESRI Conservation Program granted us ArcView GIS 3.1 software, training, and data for this purpose. ArcView GIS has already clearly demonstrated its value to our work, and we expect to see even greater benefits as we develop our technical skills and as our local government partners continue to increase their investments in GIS. The following examples show how we have applied ArcView GIS technology to analyze demographics, urban design, and community resources in a prospective urban redevelopment area in Cincinnati, Ohio.

Demographic Mapping

As can be seen in the featured maps, the urban neighborhoods in which we work are often marked by extreme concentrations of poverty when we begin. We address the poverty of a neighborhood by providing job training and supportive services for existing residents to increase their incomes and by building high-quality housing to attract new households with a range of incomes. Typically, we seek an income mix of approximately one-third low-income, one-third moderate-income, and one-third market-rate households for our

neighborhood-scale projects. The map on the following page shows the high rate of poverty in Cincinnati's West End before revitalization has begun.

Demographic mapping is also important to our marketing efforts to attract new households to a neighborhood. For our largest redevelopment projects we use a three-stage approach to marketing: (1) market research on the demographics of a neighborhood and likely target markets, (2) focus group interviews of prospective residents to determine key design and pricing preferences, and (3) a marketing campaign tailored to attract renters or homebuyers to create an economically and socially diverse neighborhood.

Street Patterns in Cincinnati's West End Neighborhood

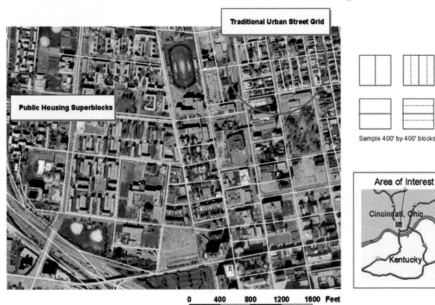

Analyzing Existing Urban Design and Master Planning

The urban design of a neighborhood can make a significant contribution to its desirability as a place to live and to its long-term success. When The Community Builders begins a neighborhood revitalization project, we convene residents and other local stakeholders to gather input for an overall master plan. By allowing our designers and local residents to work from a common base of information, ArcView GIS offers a powerful tool for consensus building and master plan development.

The map on this page compares the street network of a public housing site to the street network of the surrounding neighborhood. When vast numbers of public housing units were built in the 1940s through the 1960s, they were built in the 1940s through the 1960s, they were often placed in "superblocks" that altered the historic street networks of neighborhoods. At the time, the superblock model was used because it was thought to be a more efficient use of land. Unfortunately, the lack of access and regular street activity on these sites severely isolated public housing residents and provided attractive settings for criminal activity. Today, as we redevelop neighborhoods, we seek to recreate the street networks and design patterns found in nearby successful urban neighborhoods.

Identifying Affordable Housing Resources

The availability and condition of existing affordable housing resources are important considerations in planning a neighborhood redevelopment project. Well-maintained and

well-managed affordable housing can be used to temporarily relocate residents while a site is being redeveloped. On the other hand, if we find poorly maintained housing—whether affordable or market-rate—we often attempt to acquire the property to improve it. Our experience is that a revitalization project must address all of the needs of a neighborhood to effect long-term change. The map below uses data drawn from a HUD database of subsidized properties. A map such as this enables us to identify the location of affordable housing in a community.

Identifying Community and Civic Resources

The Community Builders, Inc., uses a multi-faceted approach to redevelopment that simultaneously addresses the physical, social, and economic needs of a neighborhood. A comprehensive education, job training, and supportive service program for local residents is a critical part of this strategy. The first step in developing such a program is to identify the existing educational and civic resources of a neighborhood. Once we identify existing resources, we can work with local service providers to strengthen their programs, to develop networks of other nonprofit agencies, and to develop new programs to fill unmet needs.

Project-Based Subsidy Program

- Public Housing
- Project-Based Section 8
- Section 236 Housing
- Low Income Housing Tax Credit

Area of Interest

Cincinnati, Ohio

Kentucky

N

| 0 | 400 | 800 | 1200 | 1600 Feet |

Affordable housing in Cincinnati's West End listed by Federal and State Subsidy Program.

Environmental Defense GIS Program

Peter Black   Peter_Black@environmentaldefense.org

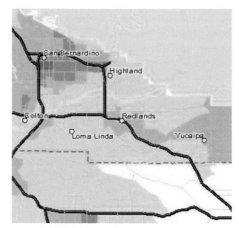

Example Scorecard output: census tract coded by cancer risk around Redlands, California.

EDF staffer verifies community environmental health conditions.

Environmental Defense is a nonprofit environmental advocacy group with four main goals: (1) stabilizing the earth's climate, (2) safeguarding the world's oceans, (3) protecting human health, and (4) defending and restoring biodiversity. Our GIS work has been successful in many areas and highlighted in the "Scorecard Project."

The Scorecard Project

Scorecard is the ultimate source for free and easily accessible local environmental information on the Internet. Simply type in a ZIP Code to find out about local air pollution and explore state-of-the-art interactive maps. Scorecard delivers accurate information on the toxic chemicals released by manufacturing facilities, the health risks of air pollution, water quality, and many other environmental topics. It can rank and compare the pollution situation in areas across the United States. Scorecard also profiles six thousand eight hundred chemicals, making it easy to find out where they are used and how hazardous they are. Using authoritative scientific and government data, Scorecard provides the most up-to-date and extensive collection of environmental information available on the Web today. Information is power—once you learn about an environmental problem, Scorecard encourages and enables you to take action: you can fax a polluting company, contact your elected representatives, or volunteer with environmental organizations working in your community.

The Scorecard received more than one million hits during its first twenty-four hours on the Web, demonstrating there is widespread interest in gaining easy access to the essential facts about pollution. Press coverage of the Scorecard has been universally enthusiastic, and the site continues to generate local stories in communities across the United States.

The Scorecard could not possibly work without spatial data and GIS software. On the

Scorecard, we use GIS in two main ways: (1) cartography and (2) spatial relations and geocoding. The maps on Scorecard are rendered graphically by Karl Goldstein's practical map server, a freeware Java™ applet that reads shapefiles and associates them with Oracle® tables. Goldstein was Scorecard's main GIS consultant. Scorecard uses several private data sets from ESRI grants for cartographic display. We also use ESRI maps and data for the country, state, county, ZIP Code, and census tract level data. This same data is used to create indexes of spatial relations that are at the core of Scorecard's localizing capability. For instance, for each census tract, we need to know what ZIP Codes it intersects, the percentage of intersect area for each, and which county and state they are in. By doing so, we are able to create tables within Oracle that intelligently create localized reports at all geographies.

If the main purpose of the Scorecard is to show local environmental information, then the spatial data for all of the point and polygon environmental polluters has to be fairly accurate. The raw data we use from the EPA is usually locationally inaccurate. We used an elaborate geocoding ranking system that had at its core the StreetMap 2000 data, the ZIP Code data,

and the county data. The raw EPA data usually has addresses for facilities, ZIP Codes, and county FIPS codes. Sometimes the data has latitude/longitude locations. In any case, all of the data on Scorecard goes through this rigorous exercise, and in the worst cases, the data is mapped to the centroid of the correct county and given a question mark map symbol.

Because of the GIS processing that we do in ArcView GIS, Scorecard users can zoom in on their hometown, see what air pollutants are coming out of specific facilities, and get summarized reports on their census tract, county, and state. Not all of Scorecard uses census-level data. For example, for our clean water layer, we map watersheds down to the HUC 8

SETTING PRIORITIES | Main Page

Want to know what the most important environmental problems in your community are? Scorecard can list the top ranking issues in your area, based on the judgment of scientists and stakeholders who participated in a comparative risk project for your area. Project reports can lead you directly to detailed local information about these problems. To get your report, just click near your community on the map below, or use the pull down menu to select your EPA region, state or local area.

Click on the Map Below to Obtain Rankings for the Following Kinds of Areas:

① Region Level Comparative Risk Report

ME Labeled States: States for which Comparative Risk Reports are available

★ County and/or local level Comparative Risk Reports are available

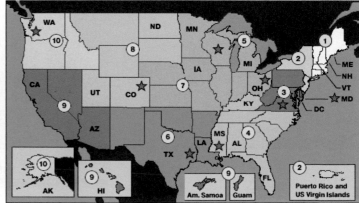

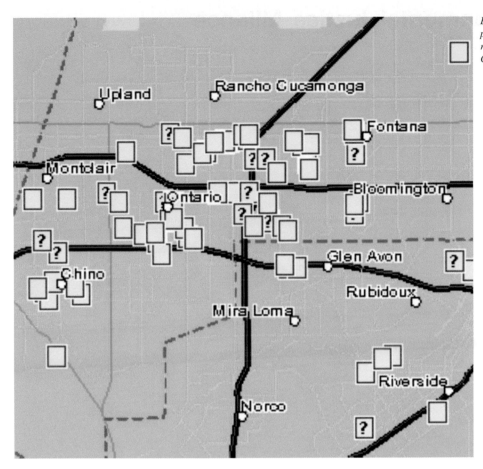

Example Scorecard output: point locations of chemical risks near San Bernardino, California.

level. GIS allows us to cross-reference watersheds with census tracts.

Scorecard has been widely publicized as a success for the right to know and environmental justice movements. It is now available in Spanish. For a complete guide to Scorecard in the news, please refer to this URL: http://www. scorecard.org/about/about-press.tcl.

Here is a specific example: Pauline Leboda lives on a modest residential street in Elyria, a small city twenty miles west of Cleveland, Ohio. Several years ago, she began suffering from periodic chest pains and other ailments. Doctors found nothing they could treat. But Leboda, who felt better when she ventured away from the neighborhood, suspected an environmental cause for her symptoms. She often smelled pungent odors from the numerous factories at the edge of her neighborhood and beyond. Her concerns were heightened when she learned about a number of children in the area being treated for asthma. Leboda decided she needed help from someone experienced with the state's environmental bureaucracy. After a few phone calls, she reached Teresa Mills. Mills, the director of the Buckeye Environmental Network, had won an environmental victory in her hometown of Columbus in the early 1990s and became a resource for others seeking help with local environmental issues. Mills helped Leboda determine that a likely source of the troubling odors was a

sponge manufacturer less than a mile from her home. A few months later, they were on the phone again when Mills went to her computer. Within seconds she was able to give Leboda a complete rundown on the company and its emissions as well as a profile of other factories in the community. "She thought I was brilliant," Mills says. "But all I did was use Scorecard." The Environmental Defense geographically indexed Web site, www.scorecard.org, is transforming local environmental activism. Now citizens can quickly gain access to a wealth of data on neighborhood pollution. It is as simple as typing in your ZIP Code, clicking "go," and being led to interactive maps and emissions data, as well as relevant background on health issues. With the help of information from Scorecard, Mills and Leboda determined that Elyria has more polluting facilities than would be typical for a community its size, many of them clustered in the poor neighborhood adjacent to Leboda's. Moreover, they found that the sponge manufacturer, Nylonge, did not have the proper permit for its emissions of carbon disulfide, a toxic chemical that contributes to the formation of smog.

Mills contacted the regional office of EPA, which investigated the matter and brought an enforcement action. The result: Nylonge reached a settlement with EPA in which the company paid a fine and agreed to start using a chemical scrubber during the summer smog season. Denny Dart, an EPA engineer who

did the on-site investigation, lauded Scorecard's role in helping people learn about health issues and take action. "My experience is that companies are quicker to settle and more willing to include citizen-friendly provisions in their settlements if they know that citizens are watching," Dart says. As these grassroots efforts demonstrate, Scorecard can be an effective tool that not only informs citizens but also empowers them to act. Pauline Leboda is breathing a little easier these days, and she and Mills are continuing their work to make sure that industrial facilities in the community are doing all they can to prevent pollution. It is difficult to precisely gauge the effectiveness of Scorecard. We do count our page hits, of course, and as of fall 2000, we averaged 750,000 page hits per month. Based on the success of Scorecard, Environmental Defense has been able to secure additional funding to expand the Scorecard project and further increase our presence on the Web with the ForMyWorld Project.

New York City Environmental Justice Alliance

Emily Chan   mapping@nyceja.org
Leslie H. Lowe   lhlowe@nyceja.org

The New York City Environmental Justice Alliance (NYCEJA) is an umbrella organization composed of fifteen grassroots member groups in low-income communities throughout New York City. Founded in 1991 by environmental activists in New York's low-income communities of color, NYCEJA became a 501(c)3 corporation in 1995.

NYCEJA's mission is to empower its member organizations to fight against environmental racism by helping them organize their communities and by providing legal analysis and technical support that they need to engage in effective advocacy and wage long-term campaigns.

NYCEJA's constituent communities are among the poorest, the sickest, and the most heavily polluted in New York City. The residents are primarily African–American and Latino. Their neighborhoods are saturated with pollution from industrial plants, sludge treatment and waste processing facilities, toxic release sites,

Percent of unemployment in the Bronx.

Percent of workers employed in manufacturing jobs in the Bronx.

and truck traffic. Many of them encompass the city's waterfront manufacturing zones, and most of them are intersected by major transportation corridors that bring large volumes of diesel truck traffic through the community.

The lack of resources and political influence in NYCEJA's constituent communities makes them particularly vulnerable to the siting of noxious facilities, as well as to the lack of natural resources in these areas. The complexity of the problems facing these communities and of the political system that must be negotiated also leaves residents disempowered and in despair. As a result, New York's low-income communities of color end up with most of the region's environmental burdens and few, if any, amenities (such as street trees, parks, and open space) to mitigate the harm.

Consistent with its mission and the principles of environmental justice, NYCEJA supports community-led initiatives by leveraging resources, coordinating strategies and advocacy efforts, tracking issues through the regulatory and legislative process, making contacts and networking with regional and national organizations, and helping to replicate projects and activities that have proven successful in one of the member communities. Through its network of attorneys, scientists, health specialists, and consultants NYCEJA provides the resources that low-income communities need to sustain long-term campaigns for environmental equity.

NYCEJA currently provides GIS technical assistance to strengthen our member organizations in doing community organizing, research, and advocacy for environmental justice. Through GIS and computer mapping, training in data collection and documentation, and legal research support, we have documented the land-use, public health, and environmental conditions of low-income neighborhoods and communities of color. Our vision of technical assistance is not an end, but a means for building our collective capacity to fight against environmental discrimination.

In NYCEJA's member communities, disparate environmental conditions are often disregarded by policy makers. Documenting the environmental conditions in underserved communities is important because data is often out of date, inaccessible, or not available for these areas. NYCEJA strives to not only assist its members collect and analyze data, but it also ensures that grassroots communities are directly involved in the entire process and to advocate for themselves.

NYCEJA's phases of capacity building through technical assistance (including GIS) are:

- Project assistance.

- Technical training.

- Community-based data collection and research (e.g., in asset mapping and hazard mapping) that will culminate in a citywide GIS of environmental and health conditions in underserved communities.

- Establish community education and training centers that will continue the research and data collection and publish research results.

NYCEJA has reached the second and third phases of this technical capacity-building process. In 2001, we are expanding our support to include comprehensive training on government processes, environmental policies, and GIS and computer mapping, in addition to NYCEJA's existing education and training work. Almost half of our membership is already engaged in GIS work with our support. An outcome of this long-term vision is to establish a community-based, citywide GIS of environmental and health conditions in low-income neighborhoods and communities of color.

NYCEJA will also conduct a feasibility study in 2001–2002 for the last phase of our citywide capacity building plan. To this end, NYCEJA is currently working with Media Jumpstart, a technical assistance project of the Community Resource Exchange, to complete a technology assessment for our member organizations.

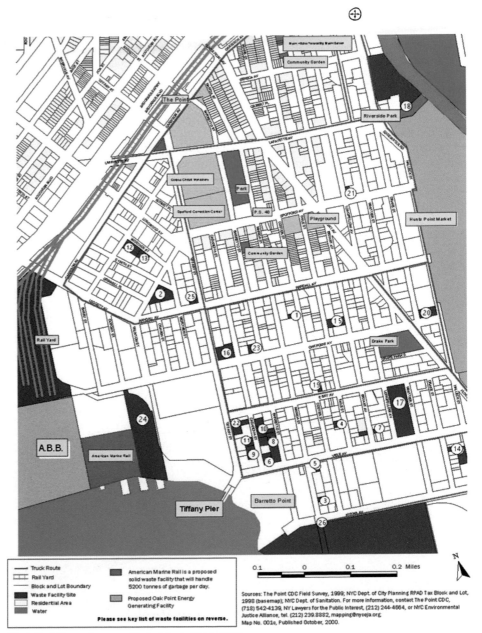

Waste facilities and proposed industries in Hunts Point.

City Scan—Empowering Urban Neighborhoods Through Technology

Richard Walker, Research Analyst

In March 2000, ESRI awarded the Connecticut Policy and Economic Council (CPEC) a conservation grant for reduced-price GIS training in Danvers, Massachusetts, to support its City Scan project. City Scan is a program that works with Hartford community groups to improve neighborhood conditions by using handheld computers, digital cameras, and GIS technology to identify, monitor, and report street-level quality of life improvement needs. CPEC has completed two City Scan projects and has received funding from the Annie E. Casey Foundation to implement it in four additional Hartford neighborhoods. The program has also received widespread attention from local and national news media, and there is great potential that the program will continue to grow for years to come.

In this era of change, information is power. New phrases like "information rich" and "information poor" have sprouted up in the media and demonstrate the new importance of information in our society. Those who have access to information and know how to use it can make change. At CPEC, we have long believed that accurate information, effectively

communicated, can help community members improve the quality of life in their neighborhood. With our expertise in using cutting-edge information tools in the public and nonprofit

Local students conduct City Scan survey.

sectors, we are working with community leaders to improve the quality of life across the State of Connecticut.

In Hartford, CPEC has partnered with community groups in the Parkville neighborhood and the Hartford High School Technology Academy to implement pilot programs of City Scan

during the summer and fall of 2000. A group of Hartford neighborhood volunteers was initially trained to use Pocket PCs to gather data regarding street-level conditions in their neighborhood. The conditions to be monitored were determined using input from community members and included issues as diverse as abandoned buildings and sidewalk smoothness. Once the data was collected, the pocket PCs were turned over to CPEC staff where they were downloaded onto a desktop computer from which reports are created and then turned over to relevant government officials.

The most powerful datasets presented in these reports were the paper and Web-based maps that were created using ArcView GIS 3.2. Each physical condition that was collected included address information. Once the spreadsheets from the handhelds were produced, this information was geocoded in ArcView GIS and plotted as points on a map. A simple visual representation of where each condition was located was very impressive to the community groups; however, ArcView GIS software's analysis tools made this information even more useful.

CityScan Bushnell Park

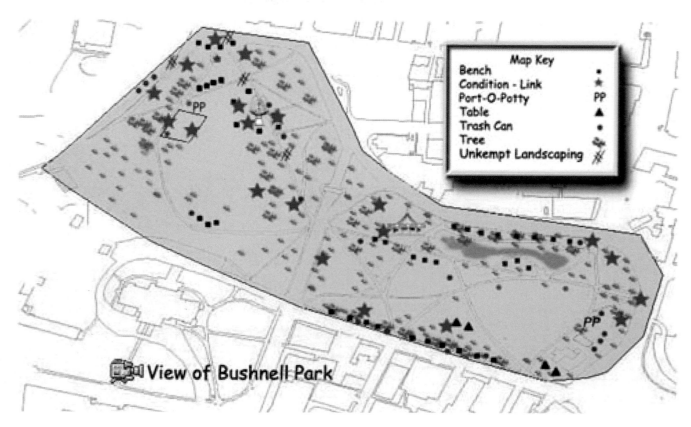

Map Key
Bench •
Condition - Link ★
Port-O-Potty PP
Table ▲
Trash Can ■
Tree
Unkempt Landscaping

View of Bushnell Park

Most of the conditions that were collected in the neighborhoods had a three-point rating scale. For example, the condition "pothole" had a rating scale of (A) less than one square foot, (B) between one square foot and one square yard, and (C) greater than one square yard. This scale allowed us to use three sizes of points on the map; small-sized points for rating A, medium-sized points for rating B, and large dots for rating C. Such maps have been powerful to community members because they indicate where the real problem areas are in certain sections of a neighborhood. Such information is also much more powerful when petitioning city government to fix these conditions.

For example, if our maps show that the neighborhood of Parkville has a pothole problem (see map below), it is not very useful to go to the city government and ask officials to fix all of them. Invariably they will complain about limited resources and the lack of ability to address all potholes in a given neighborhood. However, using the City Scan technology and our GIS software, we can show city officials a map that shows large (Level C) points are concentrated on a particular street or intersection. We can then tell city officials where exactly to concentrate their scarce resources. In Parkville, we are in the final stages of turning over the information that was collected this past summer and fall to the neighborhood groups.

Similarly, a group of students from the Hartford High School Technology Academy participated in the City Scan Parks Project. These "data analysts" were trained on how to use the equipment and surveyed the physical conditions of the five major public parks in Hartford during this past summer (see map, previous page). The list of thirteen conditions they monitored were specific to problems that would be found in a park (e.g., the quality of park benches and picnic tables and the quality of the grass and lawns). This information was then turned over to members of the Hartford Parks and Recreation Department. The City Scan Parks Project also has a Web site where the maps, pictures, and video from the data they collected are displayed. The site is located at www.city-scan.com.

This aspect of the City Scan project received widespread media attention and corporate support. Three local news outlets, including CBS and NBC, covered a press conference that was held on August 2, 2000. During the press conference the students showcased the technology they used and gave a report of some of their findings. The *Hartford Courant*, Connecticut's largest and oldest newspaper, featured two articles about the program. Similarly, the magazine *Civics.com* showcased an article on the program entitled "Citizens Wield Power Through Technology." The City Scan Parks Project received notoriety on the Web as well. The Online News Hour in its "The Buzz" and "Extra" sections featured two extensive articles on the program and displayed some of the maps that were created with ArcView GIS 3.2. Finally, Microsoft® also showcased an article on City Scan in the Press Pass section of their Web site. Microsoft has had a great interest in the project particularly because of City Scan's use of the Windows® CE operating system for the Pocket PC in its data collection.

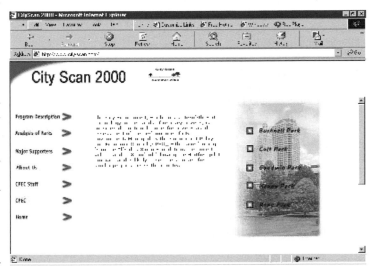

Microsoft supported the program by donating ten Pocket PCs to the program.

The Alfred P. Sloan Foundation primarily funds the City Scan project. They have provided a $435,000 Grant for Assessment of Government Performance to CPEC to support the City Scan project. The grant funds the Hartford project in Parkville and a similar project to begin in 2001 in Stamford. Also, this fall CPEC received a grant from the Annie E. Casey Foundation and the federal Youth Opportunities Grant to perform "City Scans" in four neighborhoods in Hartford. Students who live in these neighborhoods are performing the data collection as part of their community service requirement for graduation. We have also hired three of the students who worked in the City Scan Parks Project this past summer as part-time interns to process the data collected during our current City Scans. They are also being trained on the use of CPEC's GIS software. As a result of our work last year, City Scan has turned into an exciting youth development project.

In the coming year CPEC hopes to receive additional funding to continue instituting City Scan as a community service project in Hartford high schools. In addition to more GIS training in ArcView GIS, we also hope to expose students to publishing the data and maps they create on the World Wide Web.

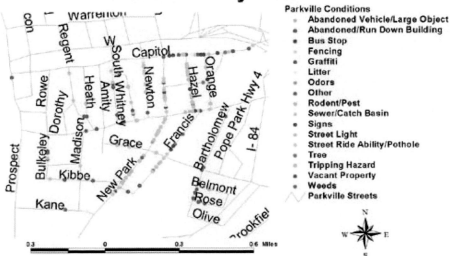

Parkville City Scan

Parkville Conditions
- Abandoned Vehicle/Large Object
- Abandoned/Run Down Building
- Bus Stop
- Fencing
- Graffiti
- Litter
- Odors
- Other
- Rodent/Pest
- Sewer/Catch Basin
- Signs
- Street Light
- Street Ride Ability/Pothole
- Tree
- Tripping Hazard
- Vacant Property
- Weeds
- Parkville Streets

Pohatcong Creek Watershed Association

Prepared by David Dempski   techsupport@pcwa.org

The Pohatcong Creek flows for approximately twenty miles in western Warren County, New Jersey. It passes through seven municipalities: Independence, Mansfield, Washington Township, Washington Borough, Franklin, Greenwich, and Pohatcong. The watershed consists of fifty-eight square miles. The land use in this area is diverse: wetlands, agriculture, urban, residential, forests, and even a Superfund site. The Pohatcong Creek's headwaters are in Independence Township, and it empties into the Delaware River in Pohatcong Township. The major tributaries of the Pohatcong Creek are the Shabbacong Creek, Roaring Rock Creek, and the Merrill Creek.

The Pohatcong Creek Watershed Association (PCWA) consists of twenty active members and meets once a month on the third Thursday of the month at Warren County Community College.

The PCWA is undertaking a three-year baseline water quality project. The project entails collecting samples of benethic macroinvertebrates (bottom-dwelling aquatic insects) and having them analyzed down to the

This screen allows the PCWA to add a site to the stream monitoring program. Sampling site photography can also be added.

order–family. An insect has a family tolerance value (FTV) that relates its ability to survive in impaired (polluted) waters. The higher the score, the less the insect is affected by pollution. The FTV is used, along with other biological indexes (see http://www.state.nj.us/dep/watershedmgt/bfbm/rbpinfo.html) to come up with the New Jersey Impairment Score

(NJIS). The NJIS indicates if the water is non-impaired, moderately impaired, or severely impaired. The habit of the sampling site is also noted. The creek water velocity, temperature, and clarity are also recorded.

The sampling is performed by trained water stewards. Sampling is generally done by a

David Dempski (right) sampling stream quality.

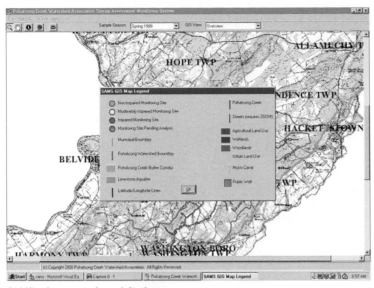

SAMS software map legend display.

five-minute run, where the sampler agitates the creek bottom and moves an 800- by 900-micron mesh net upstream to catch the insects. The raw sample is labeled and stored in alcohol, a preservative. The raw samples are then presorted by emptying the sample into a plastic tray that is labeled with 35 grids. Insects are randomly selected from the tray until 100 organisms have been removed. The sorted insects are stored in another labeled glass jar with alcohol. This 100-insect sample is then sent to a laboratory for analysis. A sample laboratory result from the spring 1999 sampling season follows. The samples are kept at the New Jersey State Museum in Trenton.

Stream Assessment Monitoring System (SAMS) Software

To collect and distribute the sampling data, the PCWA has developed a Windows-based program that incorporates GIS and database management system technology. The program was written using Microsoft Visual Basic® and includes MapObjects® for GIS functionality. MapObjects was donated to the PCWA under the ESRI Conservation Program. Microsoft Access is used for the database. The SAMS program will allow water stewards to enter in sample laboratory data and calculate the NJIS of each sample site for all the sampling seasons. The SAMS program will be distributed to municipal, county, and state officials as well as planning boards, environmental commissions, watershed organizations, and creek landowners. It will allow them to interactively view the NJIS values for each site versus various GIS themes. A GIS theme represents graphically a section of the environment. SAMS has the following features:

- Interactive graphical interface.

- Online help.

- Identify aquatic insects through questions and answers.

- Enter in sampling databases for sites, insects, and sampling results.

- N.J. Impairment Score versus Warren County map.

- N.J. Impairment Score versus N.J. DEP aerial photography (1997).

- N.J. Impairment Score versus Warren County topography.

- N.J. Impairment Score versus land use (agriculture, wetlands, urban, and forest).

- N.J. Impairment Score versus limestone aquifers and public wells.

- N.J. Impairment Score versus riparian corridor.

- GIS map legend.

- Ability to display the latitude and longitude of points in Warren County.

- Includes GIS themes for the Pohatcong Creek, Pohatcong Creek Watershed, Morris Canal, streets, sampling sites, and municipal boundaries on the NJIS views.

- Ability to print out an NJIS GIS view.

- Ability to zoom in or pan a GIS map.

- Virtual photographic tour of the watershed.

- Virtual photographic tour of the historical areas in the watershed.

- Virtual photographic tour of the sampling process.

- Calculates NJIS and prints out reports.

- Includes pictures of sampling sites.

- Data references (metadata).

The SAMS program can be requested from the PCWA Web page at http://www.pcwa.org.

SAMS software watershed slide show example.

Mount Desert Island Water Quality Coalition

Lelania Avila, Janet Redman
lalaland@acadia.net
http://www.mdiwqc.org/

Here in Maine, our conservation ethic focuses on citizen participation, stewardship, and long-term responsibility for maintenance of water quality. An excerpt from our vision statement describes our conservation mission well: ". . . The Coalition is a model for other communities of how to foster a working relationship between students and teachers with community volunteers, engaging them together in the process of solving environmental problems." We have a strong volunteer base because the MDIWQC is a fun, rewarding place for volunteers to invest their time and be effective members of the community. Through the MDIWQC programs, young (and not so young) people learn about the strength of

Phytoplankton sampling.

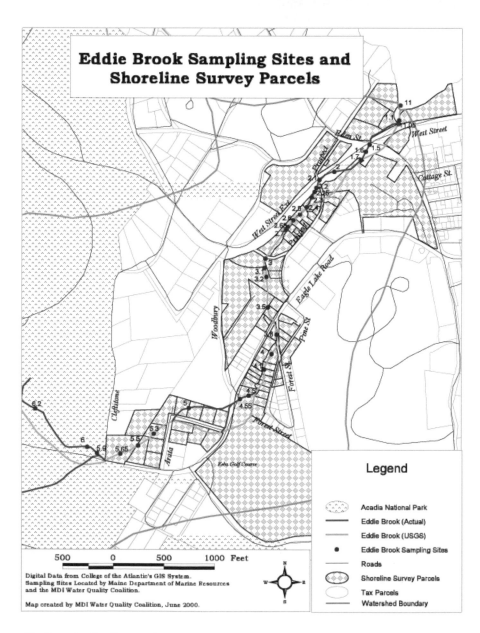

their voices and achieve a sense of ownership of MDI. This encourages them to sustain community environmental improvements and changed behaviors.

Through community education and outreach, the MDIWQC strives to help islanders voluntarily change behaviors that could harm MDI's water quality, without the time-consuming and financial drain that traditional "top-down" legislative change would involve.

We research water quality problems in our community through a systematic monitoring program and facilitate discussions with community members and town officials about these problems. In these sessions, we help brainstorm potential solutions and then provide the necessary information needed to actively pursue the solutions.

Nonpoint source pollution is the most pressing problem with water quality. It comes from everywhere, from everybody. Any person can make a difference if they know what to do. Making people aware of the problems and the actions they can take to improve situations is the only way to ensure a protected environment for the future. By using GIS, we clearly show what the problems in our community are and graphically show improvements that we are making.

We are focusing on showing how nonpoint source pollution impacts our water: clam flats and swim beaches are two important community resources that are negatively affected by nonpoint source pollution. Clam flats in our community have been closed for decades. People surveyed two years ago were unaware that storm water is untreated and washes

Water Quality Monitoring Projects 2000

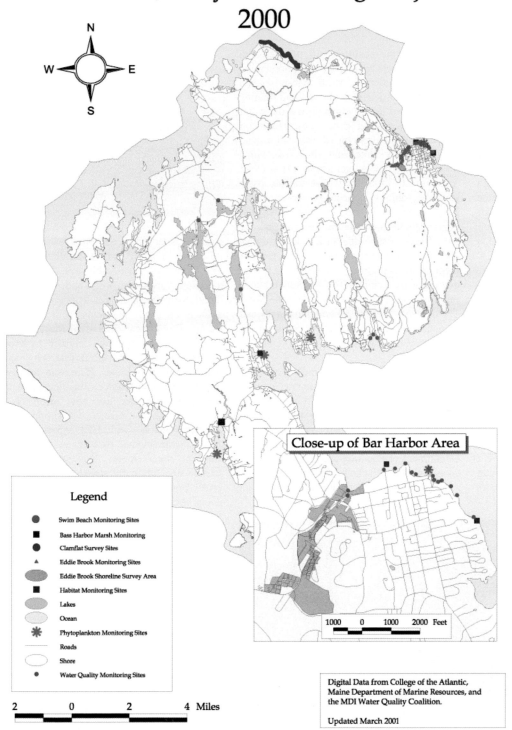

Legend

● Swim Beach Monitoring Sites
■ Bass Harbor Marsh Monitoring
● Clamflat Survey Sites
▲ Eddie Brook Monitoring Sites
⬭ Eddie Brook Shoreline Survey Area
■ Habitat Monitoring Sites
⬭ Lakes
⬭ Ocean
✳ Phytoplankton Monitoring Sites
— Roads
◯ Shore
• Water Quality Monitoring Sites

Close-up of Bar Harbor Area

1000   0   1000   2000 Feet

2   0   2   4 Miles

Digital Data from College of the Atlantic,
Maine Department of Marine Resources, and
the MDI Water Quality Coalition.

Updated March 2001

straight into the bay. People were also unaware of their individual impact and the simple things they can do to make a difference, such as planting a vegetative buffer on their property or using less fertilizer on their lawn.

At this time, the state of Maine has no protocol for monitoring swim beaches or for preventing swimmer illnesses. Storm water runoff and nonpoint source pollution are two preventable pollution sources affecting beaches. Our aim is to educate citizens about the problems and

solutions and motivate them to become stewards of where they live.

We have had success by using GIS to create data sets and maps that show others what the water quality problems are. We then facilitate community discussions generating ideas for solutions. GIS fits in with our mission by enabling us to present information to our community clearly. When people can see a map showing a pollution problem and the relationship to current land use, they see the direct

correlation. If they can see that their behavior affects the water quality directly and a simple solution is possible, they can make a difference right away. With consistent monitoring and data update displays, tangible improvements are evident. Our creative approach of engaging local students and teachers with other volunteers to solve environmental problems encourages long-term stewardship and protection of resources.

Lelania Avila: Community Service in the Schools

Laurie Schreiber

When it comes to cleaning up the marine environment in Bar Harbor, the amount of work can be overwhelming. Thanks to AmeriCorps volunteer Lelania Avila, water quality monitoring projects, carried out over the past eight years by Dr. Jane Disney at Mount Desert Island High School, have been infused with tremendous energy and have given greater scope to the term "community service." Last year, Disney was able to hire Avila through the Maine Conservation Corps to coordinate the newly formed MDI Water Quality Coalition. Since then, Avila has proven to be a boon, Disney says, in networking school programs with the community.

Disney's educational aims link students with real-life projects, while Avila views community service as her life's work. Now fulfilling the second of two yearlong AmeriCorps contracts at MDI, she has found the perfect niche that gives her a mentor in Disney while also allowing her to expand her own ideas and expertise.

In the fall, Avila returned to the sophomores, helping them collect data in the field, then taught them about the GIS system that allows them to map their information over time and track trends. She has also been instrumental in helping get more community volunteers involved in the coalition. All this research leads to concrete results, she says. Something as innocuous as a car wash fund-raiser can lead to a lot of detergent, dirt, and oil running into the storm drain, which affects the water. Certain boat-washing practices use a high concentration of chlorine bleach that can instead be significantly diluted and which will, with a little more scrubbing, get a boat just as clean. All these are little things that add up. Avila says she wants to educate people about the simple things they can do. She notes that many people come to the area because of the island's pristine qualities. A lot of things they do detract from those qualities. Government agencies are overwhelmed by demands, making it the communities' responsibility to improve things in many instances, she notes. The only way the flats are going to get opened up and the beaches made safe is if the community monitors the water.

The 1992 College of the Atlantic graduate, whose work focused on geographic information systems, has been far afield in her career. She first jaunted off to Alaska on a tip from another COA student. There, as a subsistence biology analyst for the North Slope Borough Department of Planning and Community Services in Barrow, she helped local and state fish and wildlife researchers analyze their data on caribou, polar bears, and other wild animals. "After a year and a half, I decided to move to a more sane place," she says. Seattle was a lot less wild and woolly, and so were her duties, mapping urban growth boundaries with a focus on sewer problems, with the King County Department of Natural Resources. A couple of years ago, she returned to MDI, at first just for a break. Before long, she realized this was home. Here, she felt she could make a difference. Teaming up with Disney showed her just how much of a difference. At first, Disney was simply looking for someone with GIS experience. In Avila, driven by her own ideas, she found she got much more. The position with Disney was originally slated to run for one year. "But I felt there was still more to be done," Avila says. "At the same time, I also feel I'll be doing community service for the rest of my life."

Getting the MDI Water Quality Coalition going has been a big part of the work. Over the past year, it has become a significant presence. The cooperative effort is widely viewed as a model for linking schools with the community because it both addresses community concerns and helps adults see children as valuable contributions. The coalition came about from the sheer success of Disney's work with her students, which drew an overwhelming number of requests from towns and private landowners to check various coves and pollution sources. Last year, that work resulted in a $27,500 Sea World/Busch Gardens Environmental Excellence Award. The money allowed the newly formed coalition to buy two computers and other supplies and provided a financial base for supplies and the future of the program. It also allowed the group to hire a coordinator, Avila, who, Disney says, has been instrumental in forging that future.

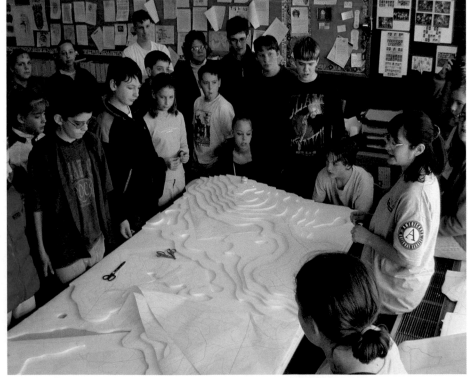

Lelania Avila (right) explains a model of the Eddie Brook Watershed.

It did not take long for the coalition to prove itself. Last summer, along with the Frenchman Bay Conservancy, Union 98, and MDI High School, they put together the island's first ever Water Quality Academy of Math and Science. Five days in July saw elementary and middle school instructors from five schools on the shore, learning about clam biology, clam flats phytoplankton tows, water quality problems, and rocky intertidal habitat; and in the high schools' water quality and GIS labs, testing water for fecal coliform, putting together data and maps, and planning projects for their students for the upcoming year. The workshop resulted in a migration of teachers and their students into the great outdoors.

Along with Disney's sophomore class, which has conducted water quality tests along the island's shore for years now, other instructors are out there teaching their students about the environmental and biological complexities of Eddie Brook, Seal Cove, and the Bass Harbor Marsh. Home schoolers are welcome as well. One family is doing phytoplankton monitoring in Bass Harbor. Lower grades are learning about habitat and have a special project—stenciling storm drains with the words "Dump no waste/Drains to bay," a logo designed to help folks remember that waste can directly affect the marine environment. Collaboration is the key. Among the group's contacts are the Frenchman Bay Conservancy, providing funds for MDI High's water monitoring projects since 1994; the University of Maine Cooperative Extension, organizing data collected from groups all over the State; the Department of Marine Resources, using the coalition's information in its own pollution monitoring; the MDI Biological Laboratory, providing supplies and financial support; the Bar Harbor Conservation Commission, acting as liaison between the coalition and the town of Bar Harbor; College of the Atlantic, providing the coalition with GIS data and support and a volunteer base; and the town of Bar Harbor, whose harbormaster, Edward Monat, has assisted coalition volunteers by allowing access to the town pier for water sampling and occasional offshore sampling.

Avila is a great example for the AmeriCorps system. Created with the support of President Clinton and Congress in 1993, AmeriCorps is one of three national service initiatives under the public–private umbrella, the Corporation for National Service. The AmeriCorps network comprises hundreds of programs nationwide including a broad array of nonprofit organizations on the national, state, and local levels.

In exchange for training and a year or two of service, recruits of all ages and backgrounds receive education awards, along with a modest living allowance and health coverage. More than 100,000 people enrolled in the program over the first four years. Among their accomplishments, members have served more than 32 million people; recruited nearly two million volunteers; taught, tutored, or mentored more than two million children; operated after-school programs for a half million at-risk youth; provided 200,000 seniors with independent living assistance; and built or rehabilitated 25,000 homes. This year, more than 40,000 people will serve in AmeriCorps. Avila's term ends next year; from the start, she found the program invaluable for the contacts she in turn used to the coalition's advantage. At this point, she has become an AmeriCorps trainer herself. "I would recommend it for anyone who has a particular interest but doesn't know where to start," she says. She and Disney have won funding for a second AmeriCorps volunteer, Janet Redman, a volunteer with the coalition from the outset, to focus on environmental education. Once her term is up, she expects to keep on with exactly what she has been doing. "I feel like I have been really blessed by everything in this community," says Avila. "I feel like I want to give back. It feels really good to know I am making a difference."

Adapted from an article that originally appeared in The Bar Harbor Times.

Watershed GIS

Heavy erosion in deforested hills in Madagascar.

Photograph courtesy of Frans Lanting, © 2001 Frans Lanting

Housatonic Valley Association, Inc.

Kirk Sinclair   housatonic@snet.net
www.hvathewatershedgroup.org

Founded in 1941, the Housatonic Valley Association (HVA) is the only nonprofit conservation organization dedicated solely to protecting the Housatonic River and its entire 2,000-square-mile watershed, from the Berkshires of western Massachusetts through western Connecticut and a small part of eastern New York State to Long Island Sound. HVA works to fulfill this mission by promoting a balance between community growth and resource protection through research, community assistance, education, and advocacy.

HVA's key achievements over the past fifty-nine years include the following:

- Permanent conservation of more than three thousand acres of land in the Housatonic watershed.

- Leadership of more than 40 grassroots groups and hundreds of HVA members to remove almost five hundred tons of garbage from the Housatonic River and its tributaries in the Annual Source-to-Sound Cleanup.

- Creation of a Watershed Environmental Resource Center, which provides environmental education and resources to more than ten thousand people each year.

- Protection of hundreds of acres of wetlands, wildlife habitat, riparian lands, and ridgelines through our direct intervention in development proposals and our assistance to towns in developing adequate zoning, wetlands, and building regulations.

- Organization and training of volunteer stream teams to perform water quality monitoring and river advocacy in the northern watershed.

- The Housatonic RiverBelt Greenway Program, which works to protect open space and improve public access along the entire Housatonic riverfront.

In 1997, HVA was honored with the Environmental Protection Agency's Environmental Merit Award for our Housatonic RiverBelt Greenway Program and with the Connecticut Environment 2000 award for our history of water stewardship. In 1998, the Connecticut Department of Environmental Protection awarded HVA its Green Circle Award for sponsoring the Annual Source-to-Sound Cleanup. In 2000, HVA received the Connecticut Greenways Council Award for resource protection in our Housatonic RiverBelt Greenway work.

Since our GIS Department was created in 1997, it has not only facilitated our mission to restore and protect the Housatonic River watershed, but it also enabled us to provide a conservation benefit to the small towns and conservation organizations that fall within the boundaries of our watershed. We primarily worked on specially funded projects such as the digitizing of conservation parcels for towns in our Connecticut watershed and establishing a Stream Team database for basins in Massachusetts. With the power of GIS as an educative and advocacy tool, there is so much more we can be doing toward establishing our own agenda for environmental protection and restoration throughout the watershed.

This past year we worked with the town of Washington, Connecticut, on a Resource Assessment project that identified and mapped their important cultural and natural resources (see maps below and on following page). This can now be used with unprotected and undeveloped parcel data, along with current zoning regulations, to establish new environmental zoning that protects valuable lands. In the coming year we will be reporting our Washington experience to other communities, as well as local conservation organizations, to initiate the process in other areas. In the future we will develop a Resource Assessment presentation and lesson plan as educative tools for schools.

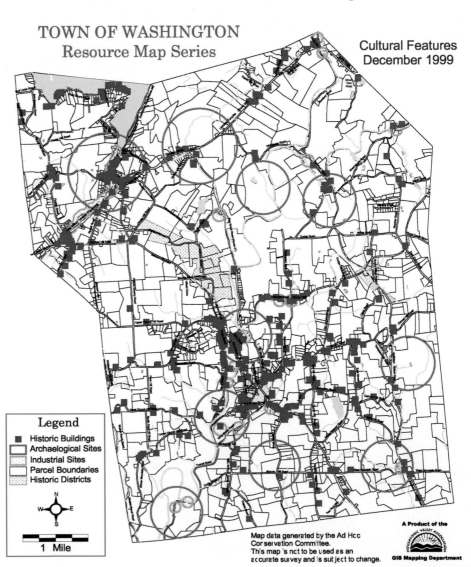

TOWN OF WASHINGTON
Resource Map Series

Cultural Features December 1999

Legend
- ■ Historic Buildings
- □ Archaeological Sites
- ⬚ Industrial Sites
- ▢ Parcel Boundaries
- ▦ Historic Districts

N W E S

1 Mile

Map data generated by the Ad Hoc Conservation Committee. This map is not to be used as an accurate survey and is subject to change.

A Product of the

GIS Mapping Department

TOWN OF WASHINGTON
Resource Map Series

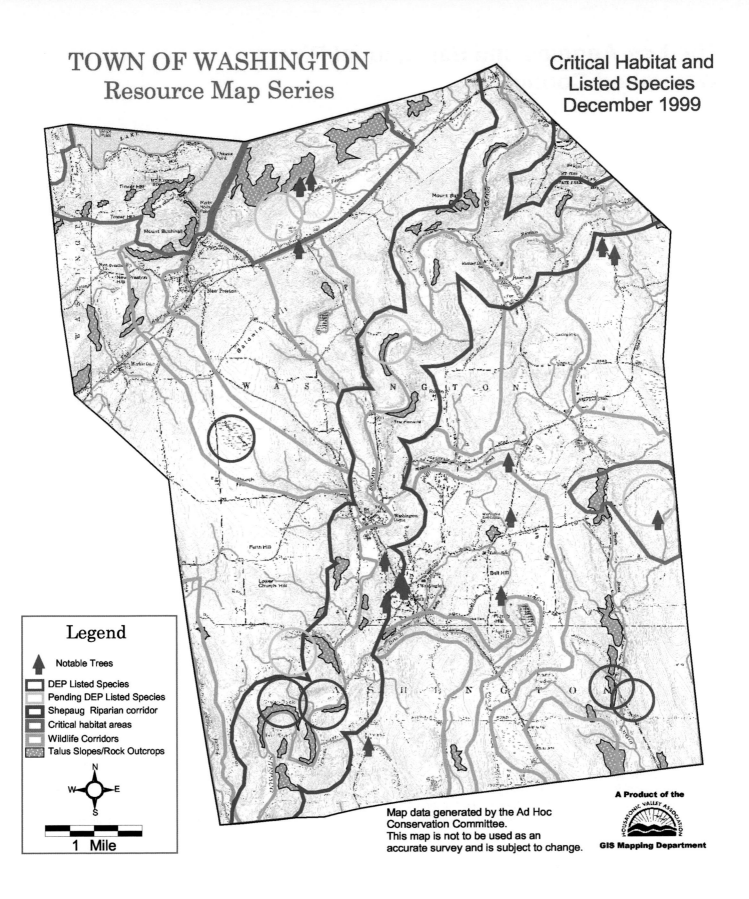

Legend

- **Notable Trees**
- DEP Listed Species
- Pending DEP Listed Species
- Shepaug Riparian corridor
- Critical habitat areas
- Wildlife Corridors
- Talus Slopes/Rock Outcrops

N
W — E
S

1 Mile

Map data generated by the Ad Hoc
Conservation Committee.
This map is not to be used as an
accurate survey and is subject to change.

A Product of the
HOUSATONIC VALLEY ASSOCIATION
GIS Mapping Department

The Los Angeles and San Gabriel Rivers Watershed Council

Rick Harter   rickharter@verizonmail.com
Rumi Yanakiev
www.lasgRiversWatershed.org/

The Los Angeles and San Gabriel Rivers Watershed Council exists "to facilitate a comprehensive, multipurpose, stakeholder-driven consensus process to preserve, restore, and enhance the many beneficial uses—economic, social, environmental, and biological—of the Los Angeles and San Gabriel Rivers watersheds ecosystem through education, research, planning, and mediation."

The three fundamental operative verbs are "preserve," "restore," and "enhance" in reference to "many beneficial uses" within the joint watershed ecosystem including "economic, social, environmental, and biological" functioning. The specifics of what constitutes preservation, restoration, and enhancement vary substantially among stakeholders, depending upon their vision and perspective. Some would like to see the entire river system returned to a naturally functioning soft-bottomed condition (all concrete removed), where steelhead could be reintroduced and riparian habitat would flourish in a greenbelt all the way between the ocean and the San Gabriel Mountains. For others, the notion of preservation means to maintain the flood control channel that has protected densely built-out communities since the 1950s. The ultimate reality will undoubtedly lie somewhere between these two limits, in a system that is intensively managed by humans but that respects the values to communities and individuals of incorporating natural environment into urban settings.

A fundamental step in preservation is to understand what still exists. Nature is highly resilient, and left unmanaged over several generations would ultimately regain the river channel. In some areas even now, substantial habitat continues to exist, albeit seriously threatened by invasion of exotic species. One of our uses for GIS is developing inventories of existing natural conditions and the quality of those conditions, especially with regard to arundo infestation. Most often, these resources are not highly visible and are known to relatively few people in the neighboring community. Once we have a clear and accurate picture of the type and location of these resources, we can begin to communicate this to decision makers both in the administrative world and the political world and to reconsider management approaches.

The idea of restoration has special meaning in an urban context in that we must identify opportunities for the reintroduction of natural landscapes in light of practical concerns that have strong constituencies. For example, day lighting urban streams and removing concrete channels are only politically feasible where it can be demonstrated that flood protection will not be compromised and deconstruction costs are reasonable. We strive to apply rigorous analysis in identifying areas that meet engineering and economic requirements, locating them within the larger context of system functioning. Another of our uses for GIS is communicating the results of technical analyses and placing those results in a geographic framework so that linkages can be better understood.

Enhancement has a wide range of meaning, but there is fundamental consensus among our stakeholders that human lives are enhanced by contact with natural environments, in part as a retreat from the intensely competitive world of urban enterprise. Therefore, we look to identify preservation and restoration opportunities in every community that has access to river/stream resources. Another use for GIS is to maintain a database of parkway planning and acquisition activities in the many jurisdictions that constitute the watersheds, so that integrated strategies can be developed comprehensively, while implemented by numerous players.

It is in representing and communicating this "bigger picture" holistic vision and perspective that the Watershed Council finds its role.

We hold monthly stakeholder meetings (average attendance of fifty to sixty-five people) on topics pertinent to watershed issues (storm water management, water quality, habitat, open space, recreational needs, and so on). We also participate in many meetings pertaining to specific projects or policy initiatives, have begun a publication series, and maintain a Web site with current activities. Our fundamental activity is *communication,* for the purpose of *influencing decisions* that affect watercourses within the double watershed as well as coastal areas that are highly impacted by runoff from the watersheds.

We will use GIS in at least two ways: as an internal tool for tracking and managing outreach and advocacy efforts; and as an analytical tool for producing mapped products in a variety of formats, most frequently for presentation support including computer projections (either directly in ArcView GIS or in PowerPoint®), foam-core mounted display boards, handouts in tabloid or page size, as well as in publications and in Web pages. Internal and analytical usages overlap and are complementary to each other in that the latter provides the message or content and the former identifies the audience. Decisions about media and format are dependent on appropriateness for the selected audience.

The internal usage will support our ability to quickly and easily identify constituencies and to formulate plans for action. At a minimum, we have in mind two sets of GIS products: jurisdictional maps/databases and subwatershed maps/databases. These are the first products we will most likely develop because of their strategic importance to our work and also because they are foundational to the analytical work in terms of who are the target audiences for analytical information. These will be completed within the first six months of our having a GIS technician onboard and will have ongoing life as information is updated and kept current. Detailed descriptions are provided below.

The analytical usage will provide the core presentation materials for projects that are currently programmed and those anticipated to be developed within the next few years. We will begin with immediate work assignments, some of which have already begun, and continue with follow-on work that we expect grant funding to support. At this time, we have five projects in mind: habitat mapping, wetlands mapping, water quality data mapping, management of the Water Augmentation Study, and maintenance of Web site materials for the Los Angeles County Task Force of the Southern California Wetlands Recovery Project (SCWRP). These efforts will be largely completed within the next three years, but also will have ongoing life as information is updated and databases continue to evolve. Detailed descriptions of these as provided below, ordered to provide a certain flow of discussion and not with regard to ranking of importance or timing.

Political Jurisdictions

The Urban Environment Initiative (UEI) Digital Database for Los Angeles County contains several maps of political jurisdictions. Two of these (congressional districts and assembly districts) were used in creating one set of our sample maps for this application. The original UEI maps are color-coded to indicate incumbents as of 1990. With the 2000 election having just occurred, it is clear that this information is out of date. We want to use the boundary information from this UEI disk (until it needs updating based on Census 2000 redistricting; updated GIS maps will be produced by SCAG), and update the database to include not only the name of the incumbent, but also contact information (address and phone number) and identification of local office staff.

The UEI disk contains city and county boundaries, but does not include council or supervisorial districts. We want to add these two maps to the project so that local politicians can also be identified. We will examine the utility of creating a map of Neighborhood Council Districts within the city of Los Angeles, which are being created under Charter Reform. Mapping of submunicipal boundaries will be available from the city of Los Angeles' GIS system and from other municipalities that maintain GIS. In instances where electronic mapping does not exist and information needs appear critical, we will have digitizing done by CSULA.

The political jurisdiction maps that we have created (see map below) indicate a feature that we want to incorporate for our use: the ability to overlay political jurisdiction on project location maps so that information can be retrieved pertinent to gaining support and communicating information about specific projects. On the sample maps, we have indicated one project example. The Rainbow Canyon Restoration/ Rehabilitation Project was one of approximately 100 projects located in GIS by CSULA for the Watershed Council last year during the campaign for support of Propositions 12 and 13. This campaign successfully resulted in statewide bond funding of nearly $1 billion for open space/park acquisitions and many kinds of watershed projects. From this point in time, we want to track nominations for specific project funding and implementation of funded projects.

Vegetation (Habitat) Mapping

The RMC has contracted with a consultant to develop GIS data for the entire San Gabriel River and Los Angeles River watersheds as part of a cooperative arrangement with the Santa Monica Mountains Conservancy. In contribution to this effort, our senior ecologist is assisting with the development of the habitat layer. Existing information available on the UEI disk, and originally from SCAG, is based on the Gap Analysis of Mainland California. The scale of this information is insufficient for the level of detail required for purposes of the RMC analyses. Therefore, more detailed mapping is being undertaken.

Digital orthorectified aerial photography in natural color flown in February 1999 has been acquired from I.K. Curtis Services on CD–ROM by the RMC, covering all of Los Angeles County. The images have a pixel size of one meter. As an initial characterization, we plan to visually scan the photography for vegetative cover within and adjacent to the watercourses, looking at a 1,000-foot-wide band, 500 feet on either side of the centerline of channel. On hard-copy printouts of photos at a scale of 1"= 400' and applying categories of the Sawyer/Keeler–Wolf habitat classification system, we will draw delineation lines based on photointerpretation. We will take these into the field to verify the characterizations. We plan to conduct at least 85 percent ground truthing, aiming at 95 percent accuracy in characterization. Field verification will focus on areas outside the Angeles National Forest, since detailed characterization has already been conducted within the ANF and digitized by U.S. Forest Service staff into GIS format. GIS data from the ANF will be integrated with the new material to cover the entire double watershed.

The results of this effort, in combination with other data layers being compiled, will be used to develop an Open Space Plan for the watersheds that will help to identify acquisition priorities and management principles. The plan is legislatively required to be in place by summer of 2001. While the mapping work will be completed before the Watershed Council has hired GIS staff or soon thereafter, the results of the effort will be made available to us and will form the basis for additional work as described here.

Assembly Districts within the Los Angeles and San Gabriel Rivers Watershed

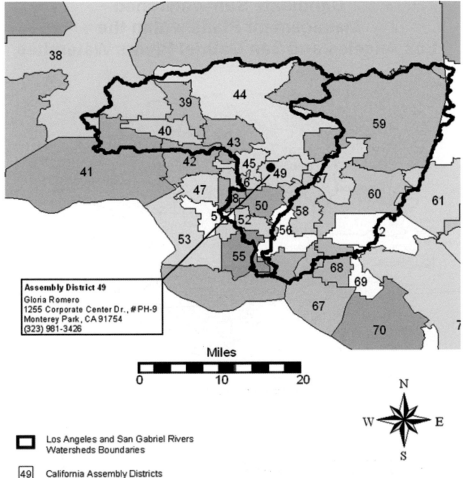

Assembly District 49
Gloria Romero
1255 Corporate Center Dr., #PH-9
Monterey Park, CA 91754
(323) 981-3426

Miles

0   10   20

N
W   E
S

☐ Los Angeles and San Gabriel Rivers Watersheds Boundaries

49 California Assembly Districts

● Rainbow Canyon Restoration/Rehabilitation Project

Produced by the Los Angeles and San Gabriel Rivers Watershed Council, Jan. 2001

Support of the Southern California Wetlands Recovery Project (WRP)

The Watershed Council is leading the Los Angeles County Task Force for the Southern California Wetlands Recovery Project (WRP), which is administered by the California Coastal Conservancy. This gives us a wider geographic purview than merely the joint Los Angeles and San Gabriel watersheds. One of the WRP objectives is to extend its current Southern California Coastal Wetlands Inventory (see Web page at www.ceres.ca.gov/wetlands/geo_info/so_cal.html) to include watershed resources. Also, one of its aims is to have Web-based information maintained at the local level for continual updating as additional information is developed. The Watershed Council intends to serve in this role as the primary Web site manager for wetlands and riparian habitats within the county. To accomplish this, we need to develop the capacity to manipulate GIS information on the Web (but not necessarily to generate every item) and to provide update services, which includes quality control.

Wetlands/Riparian Areas Mapping

Wetlands and riparian areas are a subset of the habitat mapping described above, although the area to be surveyed is surrounding watercourses, characterized habitat will be largely these types. Delineation of wetlands and riparian areas from the habitat mapping will not be based on Army Corps of Engineers jurisdictional parameters but will serve as a general indicator of sensitive resources.

As another approach to identification of extant wetland/riparian resources and in support of the WRP objectives described above, the Watershed Council has begun to develop an inventory of identified resources for the entire county, compiled from existing reports and interviews with local experts. The database is currently in Excel format and includes information on 12 attributes (hydrology, water quality, biology, history, land use, ownership, threats/pressures, and so on) and 92 subattributes (e.g., for hydrology: shape, substrate, soils, inflow, outflow, watershed area, tidal effects, seasonality, control features, dams) for approximately 140 locations. The database currently has no atlas of mapped references but includes locational description and, where available, latitude and longitude.

The next step we intend to take with the inventory is to develop it into a GIS database with mapping display. This will require developing an additional theme from the habitat mapping, so that wetland areas can be individually identified and database information can be retrieved and manipulated for analytical purposes. This step will require limited field verification within the joint watersheds because the overall habitat mapping should have captured all of the inventoried items. However, the geographic range of this work extends to all coastal watersheds within Los Angeles County and, therefore, field verification of inventoried wetland/riparian locations outside the joint watersheds will need to occur. The same methodology used for habitat mapping will be applied in this instance, with characterization of all habitat within a band along watercourses within the Santa Monica Bay watersheds (Ballona Creek and Malibu Coast) and portions of the Santa Clara River. Digitization of this information and augmentation of the habitat layer will be done under an extension of the RMC contract. Extension of the wetland/riparian theme in conjunction with the database inventory will be done by in-house staff.

Once in GIS format, we can produce maps at various scales (overview of entire joint watershed, subwatershed areas, individual wetland areas) for various purposes (publication, presentation). We can also add other dynamic characteristics to the GIS database such as land ownership, stewardship organizations, restoration efforts, and so forth. Updating such information will be done on a collaborative basis with caretaker organizations that have a stake in monitoring activities in each area. The ultimate purpose of establishing the inventory is its use as a tool in developing and managing a wetland/riparian mitigation banking system. The idea is to identify restoration and enhancement opportunities to establish implementation programs focused on identified areas and to track activities within those areas.

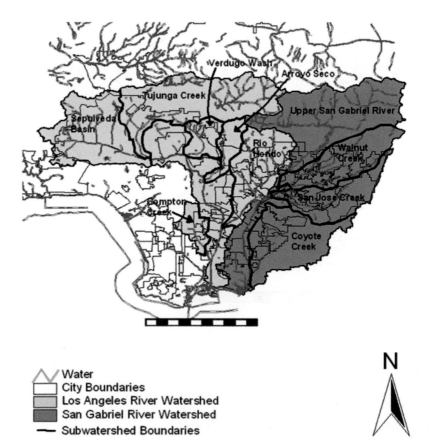

Candidate Sub-watershed Management Plans within the Los Angeles and San Gabriel Rivers Watershed

Legend:
- Water
- City Boundaries
- Los Angeles River Watershed
- San Gabriel River Watershed
- Subwatershed Boundaries

N

Produced by the Los Angeles and San Gabriel Rivers Watershed Council, Jan. 2001

Water Quality Mapping

Another of the Watershed Council's five major work plan efforts pertains to compilation of water quality data. This has several aspects. Last summer, one of our interns began to input unrecorded data at the Regional Water Quality Control Board into their existing GIS database. In early autumn, Heather Trim organized a simultaneous collection of surface water samples that were later analyzed as part of an SCWRPP study to characterize water quality both inland and offshore at a single "snapshot" in time. There are ongoing volunteer River-Watch monitoring programs by Friends of the Los Angeles River (FoLAR) and other organizations, such as Friends of the San Gabriel River, that are interested in establishing similar programs. The City of Los Angeles' Storm Water Division is mounting a new program of their own for water quality sampling. Watermaster offices and water supply agencies within the watershed have groundwater sampling programs as well. To provide a consistent overview among these efforts and to provide more opportunities for public review of surface and subsurface water quality conditions, the Watershed Council will compile water quality data from all extant sources into its own GIS system.

This effort will rely heavily on existing GIS resources available from the agencies that have established monitoring programs. Also, the LACDPW's GIS system being developed by their consultant includes information on monitoring well locations and sampling results over time.

Management of the Water Augmentation Study

The fifth major project we have initiated is a multiyear study on the ramifications of capturing storm water on-site and infiltrating it into the ground, as opposed to letting it run off-site from the "first flush." The initial pilot study will investigate two sites of a nonindustrial character. The expanded pilot study will look at seven to 10 sites, including light industrial along with residential and commercial locations, and the ultimate study, if funded, will look at several dozen sites over a multiyear period. At each site, water samples will be collected at several locations in the system at one or more surface points and at several subsurface points. Samples will be tested for more than two dozen constituents, covering both biological and chemical parameters. Managing this data will involve use of a relational database, but it would be valuable to have a geographic reference associated with the database, particularly as the number of sites become greater.

Moreover, with both existing water quality data (along with new data produced from regular periodic monitoring) and pilot study data in a GIS system, then relationships between the long-term reference information and the study information, as well as the wider geographic picture along with the specific study sites, can be analyzed. Developing the GIS for this project will be an extension of the water quality task described above, adding new points in a separate theme to represent monitoring locations associated with each study site.

Overall, the program of seven items described above would be highly ambitious if we were starting from scratch. We are fortunate to have the value of many ongoing GIS initiatives in the region, especially the two current projects sponsored by LACDPW and the RMC, along with developed databases by several of our member agencies. Our tasks will be focused largely on assembling and augmenting these resources, while gradually developing our own materials.

Conclusion

To make the changes in the ways our water resources are managed—ones have been agreed among our stakeholders as desirable—we need to reach out to others who may not fully understand the ramifications of their actions on the watershed. These include municipal decision makers—elected and appointed officials, professional staff, active citizens who are opinion leaders, and interest group leaders, along with private sector decision makers such as developers and consulting professionals. In communicating our vision to these audiences, we need what GIS provides: mapped products in a variety of media that convey the spatial relationships of issues needed to be focused on and issues that require distilling a large amount of detailed information into straightforward communication. We want to create such products in a high-quality manner that reflects the standards people have come to associate with our efforts.

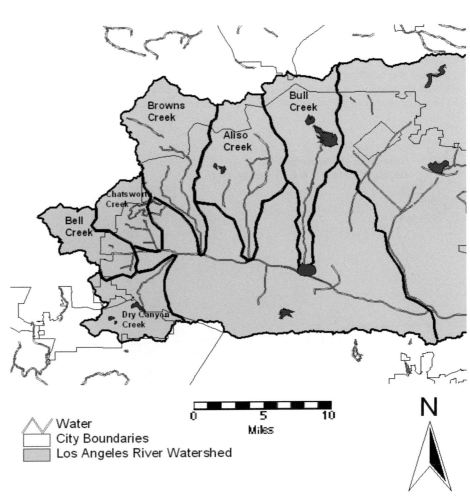

Produced by the Los Angeles and San Gabriel Rivers Watershed Council, Jan. 2001

Headwater Creeks within the Sepulveda Basin subwatershed.

Cannon River Watershed Partnership

Justin D. Watkins   crwp@means.net

In 1990, local citizens in Minnesota formed the Cannon River Watershed Partnership, a nonprofit organization. The CRWP works with many agencies, organizations, and individuals to protect and restore lakes, streams, rivers, forests, and prairies for current and future generations. The mission of the Partnership is "to protect and improve the surface and groundwater resources and natural systems of the Cannon River Watershed." This mission is accomplished through (1) education, providing information to partners and all watershed residents. We support agricultural, industrial, and business practices that minimize environmental impact and continue to encourage wise use through our River Friendly Neighbor Program; (2) coordinating efforts with organizations and local and state agencies to maximize resources, funding, and expertise on issues; (3) providing cost-share dollars to fill gaps in conservation programs; (4) promoting and advocating for local citizen initiatives by addressing local concerns; (5) monitoring water quality through citizen volunteer monitors trained and supported by CRWP; (6) collecting and tabulating all relevant geographic information to be utilized and accessed by the conservation efforts of the CRWP and any other conservation group or agency in the watershed.

Two-Year Overview

Over the next two years, the CRWP will be focusing on phosphorus and fecal coliform pollution reduction in the Cannon River and its tributaries. The EPA has assigned us a TMDL computation for both of these "pollutants" for multiple stretches of the Cannon and/or its tributaries. GIS will play a very important role in all stages of these projects. Digital data will be used to identify subwatersheds that are the greatest potential contributors of these pollutants.

Goals

The primary goals of our GIS project are as follows: (1) guiding phosphorus management efforts—coupled with a phosphorus budget generated by water sampling, geographic information will tell us which areas of our watershed need the most attention; (2) guiding fecal coliform management efforts—similarly, GIS will point out which subwatersheds may be contributing the greatest amounts of harmful bacteria to our waterways. A specific goal will be the implementation of both erosion potential (examining current possible loads of phosphorus and fecal coliform) and development suitability (suggest areas for future development) models (using ArcView Spatial Analyst),

incorporating land cover/use, soil type, slopes, and distance to stream/river. These models will be primary products of our plan and will be applicable in many future situations.

Our plans will utilize primarily International Coalition land-use data, digital elevation models, digitized streams/rivers/lakes, DNR minor watersheds, National Wetland Inventory, and the limited soils data that is currently available. Of secondary importance will be geomorphology, depth to bedrock, feedlot locations, presettlement vegetation, and others as needed (we have already acquired most of the mentioned data, most via the MN DNR and MN Land Management Information Center). The project will generate (via an inkind contributor) a residue coverage, which will estimate the percent of crop residue existing in the fields of the watershed.

Long-Term Goals

Although these tasks will absorb a majority of staff time over the next two years, the CRWP does plan to continue developing its GIS indefinitely. The long-term goal is to first collect and house the wealth of data that exists (we're well on our way already), and then make that data easily accessible to ourselves and others. As you see at the listed URL, (http://206.11.107.10/crwp/) we have already served two fairly simple maps on the Internet; the plan is to make many more available in the near future. The Cannon River Watershed is a large area, containing many agencies and conservation groups. One thing it is lacking though, is a central assemblage of information; we hope to act as that chief residence of geographic information for the entire watershed.

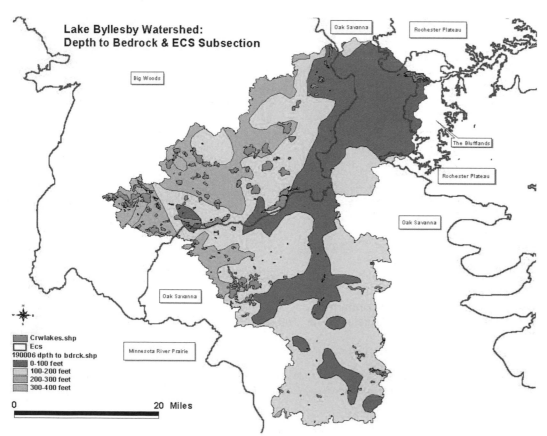

Lake Byllesby Watershed: Depth to Bedrock & ECS Subsection

Oak Savanna
Rochester Plateau
Big Woods
The Blufflands
Rochester Plateau
Oak Savanna
Oak Savanna
Minnesota River Prairie

Crwlakes.shp
Ecs
190006 dpth to bdrck.shp
0-100 feet
100-200 feet
200-300 feet
300-400 feet

0   20 Miles

Sonoma Ecology Center

Amy Goldstein, Caitlin Cornwall, and Richard Hunter
Sonoma Ecology Center, California

Abstract

This pilot study mapped and quantified current conditions and historical trends in the riparian areas and streams around the city of Sonoma. Mapping and assessment of riparian vegetation is needed to provide a credible foundation and common understanding to support discussions of riparian and salmonid restoration priorities, policy needs, and land protection and acquisition. This study covered a small central portion of Sonoma Valley, an area around the city of Sonoma approximately two and a half by three miles. We produced two vivid maps: structural channel changes between 1951 and 1999 (see map below) and the amount of riparian forest found inside the stream setback zones mandated by the city and county (see map, next page). About 19 percent of the stream mileage in the study area has been lost completely, channelized, or put underground in the last forty-eight years. The proportion of mandated setback containing riparian forest is approximately 49 percent in the city areas of the study area and 34 percent in the county areas. These maps and results will be used for educating the community, for supporting discussions with planning departments regarding upcoming

improvements in land-use policy, and for sharing as a model with other watershed groups. The study was funded by the Sonoma County Water Agency and the U.S. Fish and Wildlife Service (via CALFED).

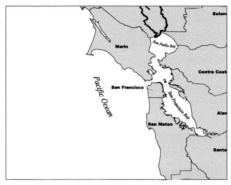

Overview map

Introduction

This pilot study mapped and quantified current conditions and historical trends in the riparian areas and streams around the city of Sonoma, in the Sonoma Creek watershed, which drains into San Pablo Bay. Human communities in Sonoma Valley have caused changes in forest cover and composition at least since the time of European settlement (1823). The presettlement extent of riparian forests is not well known, although historical ecology research is beginning. From anecdotal reports and evidence, we can generalize that presettlement riparian forests extended beyond their current extent and beyond the stream setbacks mandated by county and city planning policies. Several powerful regulation-driven processes are converging on the need to enhance riparian habitats. These processes require better data on the locations, conditions, and priorities of riparian and aquatic resources. Mapping and assessment of riparian vegetation is needed to provide a credible foundation and common understanding for discussions of riparian and salmonid restoration priorities, policy needs, land protection, and acquisition. This pilot study had three objectives: (1) to create maps to visualize the status of riparian areas and the channel network for use as tools for educating the public and policy makers, (2) to compare stream setback policies with practices, and (3) to begin to quantify some aspects of the status of riparian areas and channels.

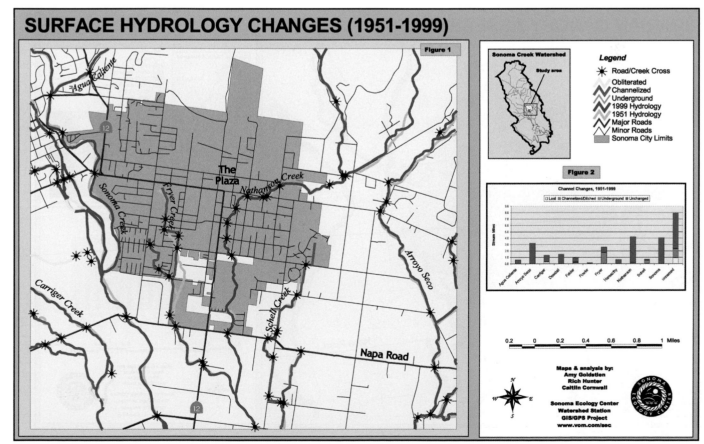

Figure 1. Channel changes between 1951 and 1999 in and around the city of Sonoma. Black stars mark road crossings.

Methods

We researched existing data sources for riparian vegetation and source imagery, both historic and current. Existing data sources, including several 30-meter Landsat TM classifications, were dismissed because they are too coarse for the scale of our analysis. We selected a black and white aerial orthophoto produced by HJW and Associates in July 1999 for the city of Sonoma. This orthophoto covers a small central portion of Sonoma Valley, a 6,600-acre area, around the city of Sonoma. Its one-foot pixel resolution provides an extremely detailed image of the city and its surrounding agricultural areas.

Analysis #1: Surface Hydrology Changes (1951–1999).

Surface hydrology from 1951 was acquired from digital line graphs (DLG) reflecting the USGS 7.5-minute Sonoma quadrangle. We created a 1999 Hydrology Theme by copying the 1951 Hydrology Theme and editing the vertexes of the creek lines to reflect the position of the creeks in the 1999 photo. This process revealed some obvious structural changes that occurred during the last forty-eight years.

Four categories of change were identified:

1 Lost: Lengths of stream that are shown on the 1951 USGS topographic map but do not appear at all on the 1999 aerial photograph.

2 Channelized/Ditched: Lengths of stream that have been rerouted and/or straightened, and may be revetted with riprap, concrete, and other man-made structures.

3 Underground: Lengths of stream that have been piped underground.

4 Unchanged: All other lengths of stream. These reaches did not appear to have obviously changed by human action between 1951 and 1999, although most of these stream miles did change position, presumably via natural migration.

Analysis #2: Riparian Canopy Assessment.

We digitized riparian vegetation as a polygon theme in ArcView GIS by interpreting the aerial photograph on-screen (heads-up digitizing). During the process, it was necessary to make interpretations of the aerial photograph that are in the process of field verification. For instance, the transition from riparian vegetation to upland or valley bottom vegetation is not always clear in the photo.

County and city riparian setback zones were modeled using a combination of field and GIS methods. Riparian vegetation was mapped as an overlay on riparian setbacks established by the Sonoma County General Plan Open Space Element and the City of Sonoma Zoning Ordinance.

The Sonoma County General Plan designates riparian corridor setbacks for four stream-type classifications, each with a different width of setback. The classifications are designated as follows: (a) Urban Riparian Corridors (50'), (b) Russian River Riparian Corridor—not applicable (200'), (c) Flatland Riparian Corridors (100'), (d) Upland Riparian Corridors (50'). (Sonoma County General Plan, Open Space Element 3.2, Policy for Riparian Corridors. Goal OS-5, Objective OS-5a & OS-5c. Page 178.) Not all streams have been officially listed in the policy.

The city of Sonoma has three creek setback overlay zones. They are measured from the toe of the creek bank outward at a slope of 2.5:1 plus setback distance, or the setback distance measured from the top of the bank, whichever is greater. The distances are specified as follows: (a) Sonoma Creek (50'), (b) Nathanson Creek (30'), and (c) Fryer Creek (30'). (City of Sonoma Creek Setback Overlay Zones City of Sonoma Zoning Ordinance. Zoning District Regulations Exhibit B, p B-6.)

The GIS line features for streams represent the centerlines of streams, but the regulations call for measuring the setback from the top of the bank. It is important to estimate the distance between the centerline and the top of the bank because in some cases this is a difference of 100 feet or more. The resolution

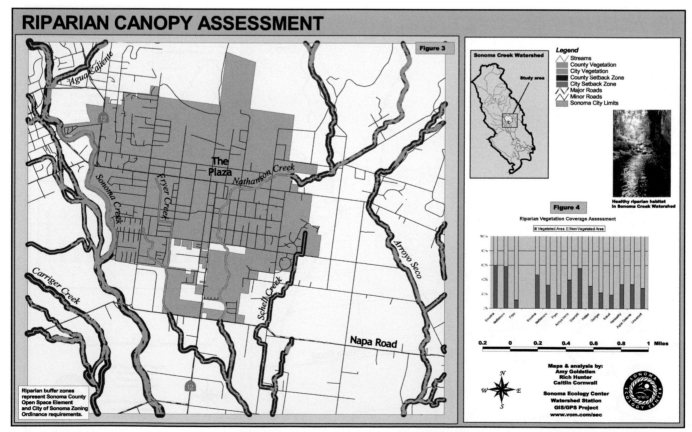

Figure 2. Riparian forest cover inside mandated stream setback zones in and around the city of Sonoma, 1999. Note: city and county setbacks have differing width.

of our 10 meter DEM topographic data was insufficient to locate the bank top. So we chose 25 representative locations where streams cross roads to sample this distance. We measured the distance between the tops of banks in the field. The measurements were divided in half and added to the county and city setback requirements to calculate the setback using a buffer command. The acreage of setback was compared to the acreage of riparian canopy within the setback to evaluate the effectiveness of the setbacks in protecting the riparian corridor.

Results

The study area contains approximately 28 stream miles on 22 creeks. Of these, 28 percent of the miles and 11 of the streams are unnamed.

Analysis #1: Surface Hydrology Changes (1951–1999). Figure 1 shows the estimated 1951 and 1999 hydrology of the study area. Some of the differences between 1951 and 1999 can be explained by natural meander patterns and some relate to the accuracy and scale of the original data source compared to the aerial photo used to create the 1999 Hydrology Theme. Out of a total of 28 miles of stream, 5.4 miles (19 percent) have been significantly altered by human land use changes: 1.4 miles (5 percent) of stream have been lost completely, 3.8 miles (13 percent) of stream have been channelized or ditched, and 0.2 miles (1 percent) have been rerouted underground.

The majority of creek miles channelized (70 percent) and underground (88 percent) since 1951 are inside the city of Sonoma. Fryer Creek is especially impacted and mostly devoid of riparian canopy. Sonoma Creek, the mainstem stream of the valley, has no apparent channelization. All the miles of stream lost since 1951 were found in agricultural areas outside the city limits. Drainage systems associated with intensive vineyard developments have replaced and simplified the drainage network that was present in 1951. However, this study does not include the extensive channel modifications that occurred in Sonoma Valley before 1951, particularly those near the city of Sonoma, which began as early as 1823.

Analysis #2: Riparian Canopy Assessment. The study area contains approximately 273 acres of riparian forest canopy and about 766 acres of setback required by city and County policy. Figure 2 illustrates the proportion of the mandated setback that contains riparian forest as of July 1999. Within City limits, there are 41 acres of riparian forest and 85 acres of setback. There are 232 acres of riparian forest and 681 acres of setback in the County portion of the study area.

For the study area, approximately 36 percent of the designated stream setback is vegetated with trees. The average riparian vegetation coverage in the setback zone along creeks regulated by the county is approximately 34 percent, which leaves about 66 percent of the county setback zones without riparian vegetation. Active agricultural operations, including livestock grazing, orchards, and vineyards, were found within county setback zones. While these land uses can be compatible with riparian conservation, setback zones have not been strictly implemented.

The city has about 49 percent vegetation coverage of its allotted setback acreage. Within city limits, urban/suburban development is the main reason for the loss of riparian vegetation. Most of the creeks in the city have been altered (channelized or rerouted underground) to maximize space for development and roadways. Houses, roads, and landscaping are found within city setback zones.

Riparian canopy is generally contiguous in the section of mainstem Sonoma Creek occurring in the study area, with some notable gaps at major road crossings (shown as black stars on figure 1). The outer margins of Sonoma Creek's setback zone are unvegetated, indicating that surrounding land uses are narrowing riparian vegetation. Close examination of the aerial photograph indicates that at most road crossings there is a break in the riparian vegetation.

Comparison of the vegetation data and channel change data reveals how altering a creek channel affects the riparian vegetation associated with it. For example, 72 percent of Schell Creek and 67 percent of Fryer Creek are channelized (figure 1). Both these creeks lack much of the riparian vegetation where they have been altered. Schell Creek riparian vegetation only covers about 19 percent of its allotted riparian setback zone. Fryer Creek riparian vegetation only covers about 11 percent of its setback zone. Generally, where the creeks still retain their natural channel, there is a substantially greater proportion of riparian vegetation in the setback zone.

Conclusions

Only changes that were clearly visible on the aerial photograph were recorded, so these results are a conservative estimate of human impacts to surface hydrology in the study area. It is likely that all the streams have changed since 1951, but some impacts, such as downcutting or spreading, are not apparent from an aerial photograph. The categories were chosen to indicate large physical alterations directly caused by humans.

An unknown but large amount of the loss of natural channel processes and riparian communities happened early in Sonoma Valley's history. We plan further work to extend studies of this type further back in time and to the entire watershed.

City regulations provide for only narrow setback zones even though some riparian functions may have elevated importance in urban settings. For instance, the buffering/filtration capability of riparian vegetation could be instrumental in cleansing polluted runoff water and slowing peak flows from runoff. Likewise, vegetated riparian buffers moderate flood flows, damages from which are particularly expensive in an urban setting.

From this study and other evidence, it appears that channel losses have particularly affected distributary channels that once carried high flows in multiple channels. Agencies and others working toward flood reduction and floodplain restoration in Sonoma Valley will be investigating whether restoring some of these distributaries would accomplish these objectives.

Although general plans for Sonoma County and city have only mandated stream setbacks for a few decades, it appears from our data and field observations that these worthy policies often go unenforced. If Sonoma Valley is to retain sensitive and cherished species and habitats such as steelhead, California freshwater shrimp, and oak-rich riparian forests, our history of bringing land uses like housing and agriculture close to streams will need to change. Future land users, and the local governments that regulate land use, will soon have to comply with stricter standards designed to protect water quality and fisheries. We will be working with planning staff and local community leaders to share our study results and provide relevant information to the process of improving land use and land-use policy affecting riparian and aquatic ecosystems.

We plan to use the maps from this study as tools for education in our community, to teach about historical ecology and the value and status of riparian areas and streams. We will be conducting more in-depth discussions about the type of information generated in this pilot study with city and county planning staff to assist them in improving practices related to permitting, public works, and enforcement. We'll share the simple methods, preliminary results, and applications of this pilot study with watershed groups around the State in hopes that our work can assist the many communities dealing with similar issues.

Chesapeake Bay Foundation

Steve Libbey, Land-use Planner

Our original GIS plan sought to bring GIS more broadly into the work of CBF organizationwide, and make GIS more readily available to community organizations and individuals via community outreach and the Internet. We originally planned to use GIS to extend the mapping work we had already begun through our Lands-at-Risk program by providing Internet access to maps to help us conduct community education and outreach. We are still working to make GIS a viable tool for CBF's conservation outreach and education goals. But to fully realize its potential, we will require a longer term than we initially envisioned.

Personnel changes and the need for training have delayed our achieving several of the goals set forth for the year. Our original GIS team leader left CBF and could not be replaced for several months. The new team leader, Steve Libbey, CBF's Headquarters Lands Planner, has not been trained in GIS, but has been learning ArcView GIS over the past few months.

Despite these changes, our GIS grants have enabled CBF to:

- Increase awareness of the potential of GIS within CBF, thereby encouraging the training of headquarters and field office staff in the program.

- Initiate, with the U.S. Environmental Protection Agency's Chesapeake Bay Program, creation of a Web site to permit community members and others to view overlays of GIS data through which they can learn more about potential development risk to their quality of life and to the Bay as a whole within the 4,000-square-mile metropolitan Washington, D.C., area.

- Begin exploring methods for creating a predictive future model of development around the Bay using GIS technology. However, because of the modeling complexity and the personnel requirements of this project, it would ideally be conducted by an independent research organization, such as a university, for CBF.

- Initiate discussions with several other NGOs on establishing a shared GIS database.

Challenges and obstacles

Personnel changes and the resulting training issues have posed the most significant challenges to our program. CBF has been addressing these challenges through the online *Introduction to ArcView GIS* training offered through ESRI.

Given that this represents introductory training only and that the trained Virginia staff member has recently moved to another organization, further training will be vital to our ability to advance the use of GIS both within and beyond CBF.

Scope and schedule

Change, as the phrase goes, is inevitable. Clearly, GIS can serve to facilitate people's understanding of the changes taking place in and around the Bay. But achieving solid proficiency in GIS within CBF's team will not be realistically achieved before summer of 2001. Once the learning curve is surmounted, however, it will open the way for advancing the use of GIS broadly throughout the organization.

Our scope of work has also been redirected to some extent, partly due to the need to rebuild in-house skills and partly due to CBF's effort to achieve some of our goals via different routes. The basic training completed by CBF

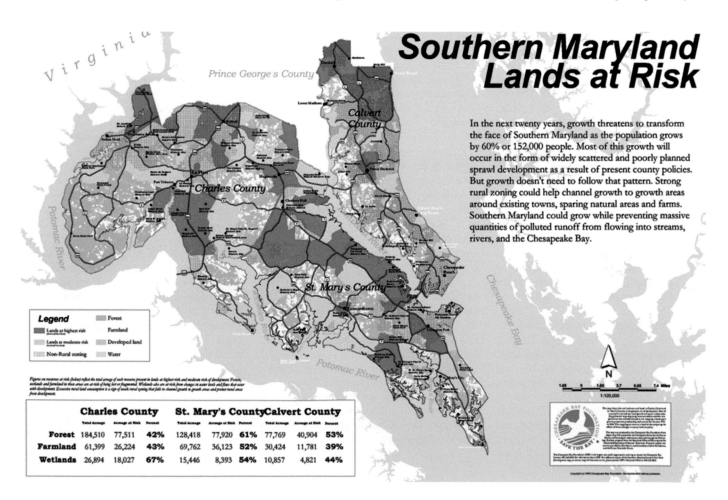

Southern Maryland Lands at Risk

In the next twenty years, growth threatens to transform the face of Southern Maryland as the population grows by 60% or 152,000 people. Most of this growth will occur in the form of widely scattered and poorly planned sprawl development as a result of present county policies. But growth doesn't need to follow that pattern. Strong rural zoning could help channel growth to growth areas around existing towns, sparing natural areas and farms. Southern Maryland could grow while preventing massive quantities of polluted runoff from flowing into streams, rivers, and the Chesapeake Bay.

Legend
- Lands at highest risk
- Lands at moderate risk
- Non-Rural zoning
- Forest
- Farmland
- Developed land
- Water

	Charles County			**St. Mary's County**			**Calvert County**		
	Total Acreage	Acreage at Risk	Percent	Total Acreage	Acreage at Risk	Percent	Total Acreage	Acreage at Risk	Percent
Forest	184,510	77,511	42%	128,418	77,920	61%	77,769	40,904	53%
Farmland	61,399	26,224	43%	69,762	36,123	52%	30,424	11,781	39%
Wetlands	26,894	18,027	67%	15,446	8,393	54%	10,857	4,821	44%

staff thus far has helped inform our discussions and collaboration with others, and enabled the continued advancement of education and conservation objectives.

For example, our goal of enhancing community members' understanding of the pace and impact of change is no longer exclusively reliant on CBF. Rather, we are pursuing this objective in conjunction with the U.S. Environmental Protection Agency's Chesapeake Bay Program (CBP). The CBF–CBP project will provide people with the ability, via the Internet, to overlay basic data layers, that is, agricultural lands, protected lands, road networks, and so on, to graphically see which ones are in the likely path of development. As initially intended, this project will take some of the decision-making power out of the hands of developers and put it back into the hands of the people. The ability to continue moving toward our initial GIS outreach goals has been facilitated by the CTSP grant, particularly through the use of ArcView GIS. We hope to see the first phase of this project online by January 2001, with revisions following the spring release of Census 2000 data.

Additionally, CBF is working on creating a predictive modeling project, currently in the initial stages of development. This project entails developing a reasonably sophisticated set of algorithms to model future development and will first be applied in the greater Washington, D.C., area. To ensure acceptance by such vested interests as business, CBF is seeking to have the project developed by an independent research organization such as a university. Such a model would provide a more precise look at future development in the region and provide a solid basis for both community outreach and informed discussions of regional goals. Here again, the CTSP grant has helped enhance our understanding of GIS capabilities and methodologies. We hope to have such a model developed by the end of 2001.

GIS Grants Impact on Our Mission:
The *Introduction to ArcView GIS* training has clearly been crucial in our efforts to communicate the conservation challenges posed by current development patterns. Further training, of course, will augment our ability to use it and to disseminate our work in GIS.

The types of maps illustrated here have been, and continue to be, useful in general purpose community outreach and education efforts. They provide solid background for discussions about rates and patterns of change. As an example, the choropleth population map demonstrates that the high-growth areas have moved decidedly away from existing cities during the 1990s.

These maps are also indicative of our own ongoing examination of the impact that growth and development patterns have on the natural environment. For example, counties around the District of Columbia have lost more than a quarter of their agricultural lands since 1982, reflecting not merely a significant change in local landscape but also the availability of local produce.

Collaborations with Others in the GIS Community
In addition to the projects discussed above, CBF initiated a collaborative project with the Trust for Public Land, the Wilderness Society, The Nature Conservancy, Piedmont Environmental Council, and others. The intention of this effort was to develop a publicly available GIS database for the broader Washington, D.C., metropolitan area to facilitate prioritizing lands for preservation and advocating smarter development patterns. The primary resources available to this effort included an experienced GIS person, on contract with the Piedmont Environmental Council (PEC), and the CTSP equipment housed at CBF.

Conclusion
The Chesapeake Bay Foundation continues to expand its ability to communicate the challenges and opportunities of conservation to citizens, elected officials, the business community, and other stakeholders regarding the long-term health of the Bay. GIS will play a significant role in that work as we move into the future. We look forward to continuing to enhance, not only our abilities, but the ability of residents throughout the watershed to take advantage of that potential. With the assistance of the Conservation Technology Support Program, we will be able to continue moving forward in informing and educating our local constituency in land-use issues.

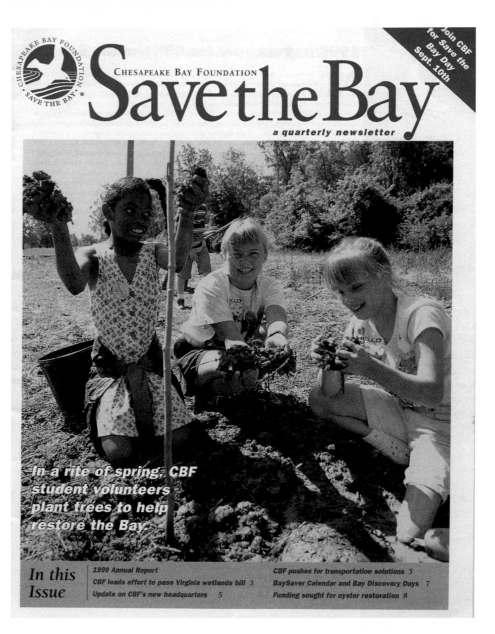

CHESAPEAKE BAY FOUNDATION

Save the Bay
a quarterly newsletter

Join CBF for Save the Bay Day Sept. 10th

In a rite of spring, CBF student volunteers plant trees to help restore the Bay.

In this Issue

1999 Annual Report
CBF leads effort to pass Virginia wetlands bill 3
Update on CBF's new headquarters 5

CBF pushes for transportation solutions 5
BaySaver Calendar and Bay Discovery Days 7
Funding sought for oyster restoration 8

Bayou Preservation Association, Inc.

Mary Ellen Whitworth, P.E.   bpa@hic.net

The mission of the Bayou Preservation Association (BPA) is to protect and restore the richness and diversity of Houston-area waterways through activism, advocacy, collaboration, and education. For more than thirty years, the BPA has worked to protect bayous from unnecessary channelization, from damage due to encroaching urbanization, and from deterioration of water quality. Our long-term goal is to assure that large, medium, and small stream corridors in the Houston region can gracefully serve the urban needs of hydrology, nature, and people well into the 21st century.

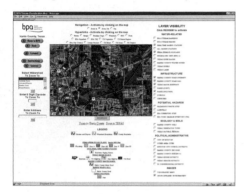

Activism
The BPA responds to threats to our bayous and streams by grassroots organizing, education, and promotion of alternatives to destructive channelization. We monitor our bayous on a monthly basis for water quality parameters and submit this data to the state and the general public. The BPA is represented on numerous water-related committees that include a broad range of stakeholders. Further, we monitor herbicide spraying on the banks of the bayous to protect critical habitat and review Corps of Engineers wetland permit applications.

Advocacy
The BPA believes that there are viable, economically responsible alternatives to channelization for our region's storm water management, such as floodplain management and buyout, regional detention basins, and natural models for streambed modification. We work closely with the Harris County Flood Control District and other city and county agencies to shape the policies that affect our bayous. We offer support to each of our city's individual watershed organizations while serving as a clearinghouse for issues affecting all the bayous and responding to threats to bayous not otherwise represented by an advocacy group.

Collaboration
Since its inception, the BPA has collaborated with other environmental organizations and governmental agencies to promote the health of our city's waterways. Examples of such collaborations include the coordination of debris and litter cleanups along bayous throughout the city, revegetation of riparian areas along the bayous, the installation of watershed signs along heavily trafficked areas of greater Houston to foster a greater sense of stewardship and ownership among our city's residents, and the installation of canoe launching points along area bayous to promote the bayous as recreational resources.

Education
The BPA is an information gathering and disseminating organization. We sponsor seminars to teach our community about bayou-threatening issues and make presentations to schools, civic clubs, and other organizations on the importance of watersheds and bayous. We are in the process of developing a plant palette of both native and introduced plant materials with growth forms and other physiological characteristics that make them useful along our waterways. A new project, Kids on the Bayou, provides hands-on opportunities for thousands of Houston-area young people to explore our city's waterways and to begin to grapple with the complex issues surrounding their management and stewardship.

Future
In the next century, most, if not all, of our remaining streams will come under intense pressure from the surrounding urbanization. It is vital that the bayous and streams that are still in their natural state be identified and classified so organizations and citizens can act intelligently and sensitively to protect these important natural assets. Hence, in addition to pursuing its ongoing objectives, the BPA is prepared to focus on disseminating such information in the years to come.

The Stream Classification Project has provided BPA with the opportunity to consolidate relevant multiscale GIS data from a variety of sources. With the assistance of its cooperator HARC, BPA developed and implemented a plan to acquire the needed data, then value add them to support the stream conservation objectives of the Stream Classification Project. The cooperators then made this data and supporting information available to interested parties using ESRI's MapObjects Internet Map Server technology (see picture above). Although very new, the Bayou Information Center Web site is already attracting tremendous attention and use in the region. The site makes it easy for users to enter a ZIP Code, address, or intersection and evaluate the resulting position relative to whatever information is of interest. Users include homeowners who want to know if their property might fall in a floodplain and developers interested in gaining information to support the development of project environmental impact statements. Since much new park land and green space area to be added in Harris County is likely to be associated with waterways and their attendant flood control zones, there is great interest using BPA's site as a model for developing an accurate and comprehensive county park polygon layer with information hyperlinks that tie to BPA's site and general park information that would be of interest to the public.

The Stream Classification Project will be completed in spring 2001. At that time, ground truthing will have been completed, and the final ranking and supporting digital imagery and stream channel informational links added to the site. BPA plans to seek funding for a second phase of the project. The original phase of the project is concentrated on ranking streams that comprise the Harris County Flood Control District's Channel Assessment Program database. There are other waterways in Harris County that are not part of this database that would be ranked in a second phase of the project if funded. Furthermore, BPA would like to strengthen the current ranking by adding additional considerations of water quality and species distribution along the higher ranked segments.

Houston's streams and bayous are a regional treasure that is rapidly disappearing. The BPA is striving to create a coherent approach to protecting what we can while exploring alternative storm water management tools. The BPA and other bayou and watershed advocacy groups can use this GIS information to prioritize the use of scarce volunteer resources. Agencies can use the information to assess the impact of existing or proposed projects along our streams and to consider possible alternatives or to evaluate the depth of community input needed in a sensitive area. Disaster response teams can use the information to evaluate response measures to spills and accidents along our bayous. Developers and land planners will be able to use the data in their design and building processes and avoid some uncertainty about whether a stream is considered environmentally sensitive. Students of all levels can use the data and maps to better understand their watersheds and bayous for school projects and research projects. With the help of the CTSP, the BPA will be able to make its Stream Classification Project a foundation block for a living body of knowledge about our bayous and streams that can be built upon and drawn from for generations to come.

Forest GIS

Forest Community Research

Jonathan Kusel   info@fcresearch.org
http://fcresearch.org

Forest Community Research is dedicated to advancing the understanding of working and wild landscapes. Our focus is understanding and sharing with others the interconnections between community well-being, natural resource management, and healthy and sustainable ecosystems. FCR pursues its mission by conducting research on rural community issues and facilitating education and dialog to support community-based approaches to sustainable ecosystem management. FCR's core principle is that research, education, and dialog must build community capacity to manage natural resources and improve ecosystem health.

FCR has coordinated and participated in studies and activities associated with large-scale ecosystem projects, been a national leader with its assessments of community well-being, developed an integrated adaptive approach to public involvement in the Sierra Nevada Ecosystem Project (SNEP), and helped to build the capacity of partner groups locally and nationally. In the SNEP, FCR pioneered ways to conceptualize community assessment research and, using GIS technology, present socioeconomic and community capacity information. FCR continues to build on this methodology to assess and link the social and environmental conditions of Pacific West communities through projects including the Klamath–Siskiyou social assessment; the Northwest Economic Adjustment Initiative (NEAI) assessment, currently underway; and a planned analysis of land-use patterns as affected by social and demographic changes over the past ten years, using 2000 census and other data. As a result of FCR's work, social assessment methodologies have changed and policies have been influenced with far more attention paid to the relationship between resource use and social and economic well-being.

Projects with community-based groups in Oregon and California include education, dialog, and participatory research. FCR started the Lead Partnership Groups (LPG). A current project, All-Party Monitoring, brings diverse stakeholders, including distant environmental groups and scientists, together to monitor community-based natural resource projects. All-Party Monitoring work entails pilot projects serving seven groups performing grassroots work in a number of communities. Similar education, dialog, and participatory research activity will increase with the recent launching of the National Community Forestry Center's Pacific West Center housed at FCR.

Activities that advance the understanding of reciprocal relationships between ecosystems and communities are critical to integrating individual, community, and societal concerns. At this intersection, between ecosystem and community, there is a need to improve the processes that facilitate learning and the implementation of practices that achieve healthy, sustainable ecosystems and community well-being.

To address this need, FCR is developing its GIS capacity. Building this capability will provide a powerful tool for integrating and presenting social, economic, and biophysical data as we work toward bridging people and groups with different and competing ideas about social and natural resources. FCR will (1) develop GIS skills that advance our understanding of rural forest communities through spatial analysis and representation of social, economic, and ecological data; (2) integrate GIS into FCR project research, analysis, and presentations; and (3) provide GIS as a service and learning tool for participatory research to more effectively link and address ecological and social issues.

FCR is generating rudimentary maps to show the numbers and extent of community-based efforts in the Pacific West (see example, next page). These maps are being used to help researchers and practitioners quickly see the work taking place in the landscape, in terms of human and capital investments, as well as the impact of these efforts. Using data provided by individual community groups and agencies, this basemap will transform into an interactive database providing information about each project or group. Within a year's time, this map will be online and interactive, and will ideally link "ecosystem workers" with stewardship opportunities.

Northwest Economic Adjustment Region Study Counties and Communities

This map portrays the distribution of counties eligible for receiving Northwest Economic Adjustment Initiative (NEAI) investment as well as counties and communities that did receive assistance and that are part of this study.

As part of the NEAI, $1.2 billion was invested between 1994 and 2000, into projects in timber-dependent communities across the Pacific Northwest.

Communities shown on this map were selected as case studies for the NEAI Assessment. Note that graduated points indicate the level of investment directly into the communities. Some investments were made at the county or regional level, these data are not reflected here. Data for three communities was not available and are therefore not represented on this map.

An online version of this map will be used by researchers to manage and access the NEAI study community database. The database includes community demographic, social and economic data as well as project specific information.

This product will assist local stakeholders as they link natural, social, and economic systems in management scenario analysis, policies, and practice. This is essential if ecosystem management is to be built around principles of sustainability, damaged ecosystems are to be restored, land and resources conserved, and the environment is to be protected and preserved for future generations.

Through the Pacific West Center and other financial support, FCR is working with the Mountain Maidu, a Northern California Native population, to build organizational capacity. The Maidu Cultural and Development Group (MCDG) received a U.S. Forest Service stewardship pilot project grant to advance their use of traditional environmental knowledge (TEK). Plans are underway to use GIS to conduct analysis on stewardship lands (i.e., determining

how many acres comprise the defensible fuel corridors). The data for this project is acquired through exchanges with the U.S. Forest Service using a GPS unit to create data points and collect data to monitor change at the stewardship site and through MCDG's own TEK collection activities. Generating basemaps and using GIS for analysis will take place over the next twelve months. Within two years, GIS will serve as a learning tool for Maidu cultural development and will be used to communicate environmental knowledge. GIS will be integrally utilized by Maidu to monitor the effect of management practices over time and, ultimately, to restore social and economic roots as well as the health and integrity of the area's degraded landscape.

The Pacific West Center's work is based on participatory research. Participatory research

brings research and scientific methods into community-based development and ecosystem management practices. The path toward sustainable development in rural communities is predicated on combining conservation efforts along with ecologically sound development. FCR is building on earlier social assessment work by enhancing its database of social, economic, and ecological variables for 310 rural communities in northern California and southern Oregon (see map, previous page). FCR will use this GIS to communicate data with communities in the region to serve them better in order to help them evaluate the relationship of community health with resource use and habitat fragmentation. This project will entail a series of workshops in communities in the center's region. At public meetings, the latest demographic, social, economic, and environmental conditions of the communities will be portrayed using large format maps. These large format maps will be used for the presentations and then left in the communities for their own uses. These workshops will tie integrally with FCR's analysis of rural land use and habitat fragmentation using 2000 census and other data.

FCR pursues its mission by conducting research on rural community issues (e.g., natural resources); facilitating education and dialog to support community-based approaches to resource use; and providing training about the use of participatory research, evaluation, and monitoring strategies to address ecological and social issues.

FCR will assist natural resource professionals and communities to study, monitor, and assess the interconnections between community well-being, participatory community development and healthy, sustainable ecosystems. GIS will facilitate integrating this data as we study and share this information within the communities and with decision makers to impact policy, increase community capacity, and restore degraded landscapes. Equally important, through center activities, providing GIS services supports the establishment of the long-term relationships with communities and community-based organizations. This ensures community participation in natural resource issues as they affect community well-being and ecological health and sustainability.

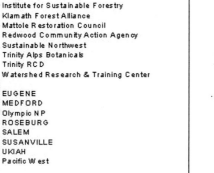

Experiments on the Land: Community Foresty Across the West

Focus on the Pacific West

USFS Stewardship Pilots
1   Antelope Pilot
2   Baker City Watershed
3   Four-mile Thinning
4   Grassy Flats
5   Maidu Stewardship
6   Pilot Creek Ecosystem Mgt.
7   Upper Glade LMSC

Ford CF Projects
A   Enterprise
B   Hayfork
C   Mount Angel

Lead Partnership Groups
▲   Alliance of Forest Workers and Harvester
▲   Applegate Partnership
▲   Applegate River Watershed Council
▲   Cherokee Watershed Group
▲   Collaborative Learning Circle
▲   Forest Community Research
▲   Forestry Action Committee
▲   Hoopa Community
▲   Jefferson Center
▲   Plumas Corporation, Feather River CRM
▲   Sacramento River Watershed Group
▲   Shasta-Tehama Bioregonal Council, QLG
▲   South Fork Trinity River CMG
▲   Sustainable Northwest
▲   Watershed Research & Training Center
▲   Willametter Valley Reforestation, Inc.

Collabative Lrng Crcle Prtps
●   Headwaters, Collaborative Learning Cirol
●   Hoopa Tribal Forestry
●   Institute for Sustainable Forestry
●   Klamath Forest Alliance
●   Mattole Restoration Council
●   Redwood Community Action Agency
●   Sustainable Northwest
●   Trinity Alps Botanicals
●   Trinity RCD
●   Watershed Research & Training Center

AMA
■   EUGENE
■   MEDFORD
■   Olympic NP
■   ROSEBURG
■   SALEM
■   SUSANVILLE
■   UKIAH
□   Pacific West

70   0   70   140   210   Miles

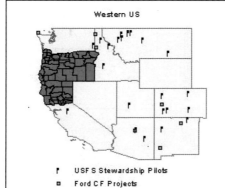

Western US

r   USFS Stewardship Pilots
□   Ford CF Projects

Forest Community Research

Created for the NCFC - Pacific West Center December 2000.

Forest Issues Group GIS Progress

Steve Benner   sbenner@jps.net

Forest Issues Group (FIG) uses GIS to analyze and, where necessary, appeal timber sales that are found to represent unsustainable forestry in the Tahoe National Forest area. Appeals of the Liberty (Sierraville District) and Bald (Downieville District) sales have been upheld, at least temporarily stopping what appears to be bad forest management. Use of herbicides on the Cottonwood Burn (Sierraville District) has been postponed for three years as a result of FIG intervention, among others. We are currently involved in advocating for protection for the areas included in the Clinton Roadless initiative, using ArcView GIS maps as a means of representing alternative scenarios for relating these areas to other sensitive areas on the Tahoe. We are following the progress of about fifteen Forest Service proposals on the five Ranger Districts of the Tahoe National Forest. Most of these projects require at least some amount of mapping to facilitate analysis and to inform our comments.

We have succeeded in halting two projects this year (Bald and Peak) based on their inconsistency with sustainable ecosystems management. We are continuing to challenge the application of herbicides to twenty thousand acres on the Cottonwood Burn based on the lack of need for chemical intervention. ArcView GIS was used to analyze and demonstrate the success of regenerating seedlings on the burn without the use of herbicides. We are currently tracking fifteen projects that may, if implemented, have negative impacts on ecological functions including loss of viability of sensitive wildlife species, degradation of water quality, conversion of vegetation species composition, or degradation of recreational opportunities. Some mapping is required for most of these projects.

The year 2000 has been a good one for those of us working toward reform of forest management throughout the Sierra Nevada Ecosystem.

The near completion of the Framework project, the submittal of petitions for federal protection for three important wildlife species (California spotted owl, Pacific fisher, mountain red-legged frog), and the agreement between the Forest Service and some California conservation groups to forego further ground disturbance pending completion of the Framework project are evidence that the public is becoming increasingly aware of forest management issues. On the Tahoe National Forest, FIG has participated in this trend by tracking all projects on our forest, deciding on a subset of these projects to pursue, participating in field trips, and supplying comments on the subset.

This year we have tracked about twenty projects. Comments have been submitted on fourteen of these. Formal appeals have been filed on nine of these projects, covering about 120,000 acres of National Forest lands. FIG appeals have been upheld (by the Forest

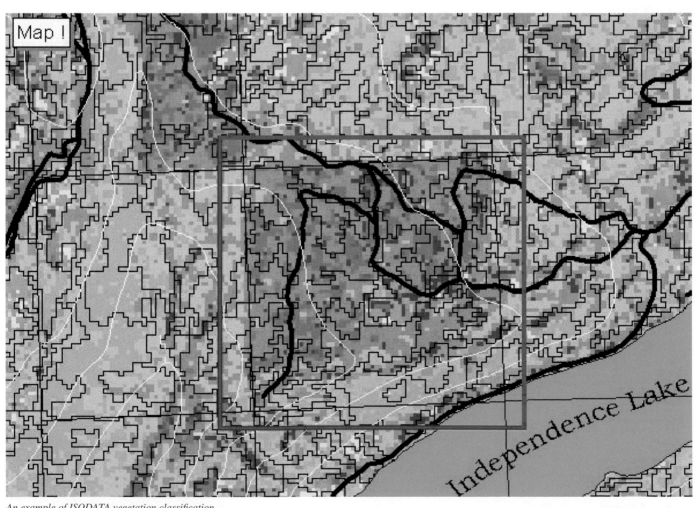

An example of ISODATA vegetation classification.

Service Reviewing Agent) in three cases. Agreement has been worked out through compromise in one other case, and one case has gone to litigation by FIG and others.

GIS has been successfully used in about half (seven) of the projects we tracked this year. We have found that resource data exists, often underutilized by the Forest Service, that can be represented on hard-copy maps to illustrate an argument for alternative management direction. An example of this is the FIG-produced map that showed polygons proposed for priority treatment with herbicides with an overlay data on conifer tree recovery for the same polygons. The data for tree recovery was "discovered" in a spreadsheet in the Forest Service planning file and would not otherwise have become public. In a situation in which project rationale was based on the need to ensure conifer success by applying herbicide, this data showed that there was no linear relationship between conifer success and proposed treatment priority. In other cases FIG has produced hard-copy maps that show spatial relationships between proposed treatments and critical habitat that would otherwise go unobserved.

FIG has found in general that well-supported data always leads to successful resolution of contested management proposals. GIS provides the tools to analyze this data, and to represent the result in hard-copy or digital maps.

As the lead GIS person for FIG I have also participated, with other FIG members, in the project of the California Wilderness Coalition to establish boundaries for potential wilderness areas in the Sierra Nevada Ecosystem in California. Areas that were identified as roadless under the RARE II program are being reconsidered for wilderness status. We have used GPS receivers to map new roads and trails that might influence wilderness boundaries, and we are in the process of redrawing suggested boundaries.

The FIG is also undertaking a change detection project utilizing Landsat TM data from a 1990 acquisition covering the Tahoe National Forest and the same coverage from Landsat 7. We are currently seeking funding for the Landsat 7 data for a year 2000 acquisition. This will provide an opportunity to study the change in vegetation (specifically conifer) cover over a ten-year period of very active timber

harvesting on the Tahoe National Forest. We have been using the ESRI extensions ArcView Spatial Analyst and ArcView Image Analysis to establish spectral signatures for several classes of canopy cover, with the help of partially ground-truthed data (from California Fish and Game Wildlife Habitat Relationship [WHR] Data and other sources) on conifer species and habitat types. The map on the previous page shows the classification results using the ISODATA function under ArcView Image Analysis overlaid with the WHR (observed/deduced) data represented by open polygons (hatch patterns would not remain transparent when converted to JPEG format). (More accurate vegetation data is being sought.) The red rectangle encloses a section (640 acres) of private land surrounded by National Forest that was heavily logged prior to 1990. We expect to see much more of this pattern in the 2000 acquisition.

Gifford Pinchot Task Force

David G. Jennings   gptf@olywa.net; 71634.127@compuserve.com
www.gptaskforce.org

The GPTF is a relatively small five-year-old grassroots forest protection organization based out of Olympia, Washington. Our focus is the forest ecosystems and biodiversity of southwest Washington. Our two major campaigns are defending the threatened roadless areas (RA) and ancient forests (AF) of the Gifford Pinchot National Forest (GPNF). The GPNF has a timber target of 62 million board feet (MMBF), with almost half of the volume to come from logging old growth stands. It is one of the four major timber producing national forests in the Pacific Northwest.

The GPTF works closely with, and provides GIS maps and/or technical support to, NW Ecosystem Alliance, Central Cascades Alliance, Wild Washington Campaign, Cascadia Defense Network, Pacific Crest Biodiversity Project, American Lands Alliance, Black Hills Audubon Society, and professors and students at the Evergreen State College.

Both campaigns—Wild Washington (Roadless Areas) and Protecting NW Ancient Forests—are gearing up to be in full swing these next two years. Educating and influencing public opinion on forest issues is vital if we are to maintain and expand our remaining landscape-level wildlife connectivity corridors and biodiversity hot spots. Easily accessible maps are key parts of larger/regional educational and advocacy outreach efforts in the campaigns to create an educated constituency for public forestlands protection in southwest Washington and with key decision makers in Olympia and Washington, D.C.

We have been tasked by the larger campaigns with generating and distributing a series of Ancient Forest and Roadless Area maps for distribution to local businesses located across the Gifford Pinchot Forest and surrounding metropolitan areas (rural service stations, bed and breakfast establishments, hiking/outdoor stores, and so on). The maps need to be large enough for someone to be able to translate the displayed Forest Road System to their own (purchased at the store) USFS brown line map. Other maps will be used for public outreach efforts and educational displays for decision makers, interested stakeholders, and the general public.

While we are a small group, thanks to a previous CTSP grant in 1997 and other support we have greatly increased our GIS capabilities in a relatively short time period. Our organizational budget is very small, and we have

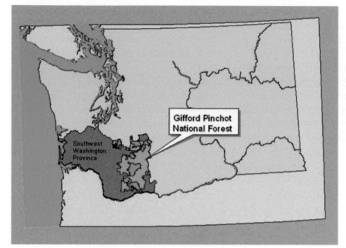

The Northwest Forest Plan established a system of "reserves" as the basis for both the continued logging of ancient forest stands elsewhere on federal public lands and as the basis for granting Habitat Conservation Plans (HCPs) to industrial timber companies in the pacific northwest.

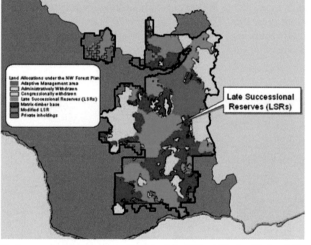

The late succesional reserves (LSRs) of the Gifford Pinchot National Forest.

This map produced by David Jennings for the Gifford Pinchot Task Force.

Data from the U.S. Forest Service and other sources

Technical & financial support provided by:

The W. Alton Jones Foundation Conservation Tech Support Program

The Gifford Pinchot Task Force is a non-profit citizen group dedicated to protecting forest ecosystems of southwest Washington.

The Gifford Pinchot Task Force can be reached at:

GPTF
PO Box 11427
Olympia WA 98508
(360) 753-4185
www.gptaskforce.org

December, 2000

recently transitioned from using part-time independent contractors to hiring full- time staff.

The GPTF also works closely with professors at the Evergreen State College to identify internship opportunities and provide subsequent mentoring of the students. One student used the work she did to develop her master's thesis.

Over the next two years we seek to improve and refine our ability to visually tell the forest fragmentation story. We see the use of large maps and posters of both maps and photos as very effective in working with rural residents and communities moving away from an extraction and into a more sustainable mode. Helping publicize GPNF's threatened areas across

the communities surrounding the forest will help us demonstrate that public opinion supports protecting roadless areas and the remaining remnants of our ancient forests.

A major role the GPTF is evolving for itself is GIS and mapping support/spatial analysis for the forest activist community of southwest Washington. This was true for the Cispus AMA victory, for the I-90 Land Exchange campaign and now again with the RA and AF campaigns.

Another role is providing the information in a live GIS presentation (laptop/LCD projector). The technology is new enough, and the subject matter is visually compelling so that almost every audience will be engaged and open.

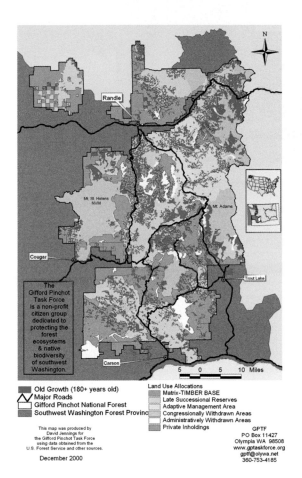

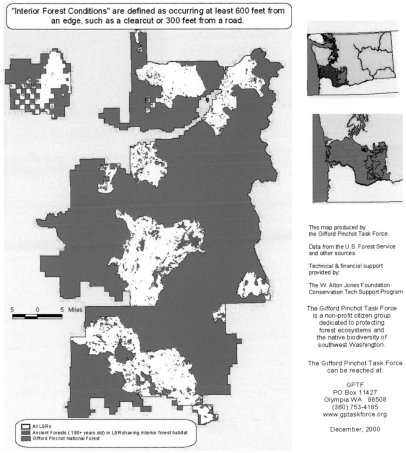

This map was produced by David Jennings for the Gifford Pinchot Task Force using data obtained from the U.S. Forest Service and other sources.

December 2000

GPTF
PO Box 11427
Olympia WA 98508
www.gptaskforce.org
gptf@olywa.net
360-753-4185

Legend (left map):
- Old Growth (180+ years old)
- Major Roads
- Gifford Pinchot National Forest
- Southwest Washington Forest Province

Land Use Allocations
- Matrix-TIMBER BASE
- Late Successional Reserves
- Adaptive Management Area
- Congressionally Withdrawn Areas
- Administratively Withdrawn Areas
- Private Inholdings

The Gifford Pinchot Task Force is a non-profit citizen group dedicated to protecting the forest ecosystems & native biodiversity of southwest Washington.

"Interior Forest Conditions" are defined as occurring at least 600 feet from an edge, such as a clearcut or 300 feet from a road.

This map produced by the Gifford Pinchot Task Force.

Data from the U.S. Forest Service and other sources

Technical & financial support provided by:

The W. Alton Jones Foundation Conservation Tech Support Program

The Gifford Pinchot Task Force is a non-profit citizen group dedicated to protecting forest ecosystems and the native biodiversity of southwest Washington.

The Gifford Pinchot Task Force can be reached at:

GPTF
PO Box 11427
Olympia WA 98508
(360) 753-4185
www.gptaskforce.org

December, 2000

Legend (right map):
- All LSRs
- Ancient Forests (180+ years old) in LSRs having interior forest habitat
- Gifford Pinchot National Forest

Old growth forest stands remaining on the Gifford Pinchot National Forest. Old growth forest stands are those with an average age of at least 180 years.

Existing ancient forest stands having interior forest habitat within the late successional reserves (LSRs) of the Gifford Pinchot National Forest.

There will be a variety of maps based on the topic, the site, and the issue at hand. We anticipate two basic forest level maps (RA, AF) similar to the hard copies submitted (but larger). Once these are finalized we will print a large quantity of each to be distributed to targeted businesses throughout southwest Washington. We will also have maps at a smaller scale that reveal where current/future logging threats exist. The ArcView GIS maps will be incorporated into a DTP format so that photos of the threatened areas can be included in the overall presentation to make the total package visually attractive.

We also see the concept of "interior forest habitat" as being a major ecological parameter/criteria for evaluating proposed management actions during future forest plan revisions. The GPTF would like to work with groups such as the Oregon Natural Resources Council and the Conservation Biology Institute to initiate similar edge impact analysis on other priority forests in the Northwest.

We think the big challenge to roadless areas will be protecting those roadless areas that fell outside of former President Clinton's Roadless Policy. On the GPNF, many of the key roadless areas from a landscape-level wildlife connectivity corridor perspective are complexes of small (1,000>×<5,000 acres) roadless areas. We have a serious challenge to keep timber sales from moving forward in those areas especially if the logging levels increase under the new administration. The maps reveal just how important those smaller roadless complexes are at the landscape level.

Mapping where ancient forest stands are part of the timber base and identifying their ecological and evolutionary values are crucial to helping tell the story of the need to end ancient forest logging.

Oregon Natural Resources Council

Erik Fernandez   ef@onrc.org, info@onrc.org
www.onrc.org
www.onrc.org/info/atlas/

The Oregon Natural Resources Council (ONRC) mission is to aggressively protect Oregon's wildlands, wildlife, and waters as an enduring legacy. Founded in 1974, ONRC's work has translated into the conservation of nearly 3.5 million acres of Oregon's most precious wildlands. We are a statewide organization headquartered in Portland with field offices in Eugene, Bend, and Chiloquin. The Oregon Wild Campaign strives to permanently protect approximately five million acres of forestlands with wilderness designation. Through GIS technology, ONRC is inventorying unprotected wildlands and working to pass wilderness legislation based on the updated inventory.

Clockwise from above: Big Bottom roadless—part of the Oregon Wilderness; Erik Fernandez,

ONRC's work has translated into protection of nearly 3.5 million acres of Oregon's most precious wildlands. In 1996, ONRC safeguarded 34,000 acres of Opal Creek Ancient Forest with wilderness designation. We also gained increased preservation of Bull Run Watershed, the drinking water source for more than one-quarter of all Oregonians. ONRC's efforts have led to letters from six city councils advocating for greater defense of their roadless areas and drinking water sources. These strategies have ushered in some of the most significant changes in forest policy in decades. Since the passage of the visionary 1964 Wilderness Act, only 13 percent of Oregon's national forestlands have been permanently protected from clear-cut logging and road building. The consequences have been devastating. Hundreds of miles of salmon streams have been damaged, driving the species that define our region to near extinction. Recovery efforts cost taxpayers millions of dollars.

Wilderness designation is the most effective means of preserving the cleanest sources of drinking water and forest habitat that nurture

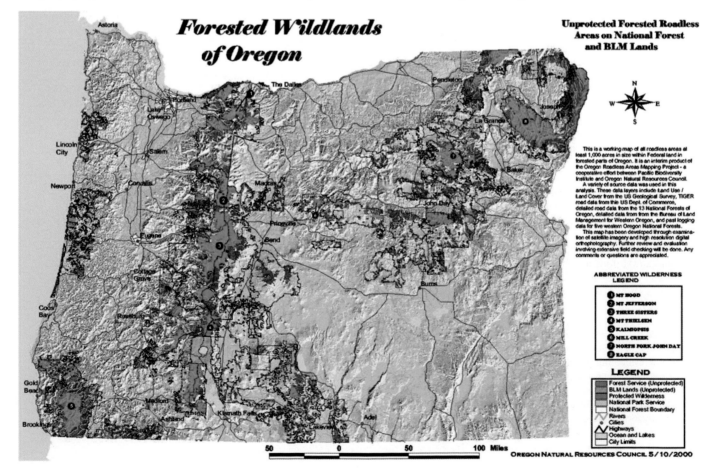

Forested Wildlands of Oregon

Unprotected Forested Roadless Areas on National Forest and BLM Lands

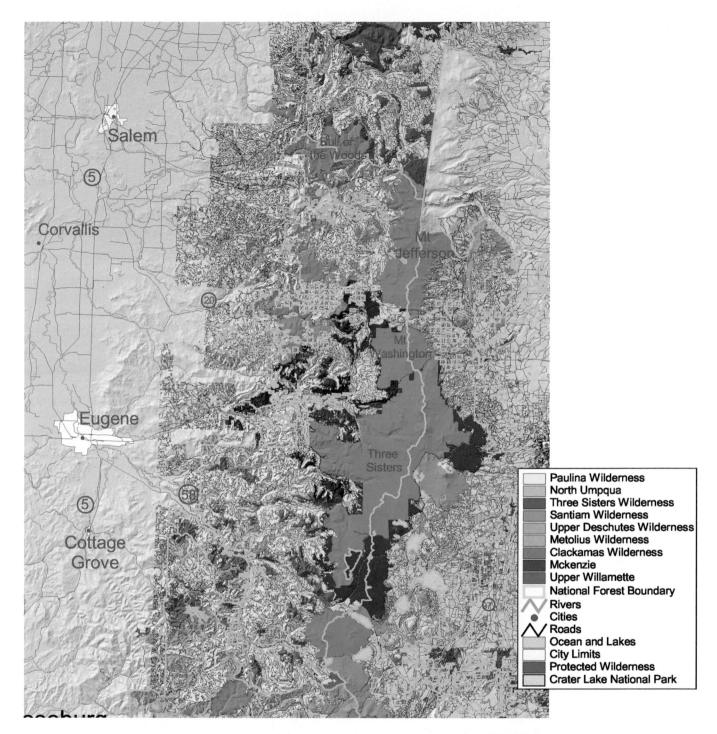

	Paulina Wilderness
	North Umpqua
	Three Sisters Wilderness
	Santiam Wilderness
	Upper Deschutes Wilderness
	Metolius Wilderness
	Clackamas Wilderness
	Mckenzie
	Upper Willamette
	National Forest Boundary
	Rivers
	Cities
	Roads
	Ocean and Lakes
	City Limits
	Protected Wilderness
	Crater Lake National Park

salmon, other forest-dependent species, and the human spirit. Wilderness is a straightforward way of permanently protecting our last pristine forests and can inspire citizens in a true grassroots campaign. In response to citizens' desire to leave a lasting legacy of clean water, abundant salmon, and ancient forests, ONRC is focused on permanent protection of forests in Oregon. Our dominant strategy is the Oregon Wild Campaign. The goal of the Oregon Wild Campaign is to protect five million acres in Oregon with wilderness designation and to permanently prevent clear-cut logging, road building, and other destructive activities on those lands. Many of these places are ancient forests providing salmon habitat

and drinking water. ONRC is working to inventory unprotected forested wildlands, build grassroots support for roadless areas, closely monitor remaining roadless areas, publicize existing threats to wilderness to build the case for permanent protection, and create a legislative proposal for wilderness protection. The Oregon Wild Campaign mobilizes grassroots support, fights for the long-term conservation of roadless areas, and promotes strong public support for wildlands preservation.

ONRC is an acknowledged leader in effecting changes in forest management policies, and we have built unprecedented support for protection of roadless areas 1,000 acres and greater.

More than 110 Oregon environmental organizations and 110 Oregon businesses endorse our Oregon Wild Campaign. More than 90 Oregon elected officials support the permanent protection of roadless areas. Our staff has an extensive and successful history as advocates for wilderness protection. Most recently, ONRC led the grassroots charge and helped lay the political groundwork for former President Clinton's decision to conserve more than 58 million acres of roadless areas including nearly two million acres in Oregon. Our challenge this year is to defend the roadless protection policy from attacks by the new administration while building broad support for wilderness.

GIS Program

ONRC has several GIS goals for our organization over the next two years. The most significant is the development of a wilderness proposal and subsequent wilderness bill. Once we have mapped the wilderness proposal boundaries, ONRC will begin work on an Oregon Wild Campaign book. Other goals include continued field verification and subsequent updating of wilderness boundaries, maps for presentation, and a PowerPoint presentation using GIS.

We are working with more than 100 organizations through the statewide Oregon Wilderness Coalition to create a wilderness bill. Local knowledge of the landscape is integral to having accurate boundary proposals. For this reason, we will travel throughout the state to meet with numerous other environmental groups, local activists, Forest Service representatives, and Bureau of Land Management employees to refine our wilderness boundary maps.

Once the wilderness proposal has been completed, we will work on a campaign book. The campaign book will highlight all of the proposed wilderness areas including maps of each area. In its current form, the book will include close to forty-five maps. The campaign book will require approximately two months of GIS work.

Staff planning meeting adjacent to the Metolius Wilderness Proposal Roadless area.

As a statewide organization involved in numerous facets of forest protection, ONRC is directly involved in extensive advocacy and outreach. GIS maps play a critical role in our advocacy and outreach. We use map displays at conferences, tabling events, press conferences, and meetings. We plan to use GIS maps and other visual tools in ONRC's outreach work in perpetuity.

In the past, we have used data from numerous sources for our projects and anticipate doing the same in the future. In addition, we have developed roadless area coverage for the state of Oregon, which ONRC will continue to refine into a wilderness proposal. We are also currently working on data sets for clear-cuts on federal land in Oregon and for old growth on federal land in Oregon. Our main sources of GIS data are from the U.S. Forest Service, the Bureau of Land Management, the Regional Ecosystem Office, Defenders of Wildlife, The Nature Conservancy, U.S. Geological Survey, GAP, U.S. Fish and Wildlife, Oregon Spatial Data Library, U.S. Census Bureau, Environmental Protection Agency, and many others.

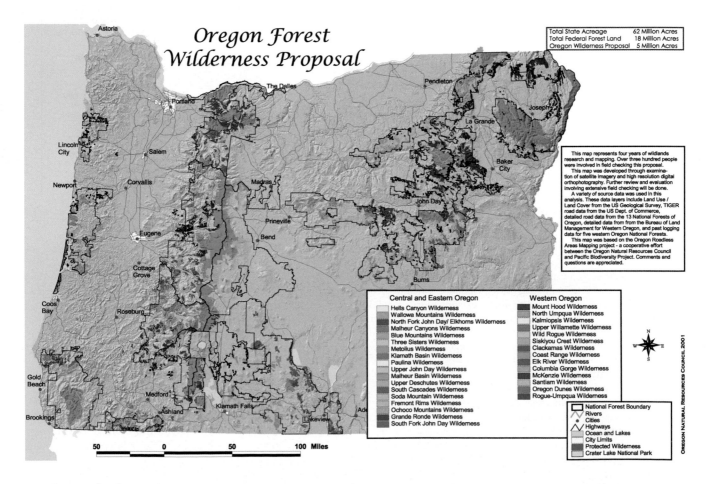

Conservation Design

Craighead Environmental Research Institute

Frank L. Craighead lance@grizzlybear.org
www.grizzlybear.org

Lance Craighead.

In the year 2000, the Craighead Environmental Research Institute continued work on several key aspects in the field of conservation planning. Our work continues to focus on the conservation of carnivores and their habitats. Conservation plans (or reserve designs) are a combination of art and science applied to landscapes and human activities. The goal of a conservation plan is to maintain viable populations of native species, in intact ecosystems, on a spatial scale large enough to maintain large carnivores, as human populations and activities increase. Human populations need space to grow and develop, but humans are extremely adaptable. We can direct our own growth, and in so doing we can conserve sufficient habitat for other species to coexist. Otherwise, we can continue to make the mistakes of the past: driving species to extinction and making the planet less inhabitable for people and wildlife.

In 2000 CERI completed one project: a reserve design for Ted Turner's Flying D Ranch, which was a pilot project for conservation planning on private lands, and we continued work on ten other conservation projects: (1) the Bozeman Pass–Interstate 90 Wildlife Crossing Study, (2) the Transboundary Reserve Design for Southeast Alaska and adjacent Canada, (3) the Coastal British Columbia Conservation Plan, (4) the Kakwa Park Grizzly Habitat and Movement Study, (5) the Conservation Area Design for the Wyoming Great Divide, (6) the Illegal Bear Mortality Investigation, (7) the Northern Rockies Conservation Area Design, (8) the B-Bar Ranch Wildlife Habitat Study, (9) the Conservation Area Design for the British Columbia Inland Rainforest, and (10) Predicting Impacts of Global Warming on Focal Species in Montane Ecosystems and designing adaptive conservation strategies. These are important pieces of larger conservation landscapes—the Yukon-to-Yellowstone and the western rain forest. To disseminate information, CERI continued its outreach, advocacy, and educational efforts.

At CERI we feel that the Yellowstone-to-Yukon region and the Pacific Northwest Coast are the two areas for highest conservation priority in North America at this time, in terms of the possibility of designing reserves large enough to maintain large carnivores and exceptional biodiversity and in the magnitude and immediacy of threats that these landscapes are facing. We still have time to make a difference in these areas.

At CERI we are also becoming more involved on a global scale. Lance Craighead was selected to be a member of the International Union for the Conservation of Nature (IUCN) World Committee on Protected Areas (WCPA) in 2000. This appointment should lead to a greater involvement and influence in protecting critical habitats for native plants and animals in North America and perhaps in other parts of the world. Lance completed his first book, *Bears of the World,* published by Colin Baxter Press (Europe) and Voyageur Press (North America). Lance and April Craighead spent seven weeks teaching conservation biology and related subjects to university students in Namibia (Southern Africa) in February and March 2001 in conjunction with the Round River Conservation Studies and the Cheetah Conservancy.

Project Summaries

The Bozeman Pass–Interstate 90 Wildlife Crossing Study

We have been planning this study for more than a year and had hoped to secure funding from the Montana Department of Transportation (MDOT). Although this funding has been further delayed, we need to begin fieldwork at this time to be able to provide timely input into the highway maintenance and construction process. The purpose of this project (see map below) is to determine where animals are crossing or attempting to cross Interstate 90 over Bozeman Pass. We hope to identify important crossing routes and develop strategies to make the highway less of a barrier

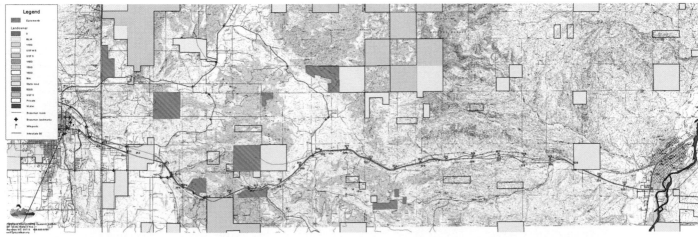

Bozeman Pass.

to wildlife movement. If MDOT can be convinced that certain areas are key movement routes for wildlife, they can allocate funds and plan to build crossing structures such as underpasses, overpasses, or elevated spans. To implement such design features requires that MDOT begin planning at least three years before construction begins.

Accordingly, CERI has begun a roadkill study along I–90 between Bozeman and Livingston, Montana. Although MDOT and the Montana Department of Fish, Wildlife, and Parks (FWP) collect some roadkill information, it is sporadic and location data is less than exact. To collect better data for this area CERI has assembled a team of volunteers who commute regularly between Bozeman and Livingston. These people will observe and record species and locations (by milepost) of animals killed along the highway. Locations of carnivores and other unusual animals will be verified by CERI personnel and pinpointed with GPS instruments. After snowstorms, winter track surveys will be done along the highway edges. All locations will be entered into a GIS mapping project. As the winter progresses a clear

picture should emerge as to which areas are most dangerous to both wildlife and motorists alike. During the summer, roadkill surveys will continue, and remote cameras will be used at important crossing sites to document animals crossing or attempting to cross the highway.

Transboundary Reserve Design for Southeast Alaska and Adjacent Canada

In partnership with Round River Conservation Studies and regional conservationists and scientists, CERI began work to complete a regional conservation area design (CAD) for the major watersheds of the greater Transboundary area—from Prince Rupert, British Columbia, to the Tatshenshini River. This regional design employs the principles of conservation biology and is complementary to designs being developed for other coastal and interior regions. The final conservation design will identify core areas that are critical to protect the resident grizzly bear and other carnivore populations, the ecological viability of the region, and the connective areas necessary to link the core areas and buffer areas surrounding the core and connective areas, so

that wildlife can continue to move freely without human barriers.

The conservation strategy will be based on the unique biological characteristics of each watershed in the region and will combine plans for each watershed into a final overall conservation area design. Recommendations within the strategy will identify significant ecological sensitivities and appropriate economic activities within each level of the conservation design (i.e., core, corridor, buffer areas). A major goal of the conservation strategy will be to provide ecological information, as appropriate, in support of short- and long-term legal, political, and media campaign efforts of the Transboundary Council.

One of the critical core areas for conservation is the Taku River watershed in northern British Columbia and southeast Alaska. Lance Craighead is helping to supervise a grizzly bear habitat and movement study on the Taku with Round River Conservation Studies. Lance visited the area in August 2000 as the study was getting started. In 2001, Lance will devote half of his time to the Taku Study and

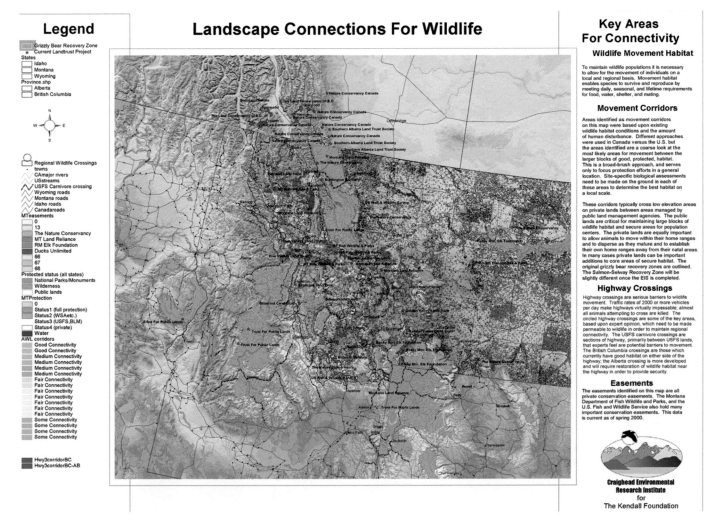

Northern Rockies Conservation Area design.

Transboundary Reserve Design. Fieldwork includes habitat mapping, remote camera and DNA sampling, and interviews and discussions to incorporate traditional knowledge of the Taku Tlingit people. The Taku drainage will be the focus of a more detailed conservation area design as part of the broader Transboundary CAD.

The Midcoastal British Columbia Conservation Plan

Similarly, CERI is collaborating with British Columbia conservation and research organizations to apply the techniques of GIS-based reserve design that we are developing in the Rocky Mountains to the problem of maintaining viable populations of grizzly bears and black bears (and thus, biodiversity) in coastal British Columbia. This analysis is based on vegetation; salmon spawning habitat; topography; and known distributions of grizzly bears, bear trails, and scent-rubbing trees. In coastal British Columbia, using a conservation area design developed by Round River Conservation Studies and the Canadian Rainforest Network as a basis, we are working with other conservation groups to further identify and protect habitat. We are cooperating with the Wildlands Project, the Raincoast Conservation Foundation, the Western Canada Wilderness Committee, the Valhalla Wilderness Society, First Nations (Native Americans), and the British Columbia Ministry of the Environment in

developing data layers, collecting field data, and designing a habitat suitability model for coastal bears. CERI assisted in investigations of two sections of the central British Columbia coast in 1998 and participated in a biological field camp on the Ecstall River along the northern British Columbia coast in 1999. Reserve design for coastal British Columbia includes such areas as the Spirit Bear Provincial Park and Wilderness, the Great Bear Raincoast, the Kitlope Wilderness, and others. Keeping these areas intact and maintaining habitat connections among them will also require protecting much of the remaining intact forest that is slated for industrial-scale logging. Due to the efforts of the Canadian groups, many of the intact watersheds were protected in early 2001; many others have had logging deferred for the next few years. In addition, a three-year moratorium was placed on grizzly bear hunting in British Columbia in part because accurate estimates of population size and trends are needed, especially along the coast.

In June 2000 Lance Craighead spent two weeks beginning a grizzly bear habitat and monitoring study on the Kowesas River for the NaNaKila Institute, a First Nations nonprofit conservation group located in Kitamaat Village, British Columbia. The Kitlope Wilderness Area and Kowesas drainage in British Columbia and surrounding coastal rain forest areas once supported highly productive bear

populations. However, due to largely unrestricted hunting, grizzly bears were almost completely extirpated from the Kitlope and surrounding areas. There is evidence that the long-term impacts of boat and aircraft assisted hunting, even in an unroaded drainage such as the Kitlope, leads to grizzly bear population decline and prolonged population depression.

At the present time, grizzly bears are beginning to return to the Kitlope. This is due primarily to the closure of the hunting season and management of the area by the Haisla band to prevent poaching and allow wildlife populations to come back into balance. Grizzlies are important to the Haisla as well as to the coastal ecosystem, and the Kitlope will continue to be managed to protect them. At the same time, pressures from the logging industry are threatening to damage and alienate grizzly habitat in surrounding areas such as the Kowesas watershed. The Kitlope alone does not have enough habitat to maintain a viable grizzly population over the long term of several hundred years.

Techniques to monitor bear populations, even highly invasive techniques such as radio tagging, are only marginally adequate. Therefore, the Kowesas study does not attempt to census all the grizzly bears in the area. We hope it will provide a relative index of the population trend over time and that it will document the

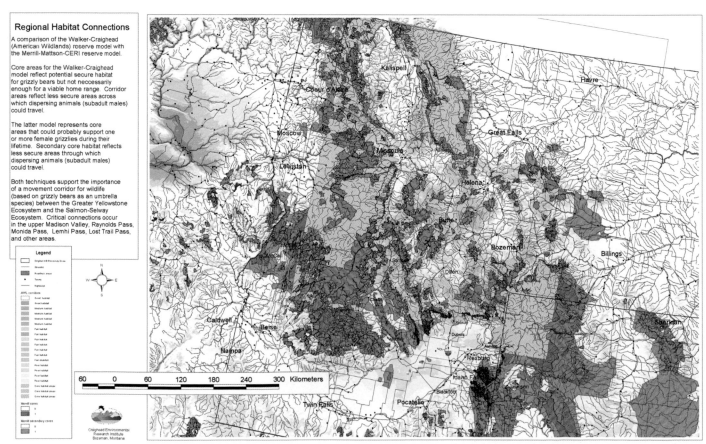

Grizzly Bear habitat model.

areas that are most important to grizzlies for food, security, and travel, at different times of the year. This information will enable the Haisla to modify their subsistence activities to minimize conflicts with the grizzly bears. The maintenance of a healthy and widespread grizzly population will also serve as an indication that the land management and resource utilization of the Haisla are being done in harmony with ecosystem functions and the maintenance of native species.

The Kakwa Park Grizzly Habitat and Movement Study

In interior British Columbia we are also cooperating with the Valhalla Wilderness Society and other groups of the Y2Y coalition on a field study to determine grizzly bear use and travel corridors in Kakwa Provincial Park. Lance and April Craighead are assisting Wayne McCrory with noninvasive fieldwork using remote cameras, DNA from hair samples, and track and sign surveys to determine the spatial context of grizzly bear use and to help plan the siting of future Kakwa Park facilities. Kakwa is adjacent to Mount Robson Provincial Park, which borders Jasper National Park. Grizzlies move throughout this region when they disperse, and bears from Kakwa may migrate seasonally over the Continental Divide to feed on salmon in the Fraser River drainage. Field data from this study will be an important part of regional conservation plans in the Canadian Rockies and will provide an empirical basis for this portion of the Y2Y Conservation Area Design. The Valhalla Society has developed maps of bear habitat classes, and CERI will assist in the incorporation of this data with movement data in the larger Y2Y conservation plan. Due

Kakwa research cabin.

to the rapid rate of industrial scale logging in this boreal forest, grizzly habitat needs to be identified and protected in a scientific manner to prevent severe fragmentation and the isolation of small populations. Future fieldwork led by Wayne McCrory will concentrate at higher elevations where female grizzlies and cubs are believed to be traveling.

Wyoming Great Divide

The Wyoming Great Divide encompasses the region south of the Greater Yellowstone Ecosystem surrounding the high basin of the Red Desert. The Wildlands Project identified this critical region as a critical link between the northern and southern Rocky Mountains in the United States (see map on opposite page). This area encompasses much of central and western Wyoming, northeastern Utah, and southeastern Idaho. Wolves, bears, elk, pronghorn antelope, and other wildlife species need to freely move from the Wyoming Tetons to the Targhee range of Idaho, through the Salt River Range and Commissary Ridge of Wyoming to the Wasatch, Uintas, and Book Cliffs of Utah; or from the Wyoming Absaroka range to the Wind Rivers, through South Pass to the Green Mountains, Shirley Basin to the Medicine Bow mountains of southern Wyoming and northern Colorado. It is at this scale that we need to plan for the long-term persistence of wide ranging species. In 2000 CERI began an important component of this effort in partnership with the Wildlands Project, Round River Conservation Studies, and Wild Utah.

Lance Craighead has been a scientific advisor during the development of an initial preliminary conservation area design for the region. At a finer scale CERI is working with Round River and Wild Utah on the vegetation classification of a central portion of the area using TM-7 imagery. This work is focusing on the delineation of riparian habitat. In collaboration with American Wildlands, CERI is developing a core corridor model of potential grizzly bear habitat using least-cost path analysis for the entire region.

Kakwa wilderness.

Illegal Bear Mortality Investigation

Solid evidence of illegal mortality of bears is very difficult to obtain. There are two primary components: (1) poaching (or deliberate killing) and (2) unreported accidental killing. In most jurisdictions, there is no penalty for accidental killing if it was done in defense of life or property, and the benefit of the doubt is usually in favor of the person reporting it, so that unreported accidental killings may be relatively low in number. Poaching, however, may be a significant source of mortality in many bear populations. Poaching may occur for several reasons. It is thought that most poaching takes place for personal economic gain, particularly for obtaining bear parts for trade in oriental medicine.

To estimate the extent and magnitude of illegal killing of bears and to try to pinpoint areas of special concern where it may have a significant effect on population stability of bears, CERI began an investigation of public and agency perceptions and a review of reported poaching in collaboration with the Valhalla Wilderness Society. As far as we know there has been no such systematic survey of local people and agency personnel to determine their perceptions.

We are designing and distributing a questionnaire that will quantify the perceived degree of illegal mortality, as seen by different constituencies who would most likely know of any evidence of illegal mortalities. Such people include hunting groups, state (or province) biologists, state (or province) game enforcement officers, federal biologists, federal game enforcement officers, fishing groups, conservation activists, wildlife photographers, ecotourism outfitters, big game outfitters, and others. Each group of people has varying biases regarding the severity of the problem and will accept different levels of evidence as convincing. We are evaluating these factors and comparing results among different groups of respondents to determine what we feel is an accurate measure of the situation in different regions of the western United States and Canada.

This initial survey is being conducted by April Craighead at CERI who has been hired to complete the project on a part-time basis. The survey focuses on brown (or grizzly) bears and black bears and will be targeted at known bear populations in Alaska, British Columbia, Montana, Idaho, and Wyoming. Results of the survey will be used to help design and implement antipoaching strategies in various areas. Compilation of available data, survey methods and implementation, and development of an effective antipoaching campaign for this region will continue throughout 2001.

The Northern Rockies Conservation Area Design and the Yukon-to-Yellowstone Initiative

A conservation area design for the U.S. Northern Rockies is a concept that has its roots among many organizations and individuals. Currently these diverse starting points have developed into strategies that are converging into a common effort under the guidance and coordination of the Yukon-to-Yellowstone Conservation Initiative (Y2Y) and the Wildlands Project. Science-based GIS concepts, legislative mandates, advocacy and educational approaches, and local or Statewide or regional interests are all integral to the realization of the vision: the scientific analysis, modeling, and implementation of a Northern Rockies Reserve Network that can maintain viable populations of native plants and animals (see map on page 43).

Advances in computer power and software have led to a new generation of ecological models. These models incorporate spatially explicit habitat information and known, individual behavior rules to predict the landscape-level dynamics of animal populations and their spatial distribution. CERI is working at the leading edge of technology to develop spatially explicit, individual-based computer simulation models for grizzly bear populations.

One approach uses object-oriented design techniques. A CERI research associate, Vickie Backus, is developing Java™ language programs to model the behavior of grizzly bears (and other species) as individual "objects" and simulate their probable movement using a robust behavioral rule base. Representing all individuals in the model allows the simulated population to be an accurate representative of an actual population, even in habitat that is currently unoccupied. This is important because even though all grizzlies share a common set of behaviors, they also possess behaviors unique to their sex and age. Furthermore, simulating individuals allows the effects of local interactions of bears with humans and bears with other bears to be included within the model. Since the model world is constructed using a habitat matrix derived from remotely sensed imagery of a known landscape, it will predict what habitat is important to the focal species and what habitat is avoided, inaccessible, or deadly. Through this simulation approach, we will produce resultant modeling techniques adaptable to a wide variety of focal species and different spatial locations and produce a powerful tool to assess the effects that changes in the landscape caused by such things as industrial development and global warming have on a given population of a focal species.

Another approach will use a Java-based model developed by Dr. Michael Gilpin, which predicts the movements of individual animals and the viability of a population in a spatially explicit landscape. This model will eventually be compatible with ArcView GIS software, the most widely used mapping application software, so that a spatially explicit population viability analysis can be done for any given population (we will help develop a grizzly bear model), and it can be viewed on a landscape map of the actual habitat. Dr. Gilpin is completing the final edits of the model this winter. CERI is seeking a graduate-level biologist and modeler who will then adapt the basic model to a grizzly bear population by defining the population parameters and the area in question. Both model approaches are expected to produce similar results; they were developed independently by different modelers, and we feel that by supporting both approaches the final results will be strengthened in areas where they agree.

The Craighead Environmental Research Institute continues work in refining a baseline reserve design map of the Northern Rockies between the Greater Yellowstone Ecosystem, the Northern Continental Divide Ecosystem, and the Selway–Bitterroot Ecosystem. This is a key piece of the Y2Y puzzle that must be solved to prevent the continued fragmentation of wildlife habitat that is resulting in small wildlife populations isolated on "islands" of habitat. If the puzzle is not solved by designing and protecting regional reserve networks, the inevitable result will be loss of species through local extinction events. Using GIS technology, we are able to store site-specific spatial habitat information about the landscape, locations of human developments, distribution of roads, and other data.

We synthesize additional data into a comprehensive design incorporating important features including undisturbed roadless areas, riparian areas and wetlands, old growth forests, and areas of local conservation interest. This analysis uses an iterative approach; results of each stage of analysis will feed back and incorporate additional data to fine-tune the original model. The general approach is to use large carnivore populations as an "umbrella" for biodiversity and add or subtract from that original model as additional data becomes available. Working with CERI, Troy Merrill completed a grizzly bear habitat suitability model for the Northern Rockies and then used a least-cost path analysis to examine the best habitat for movement between the four grizzly bear recovery areas: Greater Yellowstone, Salmon–Selway, Northern Continental Divide, and Cabinet–Yaak (see map, page 44). Lance

Craighead and coauthors were scheduled to present a paper at the 13th International Bear Conference in Jackson, Wyoming, in May that compares the results of this model with a previous least-cost path model (Craighead–Walker model) and with another model that is in the process of completion done by Helen Motter with CERI and American Wildlands.

As a part of the bigger picture, Lance Craighead has been working with the Y2Y Science Advisory Group to identify focal species to be used as the first step in developing a Conservation Area Design for the entire Y2Y region and for a smaller pilot study area. Lance received a contract from Y2Y to identify key species to be used in the CAD by combining the expert opinion of members of the advisory committee into a small list of species and habitats to be mapped. The results of this analysis will be presented at the annual meeting of the Society for Conservation Biology and will be submitted for publication to ensure peer review of the process. Results of this effort will inform our mapping and modeling efforts at finer scales, and those in turn will help refine the CAD at the total Y2Y scale.

CERI has also been involved in the Y2Y Metadata Framework Project that was funded in part by the National Biological Information Infrastructure. A comprehensive transboundary data set has been developed for the Crown of the Continent region stretching roughly from

Helena, Montana, to Banff, Alberta, Canada. This data, and metadata, will be archived at Montana State University in Bozeman, Montana. Initial data layers are:

DEM–U.S. and Canada
AVHRR
Cut Blocks
GAP
Extent of Imagery
Parcels
Reservations
Roads
Watersheds (HUC5)
Toponymy
Oil and Gas Wells
Geodetic Control

The B-Bar Ranch Wildlife Habitat Study: Integrated Ranching and Wildlife Conservation on Private Lands
For the past three years CERI has hosted a Conservation Area Design Workshop at the B-Bar Ranch in Emigrant, Montana, due to the generosity of Mary Anne Mott and Herman Warsh, the ranch owners. The B-Bar is located on the northwest border of Yellowstone Park at the head of the Tom Miner Basin in excellent grizzly bear habitat (see map below). Several resident grizzlies use the ranch at different times of the year, and dispersing wolves from Yellowstone have also visited. The ranch also helps support mountain lions and probably wolverines as well as smaller forest carnivores.

There have been no serious conflicts with predators on the B-Bar, but neighboring ranchers have lost dogs and perhaps livestock to wolves and have had unpleasant close encounters with grizzlies. CERI biologists have spent several days examining wildlife habitat on the ranch and watched a female grizzly with two cubs there for two days in May 2000. CERI prepared some preliminary maps of the ranch and the neighboring area in October and have assisted ranch personnel with their own mapping efforts. During the winter of 2001 we plan to conduct snow track counts on the ranch. Our goal is to help develop a cooperative wildlife plan with the B-Bar Ranch and its neighbors to ensure that ranching and wildlife habitat can continue to coexist. We feel that this area could be a showcase for ecologically sensitive ranching.

Tom Miner Basin lies on the northwest boundary of Yellowstone National Park. The Basin is home to a number of historic ranches and has a long agricultural tradition. It is also home to a great abundance of wildlife including grizzly bears and gray wolves. Although this could be a volatile mixture, conservation-minded landowners have a long tradition of coexistence alongside wildlife. Increasing economic challenges to ranchers, as well as increasing predator populations, could make coexistence less likely in the future.

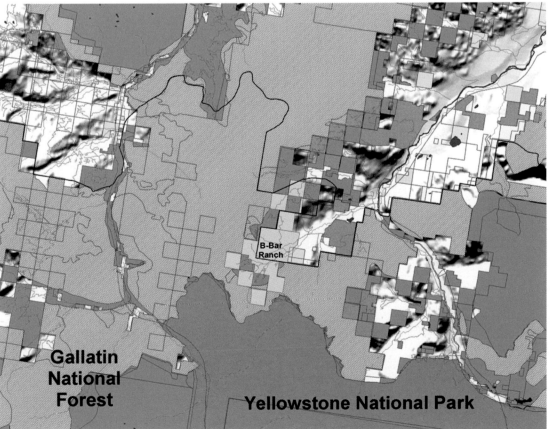

The B-Bar Ranch

The B-Bar Ranch occupies an important ecological position. It lies partly within the Grizzly Bear Recovery Zone of the Greater Yellowstone Ecosystem, and is adjacent to and includes critical roadless area habitat. It is an extremely important linkage between the protected area of Yellowstone National Park and the roadless areas of the Gallatin National Forest.

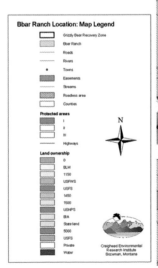

We plan to work cooperatively with Tom Miner Basin landowners on landscape-scale, long-term conservation plans that enhance wildlife habitat while ensuring the viability of individual ranches. We plan to:

- Determine presence and seasonal distribution of wildlife species of special concern (elk, grizzly bears, and wolves) in the Basin through field investigations as well as local interviews.

- Develop habitat maps for these species to improve our understanding of their seasonal requirements and movements.

- Evaluate Forest Service management plans and practices on surrounding public lands to project future spillover impacts on private lands due to habitat changes.

- Engage landowners in participatory mapping exercises to identify and ameliorate potential conflict sites between wildlife and ranch activities.

Our research methods will emphasize field surveys, habitat type sampling, and participatory mapping with landowners and land managers in Tom Miner Basin. Our methods will include:

- Survey and document presence and seasonal distribution of focal species, using track and sign surveys, automated cameras, field observation, and interviews with local people.

- Locate and map key habitat elements for focal species.

- Use geographic information systems technology to develop preliminary distribution and habitat maps.

- Convene local workshops to review preliminary maps with landowners and residents, comparing wildlife information with local expertise on land-use activities using the following techniques:

    - Overlay seasonal habitat use by focal species

    - Identify potential conflicts

    - Cooperatively design mitigation and proactive conflict avoidance strategies

- Analyze U.S. Forest Service management activities on adjacent public lands, assessing potential for cross-boundary habitat conflicts.

We are optimistic that these efforts will lead to successful long-term coexistence between people and the abundant wildlife of Tom Miner Basin. Our holistic, landscape-level approach ensures that we are addressing problems and conflicts at the right scale. We expect that this integrative, participatory effort will minimize conflicts and maximize the benefits of being neighbors to predators and other wildlife. Ultimately, this project should serve as a model for other places where agriculture and wildlife are in close proximity.

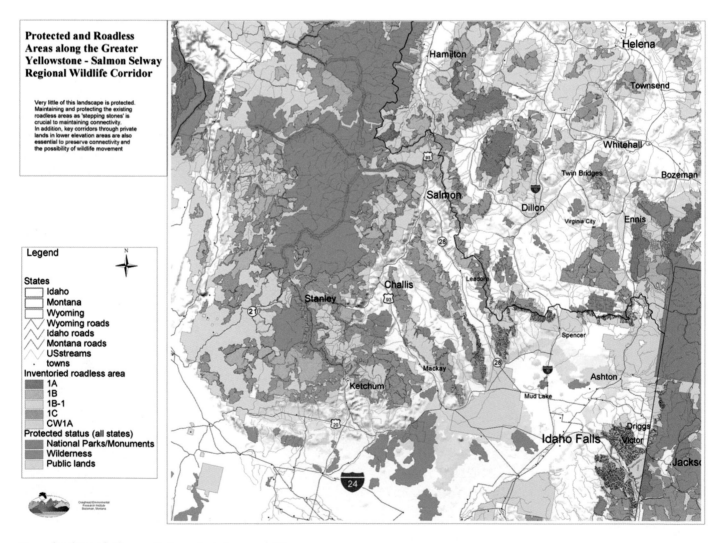

Example of corridor/connectivity analysis (see page 50).

A Conservation Area Design for the British Columbia Inland Rainforest

Because of our involvement with Wayne McCrory and the Kakwa Park grizzly study, CERI has also been asked to help develop a conservation area design for the Inland Rain Forest that overlaps Kakwa Park and extends south as far as northern Idaho and northwestern Montana. This area comprises a large portion of the Yukon-to-Yellowstone region but has received far less attention from biologists and conservationists. Conversely, because of greater conservation efforts in other regions, particularly coastal British Columbia, logging companies have concentrated on timber harvest at unsustainable levels in this region. This area contains the world's largest clearcut. A smaller clearcut, on the road to Kakwa Park, takes almost an hour to drive through. Conservation groups in the region are mobilizing to identify and map critical wildlife habitat and to guide development of the region so that connectivity and ecosystem functions are not greatly disrupted. CERI personnel will assist in developing data layers and analyzing wildlife habitat for incorporation into a CAD for this region.

Predicting Impacts of Global Warming on Focal Species in Montane Ecosystems and Designing Adaptive Conservation Strategies

Increases in atmospheric concentrations of greenhouse gases are altering global climate, and there is widespread agreement among scientists that this human-caused warming will continue. Global warming will have significant impacts on the earth's plant and animal communities. Changing climate will drive shifts in the distribution and diversity of plant and animal species as populations migrate to track suitable climate. If the rate of change is too rapid, some species will not be able to keep pace with climate change resulting in local extirpation of populations and species extinctions. Organisms that are sessile, such as plants, will be particularly vulnerable to rapid climate change.

Climate-driven changes in the distribution and viability of important plant taxa will also have significant implications for faunas that are dependent on specific plants for food or habitat. Climate change will also alter competitive interactions between species, change the distributions of pests and pathogens, and affect the frequency of disturbance events such as stand-replacing fires. As a result of these and other climate-driven changes, species, communities, and whole ecosystems once considered protected inside national parks and other reserves now face major threats to their survival. Changing climate may drastically reduce the carrying capacity of the Yellowstone region for these and other species. If these species can no longer rely on Greater Yellowstone as protected range, it is not clear that they can survive in the Rocky Mountains.

To try to maintain species and habitat, we must anticipate the potential impacts of climate change on Greater Yellowstone. During past episodes of rapid climate change, both plant and animal species responded by moving up or down in elevation, or by moving south or north in latitude, to keep pace with optimal climatic conditions. This response will be greatly hindered or completely blocked by current land-use activities of humans and by continued human alteration of habitat as the climate changes. To allow species to adapt to changing climate, we need to provide natural habitat connections for both altitudinal and latitudinal movements of plants and animals in response to changes in temperature, precipitation, and other factors. We need to leave these connecting habitats free of human alteration for long periods of time (thousands of years) to maintain adaptive options for native species. To do this we need to make land-use decisions now in order to protect regional "corridors." In particular, we need to maintain corridors to higher elevation areas and along the spine of the Rocky Mountains to allow such long-term movements. We need to identify spatially specific locations where a connected matrix of lands are possible.

To begin this work, we hope to model the potential impacts of climate change on three key members of Greater Yellowstone's fauna by predicting changes in vegetation, temperature, and precipitation. We will then use the results to identify critical "conservation corridors" and design conservation strategies and monitoring projects that can form the basis of a resilient, long-term conservation program for the region. The initial approach may entail spatially explicit models of potential changes in the distributions of three mammal species and their habitat: grizzly bears, pika, and bighorn sheep. Collectively, these three species occupy a range of habitat types threatened by climate change. Each represents unique conservation challenges and adaptive potential. Subsequent work will predict the effects of climate change on other key focal species. A workshop is being planned for the fall of 2001 in Bozeman to determine the best approaches to take.

General circulation models (GCMs) are a primary source for data on future climate change. Many of these climate scenarios are now readily available from the Data Distribution Center of the Intergovernmental Panel on Climate Change. In addition to GCM data, climate change studies have often used simple sensitivity tests to evaluate the responses of species to specific changes in climate variables. A range of probable future climates will then be used to develop correlative models and terrestrial biogeochemical models. Correlative models are based on relationships between a species and key environmental variables known to influence its distribution such as the relationship between absolute minimum temperatures and species mortality. They simulate a range of potential habitat scenarios as the response of

species and their habitat to climate change. Terrestrial biogeochemical models, such as MAPSS and BIOME3, are another type of model that has been used to simulate changes in the distributions of plant types in response to climate change. These models are more mechanistic than correlative models and include the physiological response of vegetation to increased atmospheric CO_2 concentrations.

These models can be used to predict responses of plants to climate change and the resultant response of focal wildlife species in the Greater Yellowstone ecosystem. Areas currently maintained by humans in an altered condition (e.g., agriculture, subdivisions, roads) will be assumed to remain in an altered state. Additional alterations will be predicted based on current trends in human population growth. Areas of high probability for maintaining native plant and animal communities will be identified that traverse complete altitudinal and latitudinal gradients. This information will be used to implement habitat protections in areas that will have the greatest long-term benefits for maintaining biodiversity.

Outreach, Advocacy, and Educational Efforts

Outreach and educational efforts have been struggling to keep pace with our research and analysis activities. We feel that this is a vital part of our conservation efforts. Scientific studies are of limited use when the results are confined to technical journals and professional meetings; we plan to increase our efforts to inform the public and agency managers of the needs and the threats of grizzly bear populations. In 1999 and 2000 we have acquired video film and editing equipment to be used to help explain how large-scale conservation area design is the only long-term solution for protecting populations of threatened grizzly bears and other wide-ranging wildlife species.

The CERI Web site is a valuable resource for education that was updated in December 2000. We are beginning a series of Web pages with information on all eight species of bears and their worldwide status and distribution. The basis for this work is the book, *Bears of the World,* that Lance Craighead completed in August and that was published in November 2000 as part of a series of wildlife books by Colin Baxter Press (in Europe) and Voyageur Press (in North America). The CERI Web site will soon contain links to groups and individuals that are actively involved in conservation of all these bear species.

In 2000 we purchased a video projector to use for presentations and workshops. This equipment received a great deal of use throughout the year. Much of this consisted of public presentations designed to educate the public about the behavior and ecology of grizzly bears and other wildlife, to inform people of their needs, and how to peacefully coexist with them. In addition to CERI presentations we loaned the projector to the Sierra Club, EarthJustice Legal Defense, American Wildlands, Yellowstone to Yukon, and other conservation groups.

In October of 1998, 1999, and 2000 we convened workshops at the B-Bar Ranch near Gardiner, Montana. Leading scientists and conservation advocates attended. The art and science of maintaining viable populations of native species, as human populations increase and habitat alteration accelerates, is a constantly changing field of expertise. New tools and information are being constantly developed. These meetings provide a forum for the exchange of ideas and technical approaches,

primarily designs using carnivores as umbrella species to determine the amount of land necessary and the available habitat that needs to be protected in order to maintain populations of species, including grizzly bears, wolverines, mountain lions, and so forth, and for implementation of those ideas through public outreach. The two goals of this workshop are to keep scientists abreast of new developments to increase their effectiveness, and make these critical efforts to preserve biodiversity accessible and understandable to the concerned public.

In 2001 we plan more public presentations, the preparation of written materials, the completion and distribution of video materials, development of additional Web-based educational

materials, magazine and television interviews, popular articles and books, and other educational activities. Currently, Lance Craighead is scheduled to give presentations and workshops for the Yellowstone Institute; the annual Sierra Club Writer's Workshop in Montana; the Yukon-to-Yellowstone Conservation Initiative Conservation Science Consortium; the International Bear Association Symposium in Jackson, Wyoming; the Society for Conservation Biology Annual Meeting; and many others.

Charles Craighead and Bonnie Kreps completed their documentary film on Mardy Murie, Arctic Dance, in the spring of 2000. Currently Charles is beginning work on a book version of the documentary.

Products and Implementation

Presently, corridor or connectivity analysis has been mandated in many state and federal

land management documents (see map on page 48). Corridor analysis is also a logical tool for compliance with the letter and spirit of the National Forest Management Act of 1976, the Endangered Species Act of 1973, and the National Environmental Policy Act of 1972. Wildlife movement corridors have been mentioned or broadly outlined in Environmental Assessments, Conservation Easements, and Forest Oil and Gas Leasing documents. The Transportation Equity Act for the 21st Century (TEA-21) contains language of "corridor preservation activities" that are integrated into joint community and transportation plans to reduce the environmental impacts of transportation. One of our objectives is to help standardize and coordinate techniques for identifying and protecting wildlife habitat on public lands

and to make the results of this analysis available and understandable to agency managers. Another objective is to facilitate wise stewardship on private lands with the cooperation of landowners, the use of conservation easements and other incentives, and the design of biodiversity reserves on private lands.

The Northern Rockies Conservation Area Design analysis forms the basis of outreach efforts and inputs into agency decision making processes. Important core and corridor areas are being identified and need to be maintained. As our level of knowledge increases, conservationists may be able to determine trade-offs and alternatives that can still maintain connectivity for wildlife throughout the region, but for now the most important consideration is to maintain options. The results of this project are guiding conservationists and land managers where to focus their efforts. We need to protect the areas we have identified as important linkages so their use is not precluded by unwise land management decisions. We need to educate the public and our civil servants as to their importance to wildlife and biodiversity. We need to keep them as biologically intact as possible until we can make wiser choices than we have in the past regarding human uses and wildlife needs.

Our Northern Rockies Conservation Area Design approach is based on habitat suitability models of key carnivore species, particularly grizzly bears. We are beginning to incorporate data on known routes of animal movements. We are communicating with, and planning to coordinate our analysis with, proposed work on historic ungulate migration routes in the southern GYE being developed by the Greater Yellowstone Coalition, the Wyoming Wildlife Federation, and the Jackson Hole Conservation. We are also in frequent communication with other grassroots conservation groups throughout the region such as American Wildlands, the Northern Rockies Conservation Cooperative, the Ecology Center, and others. The results of our analysis will be given to these groups as they become available for timely incorporation into activism, advocacy, and outreach efforts. Similar strategies, but with different advocacy groups, government agencies, and media sources are being developed for reserves on the Canadian side of the border. Key participants, in coastal reserves particularly, are Native Americans (or First Nations).

Implementation Accomplishments
Our educational and advocacy efforts are helping raise the Craighead Environmental Research Institute from a quiet, narrow-focused wildlife research group to a widely recognized authority on bears and an important player in conservation decisions. As a direct result of the advances we have made in our various conservation projects and the capacity building we have accomplished in our organization, we have attained a much higher profile and have become much more effective in our educational and advocacy efforts. Whereas previously we conducted research and presented our results primarily in scientific meetings and journals, we are now consulted by both the conservation community and the government land management agencies to help them make decisions concerning bears.

By 1999, with the project advances, support, and exposure afforded by unrestricted grants, we were invited to speak to a symposium of high school teachers in California by the prestigious California Academy of Sciences and to write a book on bear ecology and conservation. We contributed a book chapter in *Carnivores in Ecosystems,* published by Yale University Press. Our presentations at scientific and agency meetings were well received, and many of our ideas and recommendations

began to find their way into government documents and planning decisions. We were interviewed on television, radio, and by magazine and newspaper journalists. We testified in court cases and before Montana legislative committees. CERI scientists were asked to join scientific committees of the World Wildlife Fund, the Yukon-to-Yellowstone Conservation Committee, and the Wildlands Project. We were able to collaborate on important research projects with the Valhalla Wilderness Society and the Wildlands Project.

In 2000 we were asked to make recommendations to the Montana Department of Highways to make highways safer for bears. We attended several meetings scheduled with land trusts and their funders to help them decide which private lands have the highest priority for protecting bears and other wildlife. The U.S. Fish and Wildlife Service; the Montana Department of Fish, Wildlife and Parks; and the U.S. Forest Service invited us to participate in agency meetings to plan for protecting movement habitat so that bears can move to new areas. The Canadian Wildlife Service and the British Columbia Ministry of Environment, Wildlife Branch asked Lance Craighead to review the decisions regarding the hunting of grizzly bears in British Columbia. We continue to play an important role in planning the future of bear habitat and bear populations from the Rocky Mountains to the Pacific Coast Rainforest.

We plan to widen the scope of our cooperative efforts and to increase our office staff in the year 2001 so we can work more efficiently and scientists will have more time to devote to education and advocacy. We have acquired some key equipment to help us get the message out including a state-of-the-art video projector for presentations and workshops. We have completed the acquisition of a video editing system and will soon acquire a better digital video camera. We are also editing a film on conservation area design to help others understand the importance of these projects and the approach we are taking. We plan on increasing our expertise and our influence with the goal of ensuring that bears will be able to coexist, in perpetuity, with humans. We owe our present success in large part to the generosity and support of a key group of foundations to whom we give our heartfelt thanks.

How Our Efforts Will Be Measured
The successful completion of the CERI program will be measured directly by the development of conservation plan (reserve design) maps and data layers and the dissemination of this information to conservation groups, government agencies, local governments, tribal governments, and land trusts throughout the region. Success can also be measured indirectly by the degree of cooperation and communication that is established among the

participants as the project proceeds. Subsequently, the success of the project will be measured by how thoroughly the resultant reserve design is incorporated into agency land management policy, private land stewardship and easements, and public attitudes toward maintaining regional reserves on this grand scale. The ultimate test will be how many of the existing native species are still extant in these regions (the Rocky Mountains and the Pacific Northwest coast) after 500 years or more.

Recent Publications
Lance Craighead. 2000. *Bears of the World.* Colin Baxter Press, Edinburgh, Scotland. 132 pp.

F. Lance Craighead, M. E. Gilpin, and E. R. Vyse. 1999. Genetic Considerations for Conservation of Carnivores in the Greater Yellowstone Ecosystem. Chapter 11; In *Carnivores in Ecosystems,* ed. Clark, T., S. Minta, and P. Karieva. Yale University Press.

Lance Craighead. 1999. Contributor In: *Terrestrial Ecoregions of North America; A Conservation Assessment.* Rickets, T. H., E. Dinerstein, D. M. Olson, and C. J. Loucks, et.al. Island Press. 368 pp.

Horesji, Brian, Barrie Gilbert, and Lance Craighead. 1998. *British Columbia's Grizzly Bear Conservation Strategy: An Independent Review of Science and Policy.* Western Wildlife Environments Consulting Ltd. Calgary, Alberta. 64 pp.

Project Support
For our current projects we have been funded by the Turner Endangered Species Fund, the Maki Foundation, the Jackson Hole Community Foundation, the Wiancko Foundation, the Gilman Ordway Foundation, the Great Bear Foundation, Patagonia Inc., Hewlett–Packard Company, Environmental Systems Research Institute (ESRI), Jean Craighead George, Roland Dixon, Mary Anne Mott, the Scott Opler Fund, the Yukon-to-Yellowstone Conservation Initiative, the LaSalle Adams Fund, the Wilburforce Foundation, the Norcross Wildlife Foundation, Inc., the Henry P. Kendall Foundation, a private family foundation that wishes to remain anonymous, the New-Land Foundation, the Fanwood Foundation, and numerous individual contributors. For other cooperative projects in Canada for which CERI acts as a fiscal sponsor we have been funded by the Richard and Rhoda Goldman Fund, Sun Microsystems, Clayoquot Island Preserve, the Fanwood Foundation, the Tides Foundation, the Lynchpin Foundation, an anonymous family foundation, and numerous individual contributors. We often receive logistical support from Lighthawk.

LEGACY—The Landscape Connection

Robert Brothers, Ph.D.
Project Manager, LEGACY—The Landscape Connection
www.legacy-tlc.org

Left to right: Chris Trudel, Robert Brothers, Curtice Jacoby.

Our mission is to provide information for the protection and restoration of ecological integrity in northwestern California and southwest Oregon. We believe that this can best be done by integrating local knowledge and science. GIS is our means to do this.

Our GIS products are based on data gathered from scientists working for public agencies and universities or as private consultants. They are requested, refined, corrected, and added to by local citizens, landowners, and conservation groups.

Three fundamental kinds of GIS applications include the following:

- The creation of individual coverages (old growth forests and protected areas on federal land).

- Combining these into maps that illustrate a particular point (amount and percent of old growth protected).

- Analyzing the results of these combinations in terms of measured variables (acres of old growth protected on federal land versus total acres of old growth in all ownerships).

We are frequently called on by conservation groups to produce these kinds of issue-specific products for a particular watershed or larger area to provide support for a particular conservation and restoration project. This branch of our work we refer to as "GIS services."

However, there is another way of combining individual coverages that reaches beyond specific issues to larger questions that involve the interaction between many terrestrial and aquatic variables at landscape and regional scales.

This work we call our "Vision Map" because it seeks to develop a broad vision of the scope of ecological processes and a long-term vision of how the most crucial aspects of our region can be conserved and restored for the benefit of future generations.

We are developing the Vision Map at the request of conservation groups who are seeking ways to prioritize their current activities and integrate them into a long-term plan. For example, LEGACY's mapping of old growth and low road density revealed a crucial connection for wildlife moving between the Pacific Coast and redwood forests in the interior. This led other groups to help in the acquisition of corporate timberland in the Gilham Butte area and the development of conservation easements by adjacent landowners. Future Vision Map work will help establish similar conservation priorities in other areas so that they can be used to focus activities ranging from timber harvest plan appeals to road rehabilitation projects.

The ultimate success of this work is dependent on dialog and mutual learning between scientists and local citizens. The best scientific information must be presented clearly so its relevance can be seen, and so it can be changed or updated by local people who know the latest facts that are not revealed by remote sensing data.

Our region is blessed by an unusually high degree of citizen activism, involvement, and expertise in conservation and restoration, and it is still under severe threat by large-scale logging and other misguided development. By making science available to citizens, and citizens available to science, we will be able to facilitate the most effective use of our limited time and energy.

To best serve the active campaigns and long-term needs of conservation groups in northwest California and southwest Oregon, we have two primary programs between which our time is divided equally.

GIS Services

For the past five years LEGACY has been supplying GIS technical support and products to local citizens, landowners, public agencies, and conservation groups. We have developed a reputation as a friendly source of clear technical advice and assistance in the use of computers and GIS software, and as an accessible,

affordable producer of GIS coverages, maps, and analysis.

Technical Support

This includes GIS and computer technical support, the processing of data brought to us by others (clipping, projecting, file format conversions, statistical calculations), and printing poster-size 24- by 36-inch maps on our plotter. Our expertise in ARC/INFO® has helped many other groups and individuals make the best use of their ArcView GIS skills, and we are the only group with an ARC/INFO license in the North Coast region of California that makes its staff available to serve this function. We are active in a local user's group, Mattole Area GIS (MAGIS), along with GIS staffers from Ancient Forest International, Institute for Sustainable Forestry, Mattole Restoration Council, and Trees Foundation, among others, and plan to serve as a data hub for this group.

Our most advanced computer (a Hewlett-Packard Kayak xu) allows us the technological capacity to operate complex processes on fairly large regional data sets and images. In 1999, we were able to expand our previous 1:100,000 data by acquiring and assembling important 1:24,000-scale data (e.g., roads, streams, and ownership) within our primary area of focus—23.5 million acres in northwest California and southwest Oregon (from Point Reyes to Port Orford, and east to Klamath Lake). Currently we have more than 25 GB of data covering California and Oregon with scales ranging from 1:2 million to 1:24,000. Our other two GIS-capable computers can handle the smaller data sets that a few local mapping projects require.

GIS Products

These are based on data gathered from scientists working for public agencies, universities, or as private consultants. They are requested and may be refined, corrected, and added to by local citizens, landowners, and conservation groups.

Examples of fundamental kinds of GIS products and services include the following:

- The creation of individual coverages (old growth forests and protected areas on federal land).

- Combining these coverages into maps that illustrate a particular point (amount and percent of old growth protected).

- Analyzing the results of these combinations in terms of measured variables (acres of old growth protected on federal land vs. total acres of old growth in all ownerships).

We are frequently called on by conservation groups to produce these kinds of issue-specific products for a particular watershed or larger area in order to provide support for a particular conservation and restoration project. These products are also useful in our Vision Map project.

Arcata Office—Curtice Jacoby, Chris Trudel, and Robert Brothers

Ongoing Projects in 2001
- Digitizing Potential Wilderness and Wild Rivers Boundaries in Northern California (see map at right)

This work for the California Wild Heritage Campaign office in Davis, California, involves heads-up digitizing from ground truthed USGS quadrangle maps for the California Wild Heritage Campaign to support legislation sponsored by Senator Barbara Boxer (D–CA) in the spring of 2001. Following through on previous work, we will digitize at least two additional national forests.

- Eel River Watershed Maps

Californians for Alternatives to Toxins (CAT) in Arcata, California, requested LEGACY to compile a set of basic maps to guide their campaign to track the locations of toxic sources of water pollution.

Planned Projects in 2001
- Compiling a statewide Coverage of Potential Wilderness and Wild Rivers, and Analyzing Their Ecological and Economic Impacts

Project Manager Robert Brothers is coordinating the work of seven GIS specialists (independent contractors or staff of other groups) to digitize potential wilderness areas and wild rivers throughout the state of California for the California Wild Heritage Campaign. This information will then be assembled by LEGACY into one statewide coverage, and maps of the whole state and particular regions or watersheds will also be produced, as needed. Analysis of impacts

will be essential to support the pending legislation. At least five person-days per month from April to August 2001 are expected to be spent. If possible, we hope to be able to hire an additional staff member to assist with this project, which may also extend beyond August, depending on the progress of the legislation and other factors.

- Mapping Water Diversions in Napa County

To protect stream flows for fish and other aquatic organisms, we have received a grant to map the location of each site where water is diverted from streams for use by vineyards and other agricultural uses. This work is based on data provided by the Watershed Associate and will be used at public meetings to develop action plans to restrict additional development in heavily impacted subwatersheds.

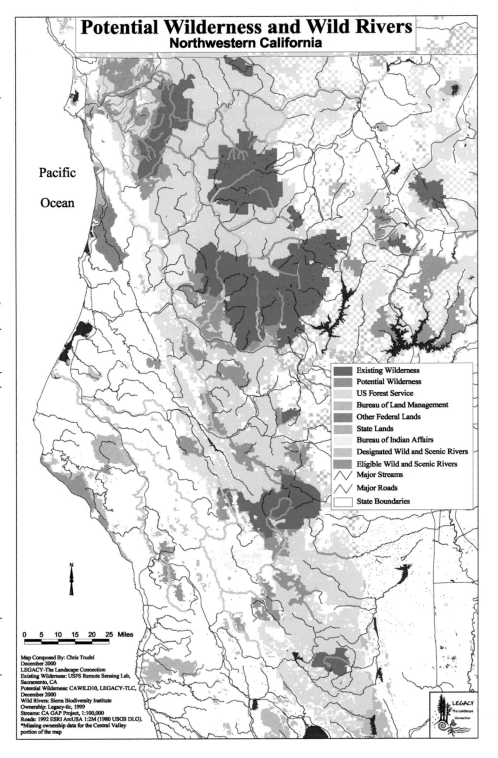

Potential Wilderness and Wild Rivers
Northwestern California

Pacific
Ocean

0 5 10 15 20 25 Miles

N

Legend:
- Existing Wilderness
- Potential Wilderness
- US Forest Service
- Bureau of Land Management
- Other Federal Lands
- State Lands
- Bureau of Indian Affairs
- Designated Wild and Scenic Rivers
- Eligible Wild and Scenic Rivers
- Major Streams
- Major Roads
- State Boundaries

Map Composed By: Chris Trudel
December 2000
LEGACY-The Landscape Connection
Existing Wilderness: USFS Remote Sensing Lab,
Sacramento, CA
Potential Wilderness: CAWILD10, LEGACY-TLC,
December 2000
Wild Rivers: Sierra Biodiversity Institute
Ownership: Legacy-tlc, 1999
Streams: CA GAP Project, 1:100,000
Roads: 1992 ESRI ArcUSA 1:2M (1980 USGS DLG).
*Missing ownership data for the Central Valley
portion of the map

Mendocino Office—Linda Gray

Ongoing Projects in 2001
- Road Assessment and Erosion Control for Greenfield Ranch

Greenfield Ranch
Greenfield Ranch Association received $24,000 of SB271 funding from the California Department of Fish and Game (CDF&G) to develop an erosion control plan on a total of 35 miles of roads (active roads and 30-year-old abandoned logging roads) within the Eldridge Creek watershed. Pacific Watersheds Associates (PWA) and New Growth Forestry are working with the Greenfield Ranch Association to accomplish this task and submit the plan to CDF&G for implementation funding. Maps with DOQQs and coverages (roads, fish-bearing and seasonal streams, culverts, the Eldridge Creek watershed boundary, and Greenfield boundary) were created as part of the Road Assessment and Erosion Control Plan to improve the salmonid habitat within the 5,400 acre Greenfield Ranch.

- Planning

Fourteen landowners of private inholdings within the Mendocino National Forest are working with the Institute for Sustainable Forestry to certify each ownership for sustainable timber harvests. About 250 coverages have been created to accomplish this work. Funding is by GIS service payments from landowners upon periodic delivery of completed maps.

- Conservation Easement Planning for Wildlife Connectivity

Members of the Church of the Golden Rule, owners of the 5,000-acre Ridgewood Ranch, are exploring the possibility of acquiring a conservation easement to protect their land for future generations. Implications for wildlife connectivity make this project especially significant since a ten-foot diameter culvert on this property allows large animals to move under the major barrier of Highway 101. Sixteen digital orthophoto quadrangles will be combined to show the relationship

of the ranch to surrounding development. Funding is by a grant from the Mendocino Land Trust, which is working actively with the owners.

- Pacific Coast to Mendocino National Forest Wildlife Corridor

Twenty-five coverages are being developed for working with private landowners, agencies, land trusts, and other conservation organizations in Mendocino County to raise awareness of the existing potential for an east–west running wildlife corridor from the Mendocino coast to the Mendocino National Forest. It is funded in part by a grant from the Mendocino Land Trust for the Ridgewood Ranch conservation easement, but mostly compiled from volunteer labor. Twenty days have been worked so far, but no completion date is expected any time soon as this project will be continuously refined over time.

- Linkage Potential for Central Mendocino County (see map below)

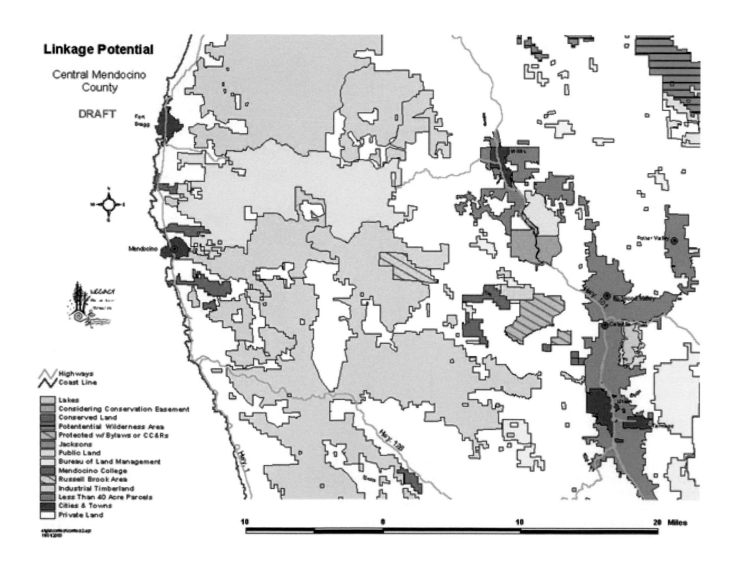

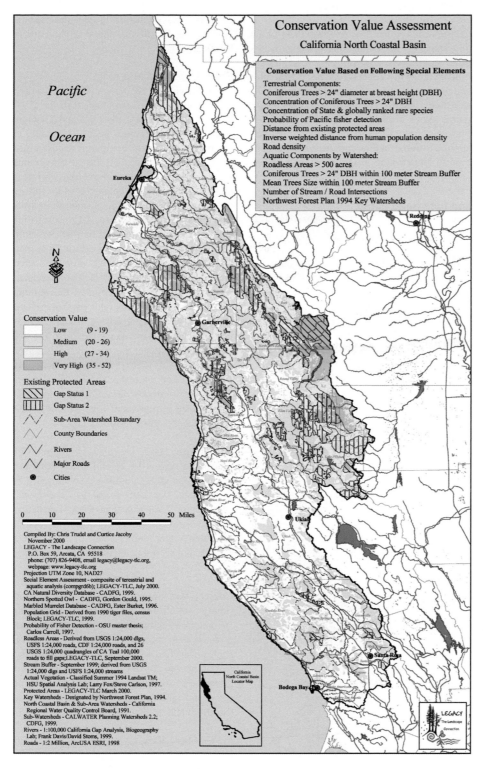

Conservation Value Assessment

California North Coastal Basin

Conservation Value Based on Following Special Elements

Terrestrial Components:
Coniferous Trees > 24" diameter at breast height (DBH)
Concentration of Coniferous Trees > 24" DBH
Concentration of State & globally ranked rare species
Probability of Pacific fisher detection
Distance from existing protected areas
Inverse weighted distance from human population density
Road density
Aquatic Components by Watershed:
Roadless Areas > 500 acres
Coniferous Trees > 24" DBH within 100 meter Stream Buffer
Mean Trees Size within 100 meter Stream Buffer
Number of Stream / Road Intersections
Northwest Forest Plan 1994 Key Watersheds

Pacific

Ocean

N

Conservation Value

Low   (9 - 19)
Medium   (20 - 26)
High   (27 - 34)
Very High (35 - 52)

Existing Protected Areas

Gap Status 1
Gap Status 2
Sub-Area Watershed Boundary
County Boundaries
Rivers
Major Roads
Cities

Eureka
Garberville
Ukiah
Santa Rosa
Bodega Bay
Redding

0   10   20   30   40   50 Miles

Compiled By: Chris Trudel and Curtice Jacoby
November 2000
LEGACY - The Landscape Connection
P.O. Box 59, Arcata, CA 95518
phone: (707) 826-9408, email legacy@legacy-tlc.org,
webpage: www.legacy-tlc.org
Projection UTM Zone 10, NAD27
Social Element Assessment - composite of terestrial and
  aquatic analysis (compgrd6b); LEGACY-TLC, July 2000.
CA Natural Diversity Database - CADFG, 1999.
Northern Spotted Owl - CADFG, Gordon Gould, 1995.
Marbled Murrelet Database - CADFG, Ester Burket, 1996.
Population Grid - Derived from 1990 tiger files, census
  Block; LEGACY-TLC, 1999.
Probability of Fisher Detection - OSU master thesis;
  Carlos Carroll, 1997.
Roadless Areas - Derived from USGS 1:24,000 dlgs,
  USFS 1:24,000 roads, CDF 1:24,000 roads, and 26
  USGS 1:24,000 quadrangles of CA Teal 100,000
  roads to fill gaps;LEGACY-TLC, September 2000.
Stream Buffer - September 1999; derived from USGS
  1:24,000 dlgs and USFS 1:24,000 streams
Actual Vegetation - Classified Summer 1994 Landsat TM;
  HSU Spatial Analysis Lab; Larry Fox/Steve Carlson, 1997.
Protected Areas - LEGACY-TLC March 2000.
Key Watersheds - Designated by Northwest Forest Plan, 1994.
North Coastal Basin & Sub-Area Watersheds - California
  Regional Water Quality Control Board, 1991.
Sub-Watersheds - CALWATER Planning Watersheds 2.2;
  CDFG, 1999.
Rivers - 1:100,000 California Gap Analysis, Biogeography
  Lab; Frank Davis/David Stoms, 1999.
Roads - 1:2 Million, ArcUSA ESRI, 1998

California
North Coastal Basin
Locator Map

LEGACY
The Landscape
Connection

already acquired) will continue to be used for this project, including such data as streams, roads, watersheds, vegetation, DEMs, public lands, industrial timberlands, and the California Natural Diversity Data Base (NDDB), which is useful in identifying rare, threatened, and endangered species that might also be protected in the process. LEGACY has already acquired DOQQs for seven of the USGS 7.5' quadrangles in this area, but at least another 11 are needed. These seem to become available at no charge from time to time through various agencies. (This work will be ongoing, depending on funding and other support.)

Biodiversity Vision Map for the California North Coastal Basin

Introduction

The protection of ecological integrity and biodiversity requires both description of what exists and prescription for how to save it. For LEGACY, this translates into an interrelated, two-phase program that we call our Vision Map, since it seeks to develop a broad vision of the scope of ecological processes, and a long-term vision of how the most crucial aspects of our region can be conserved and restored for the benefit of future generations. Based on input from scientists and conservation groups, LEGACY's GIS analysis is guided by the principles of conservation biology.

Regional Focus

Over the last five years we have developed data sets for a 23.5 million acre region in northwest California and southwest Oregon, thus we are prepared to provide GIS services to conservation groups throughout this region. A Conservation Assessment for 10 million acres of the Klamath–Siskiyou portion of this region has been prepared by Strittholt and Noss (1999). Building on this work, we have chosen to focus our Vision Map work on the North Coast region that borders the Klamath–Siskiyou to the south and west. Our offices in Humboldt and Mendocino counties are located in this region, thus our links with the scientific and conservation communities are strongest there.

The California North Coastal Basin is a hydrologically defined region of northwestern California comprised of all westward-draining watersheds of the northern coast ranges from Bodega Bay in Sonoma County to the mouth of Redwood Creek in northern Humboldt County (see map this page). From north to south, these include Redwood Creek, Mad River, Van Duzen River, Eel River, Bear River, Mattole River, Usal Creek, Wages Creek, Ten Mile River, Noyo River, Big River, Albion River, Navarro River, Garcia River, Gualala River, and Russian River.

The main goal of this project is to continue outreach to private landowners, conservation organizations, and agency personnel to raise awareness of the potential that currently exists for an east–west trending wildlife corridor made up of protected private lands and various types of public land in central Mendocino County. As population growth from Ukiah, Redwood Valley, and Potter Valley moves north while Willits population growth moves south, there is just a small area where the two growth centers have not yet met. By identifying, on a map,

the thousands of acres of land already protected either by federal, state, or local government or protected by private ownership through conservation easements, CC&Rs, or nonprofit corporations, while at the same time identifying land already lost to subdivisions, it is clearly visible where the high-priority areas for immediate conservation are. Many of the GIS coverages needed for the linkage project are ownership dependent and must be created from Mendocino County Assessor parcel data. North Coastal Basin data (which LEGACY has

Special physiographic features of the CNCB include:

- The largest stands of old-growth redwood forest on earth.

- More than 400 miles of relatively wild coastline.

- Nine major watersheds and many of the last populations of native salmon in California.

Vision Map Process

Conservation Value Assessment (CVA) is the first stage of the Vision Map process. It has three major parts, combining ecologically significant areas from Special Elements Mapping, suitable habitat from the Focal Species Analysis, and underrepresented subseries vegetation types from Representation Analysis. These are reviewed, ground truthed, and then applied to conservation work through the Community Networking Program. As one of our partners, The Wildlands Project describes it, "Networking People to support a Network of Wildlands."

Special Element Mapping

Previous work in the year 2000 has resulted in the completion of version 1.0 of the CVA, based on a process called Special Element Mapping (see map this page). All lands in the north coast were compared according to a number of variables of crucial importance to ecological integrity. Terrestrial variables included older forests, density of older forests, occurrences of rare species (State and global rankings), secure habitat for large carnivores and ungulates (as estimated by distance from human population and road density, see map next page), closeness to existing protected areas, and the probability of Pacific fisher detection. Aquatic variables included the number of road/stream crossings (as a measure of sediment input), the percent of roadless areas larger than 500 acres per watershed, the percent of riparian zones that are forested, the average size trees in riparian zones, and federally designated key watersheds. Based on scores from these combined variables, the north coast was divided into areas of low, medium, high, and very high conservation value.

In the year 2001, we plan to submit this process to review by scientific advisors and conservation groups, adding new variables as appropriate. We also intend to develop the other major components of CVA, Focal Species Analysis, and Representation Analysis, so the three major components of a complete CVA will be completed over the next three years.

Conservation Network Design is the second stage of the Vision Map process that contains specific management recommendations about how to group areas with various conservation values together to form an interconnected system of core conservation areas (such as wilderness), stewardship zones for sustainable forestry, and landscape linkages between these (Strittholt and Noss, et al. 1999).

In October 2000, we developed a draft map of management recommendations for the North Coast region in preparation for a major conference on regional connectivity, the "Missing Linkages" conference, sponsored by The Nature Conservancy, the California Wilderness Coalition, the Zoological Society of San Diego, and USGS. This map applied the conservation biology design of core reserves, sustainable use zones, and landscape linkages between them and has been developed by Reed Noss and others.

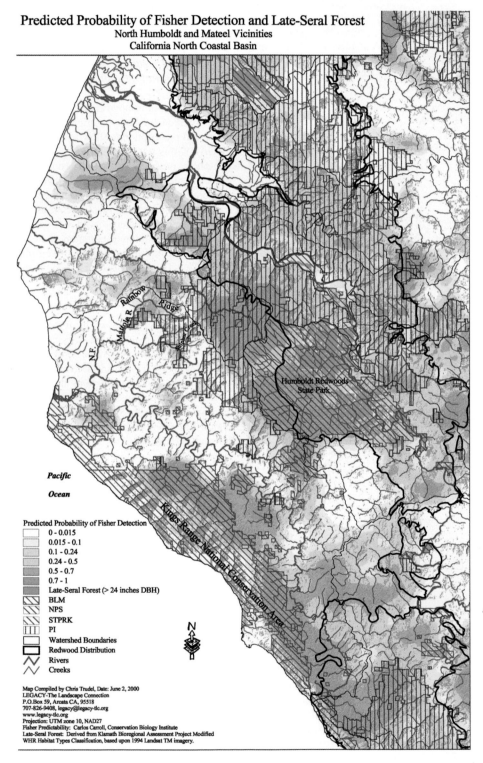

Predicted Probability of Fisher Detection and Late-Seral Forest
North Humboldt and Mateel Vicinities
California North Coastal Basin

Predicted Probability of Fisher Detection
- 0 - 0.015
- 0.015 - 0.1
- 0.1 - 0.24
- 0.24 - 0.5
- 0.5 - 0.7
- 0.7 - 1
- Late-Seral Forest (> 24 inches DBH)
- BLM
- NPS
- STPRK
- PI
- Watershed Boundaries
- Redwood Distribution
- Rivers
- Creeks

Map Compiled by Chris Trudel, Date: June 2, 2000
LEGACY-The Landscape Connection
P.O.Box 59, Arcata CA, 95518
707-826-9408, legacy@legacy-tlc.org
www.legacy-tlc.org
Projection: UTM zone 10, NAD27
Fisher Predictability: Carlos Carroll, Conservation Biology Institute
Late-Seral Forest: Derived from Klamath Bioregional Assessment Project Modified
WHR Habitat Types Classification, based upon 1994 Landsat TM imagery.

These conservation plans usually rely on publicly owned rather than private lands since dedicating them to conservation purposes is less controversial. However, since the north coast is dominated corporate timberlands, the dedication of these areas must be fully considered. As it turns out, many acres of corporate timberlands score among the highest in conservation value regionwide. Therefore, LEGACY's Conservation Network Design points to the establishment of core reserves of these lands. The relatively few acres of publicly owned lands in the region also score high, but protection of these lands alone may not be enough to fully protect the ecological integrity of the north coast. Of course the precedent for acquisition of private lands for public purposes is strong in the north coast with the Headwaters Forest Reserve being the prime example. And recently the Owl Creek grove was purchased by the state from Maxxam.

In the years ahead, a new version of the Conservation Network Design will be prepared to accompany each new version of the Conservation Value Assessment to reflect the implications of refinements in CVA for management of the conservation and restoration of the north coast.

For more information on the conservation biology theory behind our management recommendations, see "The Bioreserve Strategy for Conserving Biodiversity." 1999. (pp. 35–53 in *Practical Approaches to the Conservation of Biological Diversity*. Edited by Baydack, R., H. Campa III, and J.B. Haufler. Island Press, Washington. D.C.) This chapter was coauthored by Dr. Allen Cooperrider (LEGACY–TLC's 1995–1997 president), Steven Day (LEGACY–TLC's founder, current board member, treasurer, and secretary), and Curtice Jacoby (LEGACY–TLC's current executive director).

Application to Conservation and Restoration Work

Each level of the Vision Map process, from a coverage of old-growth forests to a system of wildlife linkages, provides an important tool for conservationists to use in their daily work. Each can help to prioritize which lands are most important to protect, for what reason, and by what means (see map this page). The wide variety of measures that the conservation groups that we work with use to conserve and restore biodiversity include enforcement of Forest Practices Act regulations; encouraging and practicing sustainable forestry, grazing, and farming; supporting voluntary measures such as conservation easements and land trusts; purchasing development rights; land acquisition; active efforts at salmon habitat restoration including landslide stabilization and road removal; and forest restoration via tree planting and thinning of overstocked brushfields for fire hazard reduction.

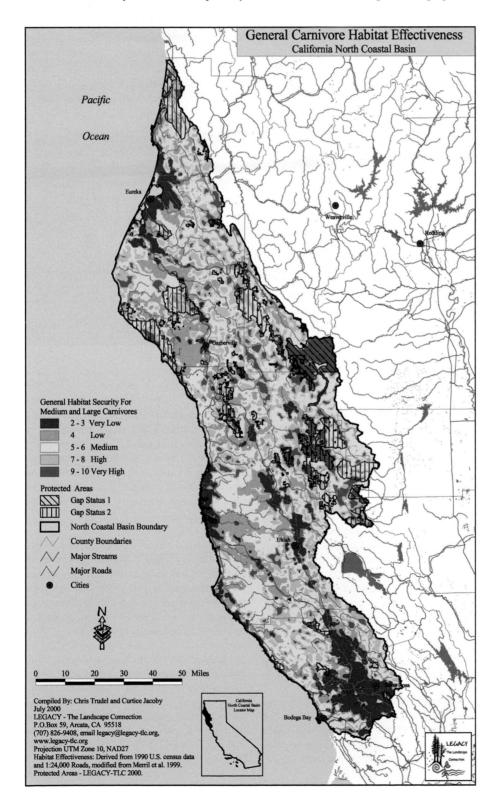

General Carnivore Habitat Effectiveness
California North Coastal Basin

Pacific

Ocean

Eureka

Weaverville

Redding

Garberville

Ukiah

General Habitat Security For Medium and Large Carnivores
- 2 - 3 Very Low
- 4 Low
- 5 - 6 Medium
- 7 - 8 High
- 9 - 10 Very High

Protected Areas
- Gap Status 1
- Gap Status 2
- North Coastal Basin Boundary
- County Boundaries
- Major Streams
- Major Roads
- ● Cities

N

0   10   20   30   40   50 Miles

Bodega Bay

Compiled By: Chris Trudel and Curtice Jacoby
July 2000
LEGACY - The Landscape Connection
P.O.Box 59, Arcata, CA 95518
(707) 826-9408, email legacy@legacy-tlc.org,
www.legacy-tlc.org
Projection UTM Zone 10, NAD27
Habitat Effectiveness: Derived from 1990 U.S. census data
and 1:24,000 Roads, modified from Merril et al. 1999.
Protected Areas - LEGACY-TLC 2000.

California
North Coastal Basin
Locator Map

LEGACY
The Landscape
Connection

The Bioreserve Strategy for Conserving Biodiversity

Allen Cooperrider, U.S. Fish and Wildlife Service
Steven Day, LEGACY, The Landscape Connection
Curtice Jacoby, Department of Forestry, Humboldt State University

Chapter for book Practical Approaches to the Conservation of Biological Diversity, *edited by Richard K. Baydack, Henry Campa III, and Jonathan B. Haufler. Island Press, Covelo, California. (In Press)*

The bioreserve strategy is a promising but largely untested approach to conserving biodiversity. The strategy involves zoning regional landscapes into areas that range from total protection (minimal human activity) to areas of intensive human use. Zoning, in this context, does not necessarily refer to a formal regulatory designation but rather to a societal agreement to limit certain human activities and uses on certain lands. This agreement may be expressed and played out in a variety of ways, from formal designation as reserves or parks to conservation easements or land-owner agreements.

In this paper we:

- Define biodiversity.

- Review the development of the bioreserve strategy.

- Describe briefly the elements of the strategy and the context in which it should be applied.

- Discuss some strengths and weaknesses of the approach.

- Describe a case history of an initiative to apply the strategy to a portion of an ecoregion in Northern California.

What Is Biodiversity?

Biodiversity has been defined as follows:

"Biodiversity is the variety of life and its processes. It includes the variety of living organisms, the genetic differences among them, the communities and ecosystems in which they occur, and the ecological and evolutionary processes that keep them functioning, yet ever changing and adapting" (Keystone Center 1991).

Detailed explanations of this definition can be found in Noss and Cooperrider (1994) as well as in many of the new books on conservation biology (Hunter 1996; Meffe and Carroll 1994; Primack 1993; Primack 1995). However, two important aspects of this definition need to be noted. First, biodiversity consists of more than just "the variety of species" or what is termed "species richness." Thus conserving biodiversity is a broader concept than just "endangered species management." Rather it involves the conservation of the full range of species, variation within species, biotic communities, and ecosystems. Second, the definition is dynamic and incorporates ecological functions or processes and explicitly recognizes that species and biotic communities change over time.

Development of the Bioreserve Strategy

The concept of "bioreserves" has been around in one form or another for hundreds if not thousands of years. Some indigenous cultures recognized areas in which human activities, such as hunting, were forbidden. More recent, the movement to create national parks, national wildlife refuges, wilderness, and natural areas is derived, at least in part, from the recognition that areas in which human activities are constrained is necessary to conserve plant and animal life.

However, this approach of setting aside areas as parks, refuges, and so on, has proven to be inadequate for at least three primary reasons (Noss and Cooperrider 1994). First, most such areas have been selected for purposes other than protection of biodiversity or with limited consideration of this as a purpose. For example, most wilderness areas have been set aside based on their fortuitous lack of roads together with high value for primitive recreation. This has resulted in a plethora of high-altitude reserve areas but few located in lowland areas. Similarly, many national parks and monuments have been selected because of their spectacular geological features rather than because they represented or conserved biotic communities.

Second, many species and plant communities are not represented or are underrepresented in any form of reserve systems (Crumpacker 1988). Given the manner in which reserves have been designated, this is not surprising.

Finally, most reserves are too small to contain fully functional biotic communities. For example, the oldest and largest national park in the coterminous United States, Yellowstone, is not large enough to support viable populations of many species (Clark and Zaumbrecher 1987). Furthermore, there is scientific evidence that national parks are losing species and that such loss is correlated with their size—with smaller ones losing more species than larger ones (Newmark 1985).

In summary, the reserve system that we have in this country today was developed in an ad hoc, piecemeal manner for a variety of purposes rather than with systematic consideration of the need to protect biodiversity. As a result, we have a system that is inadequate in terms of size and number of reserves.

To counter these limitations, conservation biologists are developing and testing ways to improve on the past approach. Three major modifications have been suggested. First and most central is the systematic design of regional reserve systems. This process includes both designation of new reserves and improved design of existing reserves. A key element of the process is a systematic effort

to ensure that all species and community types are represented in the reserve system. A second improvement is the "buffering" of core reserves (those areas where human activity is most constrained) with areas of increasing human activity and impacts. This increases the effective protection of the core reserves. Finally, the utilization of reserves can be made more effective by connecting core reserves with corridors or other forms of connectivity that allow some movement of plants and animals between core areas.

The Bioreserve Strategy

These concepts of systematic design of reserve systems, buffering of core reserves, and connectivity are the central principles of the "bioreserve strategy" as proposed by Noss (1983), Noss and Harris (1986), Noss (1992), Noss and Cooperrider (1994), and others. The application of these principles to achieve biodiversity goals is described briefly here and in more detail in Cooperrider (1994).

Goals and Objectives

The goal of a bioreserve strategy is to maintain the biodiversity of a region in perpetuity. Four fundamental objectives follow from this goal (Noss and Cooperrider 1994):

1 To represent, in a system of protected areas, all native ecosystem types and seral stages across their natural range of variation.

2 To maintain viable populations of all native species in natural patterns of abundance and distribution.

3 To maintain ecological and evolutionary processes such as disturbance regimes, hydrological processes, nutrient cycles, and biotic interactions.

4 To manage landscapes and communities to be responsive to short- and long-term environmental change and to maintain the evolutionary potential of the biota.

To these four, we have added a fifth:

5 To provide for the social, economic, and spiritual needs of the local people.

A bioreserve strategy provides a systematic approach toward achievement of all five objectives.

Design Components

Fundamental to the bioreserve strategy is the concept of zoning the regional landscape into areas of varying restrictions on human activities. These consist of four major components: core reserves, buffers, zones of connectivity, and matrix. These categories are described briefly below and in more detail in Noss and Cooperrider (1994).

Core Reserves

Core reserves are the backbone of a regional reserve system. They are areas in which the overriding goal of management or stewardship is preservation of native biodiversity and ecological integrity. They may consist of all or portions of national parks, wilderness areas, research natural areas, state parks and preserves, national wildlife refuges, or other areas in which human demands on the landscape are given a second priority. Ideally, the reserves of a region should collectively contain all native ecosystem types and seral stages as well as all native species found in the ecoregion.

Buffers

To increase the effectiveness of core reserves, buffer zones surrounding the core reserves are designated. These consist of zones in which increasing amounts of human activity and disturbance are allowed. For example, the first buffer zone might allow only light (nonmotorized recreation), and the second zone would allow for motorized travel on roads but no logging or mining.

Zones of Connectivity

A second method of increasing the effectiveness of core reserves is to ensure that there is connectivity between them. Connectivity here refers to the state of being functionally connected by movement of organisms, material, or energy. Zones of connectivity allow for the movement of plants and animals and their genes from one core reserve to another. Corridors are one form of connectivity, but not the only one.

Matrix

Finally, the matrix contains the rest of the land that does not fall into one of the above categories. In this area, human uses and demands on the landscape receive priority in the traditional manner in which land is managed. The matrix may consist of agricultural lands, timberlands, pasture land, or urban and suburban development.

Strengths and Weaknesses of the Bioreserve Strategy

The bioreserve strategy is for the most part an untested strategy—we do not know how effective such an approach will be in conserving biodiversity over the long term (one hundred years or more). Neither do we know how effective such an approach can be in resolving conflicting human demands on the regional landscape. Thus the advantages and disadvantages of the bioreserve strategy, as described below, represent a preliminary assessment of the practicality of the approach.

Strengths

An important characteristic of the approach is the simplicity of the concept. The basic concepts of core reserves, buffers, and zones of connectivity are easily understood, even though the methodologies and details of design may be quite complex. Furthermore, this landscape zoning concept can be applied at various scales ranging from the watershed to the ecoregion.

A related strength is that people are familiar with and accepting of (to varying degrees) the practice of landscape zoning. In spite of the rhetoric of the private property rights zealots, land ownership in this country has never conferred an absolute right to do anything one wants on a property. This is true under both United States law and its predecessor, English common law.

Finally, by zoning the landscape into zones of restricted human activities and areas of intensive human activity, the bioreserve approach explicitly provides for areas of concentrated human activity. Much opposition to conservation efforts appears to come from the fear that more and more human activities are being restricted. By developing a "whole landscape" strategy, provision for human needs is made at the same time that some human activities are being restricted.

Weaknesses

The strategy is not without its pitfalls and limitations. Landscape-level strategies that even hint at restricting human activity on private lands raise fears of increased government regulation and or "takings." An additional problem is the inflexibility of landscape zoning with set boundaries. Selecting, building public support, and designating a core reserve can be a contentious and time-consuming process. If future information suggests that the reserve should have been in the next watershed, it may be difficult to change the design and make the necessary adjustments.

A major difficulty with a bioreserve approach is that it requires a large knowledge base. Basic information is needed on:

- Abundance, distribution, habitat requirements, and movement patterns of ecologically important (keystone, umbrella, threatened, and endangered) species.

- Distribution, abundance, and condition of all ecosystem types (vegetation/biotic communities) and their seral stages throughout the ecoregion and an estimate of their "natural" landscape pattern.

- Natural disturbance regimes of the ecoregion—the temporal and spatial scales at which they occur, how they have been modified by humans, and how they can be mimicked in a seminatural setting.

- Area and level of protection of existing (*de facto* and *de jure*) reserves.

This is a large amount of information to acquire, synthesize, and digest, but it represents only the most basic needs for biological information. Ideally, much more information on biology and natural history, land use, and land-use impacts would be available and could be used.

Finally, as alluded to earlier, the bioreserve strategy is a largely untested approach at the

scale of the large regional ecosystem or ecoregion. We not only have little experience in applying these concepts at this spatial scale, but the relevant time frame for testing the success of such a system at the ecoregion level is decades and centuries. Thus it will be centuries before we truly know how successful such a strategy has been. However, this caveat applies equally well to all other regional biodiversity conservation strategies; we may learn pretty quickly if they are not working (by observing species extirpations and extinctions and other obvious signs) but proving success will require patience.

Applying the Bioreserve Strategy— The North Coastal Basin of the Klamath Ecoregion

We describe here a case history of an ongoing effort to apply the bioreserve strategy to a portion of the Klamath Ecoregion in northern California and south-central Oregon. This represents one attempt to apply such a strategy to a larger regional landscape. It differs from other efforts in other regions or even parallel efforts to apply a bioreserve strategy within the same ecoregion. We present it here as one example of an application that is neither better nor poorer than other applications of the bioreserve strategy but that is hopefully tailored to the region and its biological and sociohistorical context.

Background—The Klamath Ecoregion and North Coastal Basin

The Klamath Ecoregion is one of 52 ecoregions defined by the U.S. Fish and Wildlife Service. It is located in northwestern California and south-central Oregon and consists of all the watersheds or hydrobasins that drain into the Pacific Ocean from San Francisco Bay north to the Smith River. Thus the ecoregional boundaries are defined in terms of watersheds rather than some other criteria such as geology or vegetation type.

Any ecoregional delineation is bound to be arbitrary. Some interactions occur among ecoregions and thus any delimitated area is not going to encompass all species or ecological processes. However, the Klamath Ecoregion boundaries are biologically meaningful in many ways. It encompasses most of the range of coast redwoods. Mapping of potential natural vegetation of the ecoregion indicates that it contains virtually all of the potential natural sites for one vegetation type (pine–cypress forest) and most of the sites for four others (redwood forest, California mixed evergreen forest, montane chaparral, and fescue–oatgrass). Together, these four types make up more than 50 percent of the ecoregion.

Within the Klamath Ecoregion there are three relatively distinct subregions—the coast ranges, the Klamath Mountains, and the Modoc Plateau—that have relatively distinct geologic origin, climatic pattern, and vegetation.

The North Coastal Basin is a region defined by the California Water Quality control board and is generally synonymous with the coast ranges geologic province. In terms of watersheds, it is defined as all the hydrobasins draining into the Pacific Ocean south of the Klamath River and north of San Francisco Bay. Most of the region contains highly erodable soils derived from marine sediments making them highly sensitive to land use impacts such as logging.

The best known natural and scenic features are the redwood forests and the rugged coastline. And the most notorious residents of these areas are the northern spotted owl and marbled murrelet, which have received much notoriety in recent years. However, the region contains many lesser known features and species including unique coastal prairie and numerous species of endemic plants. The rivers once supported six species of anadromous salmonids as well as numerous lesser known species of fish and other vertebrates.

The redwood "rain forest" that most characterizes this region is unique in that it resides on the edge of the Pacific Ocean in a region that has a basic Mediterranean climate; that is, most of the rain comes in the winter months. This forest is a relict of more widespread rain forests that once covered much of the west. For most of the region the months of June, July, and August (at a minimum) are virtually without rainfall. Being near the ocean, however, the region is regularly covered by fog during the summer months. Much of the effective precipitation in this redwood/fog belt comes from the phenomena of fog drip—the ability of the redwood trees to capture fog from the air and transport it to the ground where it is utilized by many other life forms (Dawson 1996). When forests of the region are overcut, they lose much of their capacity to capture moisture from fog during the hot, dry months of summer. This is in addition to the well-documented ecological effects of deforestation common to most forested regions (soil exposure, accelerated erosion, sedimentation of streams, and so on).

Overall, with unstable soils and relictual forests with unique flora and fauna, the North Coastal Basin is a region that is highly sensitive to land-use impacts.

Unlike other portions of the ecoregion that contain more than 80 percent public (mostly federal) land, the North Coastal Basin consists of approximately 90 percent privately owned lands of which the major landowners are corporate timber companies. These industrial timberlands have been severely overcut and the watersheds degraded (Burkhardt 1995; U.S. Fish and Wildlife Service 1997; other). California has a Forest Practices Act that theoretically regulates forest practices on private lands to ensure that they produce "maximum sustained yield of high-quality forest products." Unfortunately, the agencies charged with implementing the act have never enforced the spirit of the law. For the most part, corporate timberlands have been cut to the point where there is little commercial timber left on them, particularly in the southern portion of the North Coastal Basin.

The net result of all of these factors and more is that the North Coastal Basin ecosystem is severely degraded. The following are key problems and evidence of such degradation.

- Redwood forests have been cut to the point where there are only a few remnants of true "old-growth" left.

- All of the anadromous fisheries of the region are in decline (Moyle 1994).

- Virtually all of the rivers of the region have been declared "impaired," primarily by sediment, under provisions of the Clean Water Act.

- Coho salmon have been declared "threatened" by the National Marine Fisheries Service in the southern portion of the region and steelhead salmon are also being considered for listing.

- Numerous other species of plants and animals of the region are either formally listed as threatened or endangered or are in some way "at risk" although not yet formally

categorized as such (U.S. Fish and Wildlife Service 1997).

Although there has been severe disruption of many of the forest and riverine ecosystems in this region, in some ways the region has more potential for ecological recovery than many other parts of California. This is primarily because of the lower human population density. The four counties that make up the total ecoregion have a population of 600,000, of which more than 60 percent is concentrated in the southernmost county (Sonoma), which sits next to the greater San Francisco Bay area.

The unique challenges of developing a bioreserve strategy in this region thus revolve around the following characteristics of the region:

- Limited amounts of public lands or existing reserves; high percentage of private lands.

- High percentage of lands owned by corporations with little incentive for long-term stewardship and virtually no regulatory enforcement of such.

- Severely degraded forest and riverine ecosystems.

- Ecosystems relatively sensitive to human disturbance because of relictual vegetation and inherently unstable soils combined with Mediterranean climate.

Obstacles to Biodiversity Conservation

The obstacles to conserving biodiversity in such a region are numerous. They include excessive demand for commodities, degraded and fragmented habitat, ineffectual regulation of forest practices and other ecosystem degrading activities, and fragmented regulatory authorities combined with a large cohort of complacent or uninformed citizenry. Problems of implementing a bioreserve strategy in such a region are numerous and are shared by many regions (Trombulak, et al. 1996).

On the other hand, some attributes of the region differ considerably from many other areas where regional biodiversity conservation strategies are being developed. Most central is the high percentage of private land and the high percentage of corporate ownership. Many of the ongoing bioreserve strategies, such as those described by Pace (1991), Noss (1993), Vance–Borland, et al. (1996), and others, are regions with a high percentage of federal or other public lands. In these cases, the strategies can rely heavily on linking up existing reserves (in the form of national parks, wilderness areas) with other federal lands currently used for multiple-use purposes.

Overcoming Obstacles

Given the nature of the region, many citizen conservationists and biologists both within and outside government have realized that conserving biodiversity within the ecoregion will require a long-term, systematic, and cooperative effort among citizens, government, academia, and nongovernmental organizations (NGOs). Furthermore, these people have realized that in a region with such fragmented land ownership, it is unlikely that a single government agency will take the lead in such an effort. Thus a need was recognized for an NGO that could provide a leadership role in developing and implementing a bioreserve strategy for the region.

In response to this need a nonprofit 501(c)3 organization, LEGACY, was formed in 1993 for such a purpose. The mission of LEGACY is to promote conservation of native biodiversity in the Klamath Ecoregion through integration of local knowledge and science. A keystone project of the group is the development and implementation of a bioreserve strategy for the North Coastal Basin, which is viewed as a long-term endeavor. As an NGO, LEGACY can play several key roles in implementing this strategy that might not be possible as a government agency. These roles include the following:

- Visionary—As a visionary, LEGACY is developing and disseminating a vision of a sustainable regional ecosystem as well as a means to move toward that goal.

- Catalyst—By collecting, analyzing, synthesizing, and disseminating information on biodiversity problems and ecosystem needs, LEGACY is serving as a catalyst to stimulate government agencies to fulfill their statutory obligations to protect elements of the environment.

- Partner—LEGACY is developing working relationships with both natural resource and regulatory agencies and also with other nongovernment organizations to pursue mutual goals. Of particular importance are the relationships with watershed groups, as will be described.

- Archivist—LEGACY is developing a means of archiving critical biodiversity information that is not being kept elsewhere because of fragmented ownership and other jurisdictional problems.

- Educator—LEGACY is developing educational programs to inform citizens, students, bureaucrats, and politicians about important concepts of conservation biology and important ecological processes of the ecoregion.

Development and Implementation of the Strategy

Johns and Soule (1996) have outlined steps for implementing a bioreserve or wildlands reserve program. LEGACY is generally following these steps; however, their program is emphasizing five aspects that may differ from programs in other regions: (1) "soft reserve design," (2) alliance with watershed groups, (3) use of geographic information systems both for data analysis and for education, (4) ever expanding partnerships, and (5) heavy emphasis on education.

Soft Reserve Design

Soft reserve design refers to a process in which the regional design is first done as a draft with "soft" lines outlining general areas where new reserves are needed and where buffers and zones of connectivity need to be established. The lines can later be modified and firmed up based on input and local knowledge of people living and working in the watersheds. This zoning procedure considers the willingness of landowners to exchange their landholdings for another location, to shift their land use goals, or to conform with the regional strategy as needed. An exploration of this willingness throughout the region is an important part of LEGACY's approach and is as important as the inventory of biodiversity in determining the ultimate reserve design. In a region consisting largely of private lands, this procedure allows LEGACY to introduce the concept of a regional reserve system in a manner that is less threatening to local residents and landowners.

The Watershed Connection

To complement this approach LEGACY is developing strong ties and communication with local watershed groups working to produce biodiversity conservation plans for the individual watersheds and river basins of the regions. Such groups are active in the region and more than twenty groups are working on such plans. This interaction with local watershed groups is a two-way process. The bioregional group, LEGACY, provides information on regional issues and problems and how the local watershed plans can be made congruent with the larger regional reserve design. This may affect placement of reserves, buffers, and corridors within the watershed. In addition, LEGACY is prepared to assist in the design of citizen sampling and monitoring programs as well as with GIS as will be discussed.

The watershed groups, being more familiar with specific areas, can provide LEGACY with more detailed information on local biodiversity "hot spots" or other critical areas for biodiversity conservation. They also provide LEGACY with information on de facto reserves (e.g., areas in which biodiversity is being protected through conservation easements, other legal instruments, or simply through landowner stewardship).

LEGACY's strategy is to encourage watershed groups to incorporate regional reserve design needs into their individual watershed plans. In this manner, we anticipate that the ultimate implementation of a bioreserve system will come from the watershed groups themselves. The watershed plans will thus be the building blocks for an integrated regional biodiversity conservation plan.

GIS as Medium and Message

Like many other conservation planning efforts, LEGACY is relying heavily on GIS technology for several purposes. GIS provides a means for efficiently manipulating spatial data to assist in analysis of biological data and in spatial design. However, generation of maps that display regional- and watershed-level conditions is also an extremely valuable communication tool (Cooperrider, et al. 1996). We have observed that the maps produced by LEGACY are some of the most effective tools for communicating with citizens, other organizations, and government agencies.

One of the key roles of LEGACY is to provide the watershed groups with similar GIS capability by either assisting them in doing local GIS work or by advising them in setting up their own GIS systems. Many of the local watershed groups are working on limited budgets and with volunteer labor, and it is not cost-effective for them to try to set up their own GIS system.

Other Partnerships

LEGACY is also working to form alliances with a variety of other organizations and agencies that share all or part of its biodiversity conservation goals. These range from state and federal agencies with broad mandates

such as the California Department of Fish and Game to nongovernment organizations with relatively specific focus such as the California Native Plant Society.

The U.S. Fish and Wildlife Service is playing a key role in developing a biodiversity conservation plan for the region. The service has developed a holistic strategy for restoration of the Klamath Ecoregion. This strategy describes the various ecological issues that need to be resolved for effective restoration of the lands, waters, and biota of the ecoregion. This Fish and Wildlife Service strategy recognizes that ecosystem restoration cannot be accomplished by one agency or organization working in isolation. Rather, it proposes that restoration of the biodiversity of the region must be accomplished by the diversity of agencies, NGOs, and private citizens working in tandem or in a coordinated fashion to accomplish this goal. LEGACY's role as an NGO pursuing a regional biodiversity conservation plan is completely congruent with the Service's ecoregion restoration strategy.

The Fish and Wildlife Service has also provided a critical service to the region by setting up a virtual GIS laboratory at Humboldt State University (Carlson, et al. 1995). This laboratory is developing seamless GIS layers of land ownership, soils, vegetation, plant and animal distributions, and more, for the entire ecoregion. These GIS layers are being made available to citizens, NGOs, and other agencies throughout the region. This allows for resolution of ecological issues with all parties having access to the same information. Having access to such a laboratory and such information has greatly facilitated many of the tasks that LEGACY is pursuing.

Ultimately, restoration of the biodiversity of the ecoregion will require participation of the corporate timber companies of the region. Unfortunately, at present there seems to be little common ground and little incentive for either conservation groups or the companies to work together. There are some signs, however, that this may be changing, particularly with some of the smaller timber companies. With the listing of coho salmon as threatened and the imminent listing of a number of other species of fish and wildlife, all parties may find that there is a stronger need to work together to achieve both common and individual goals. One approach to bringing timber companies into the conservation arena which is being tested in this region now, is to develop mutually acceptable conservation plans at the watershed level rather than at the regional or companywide level.

Education

The importance of education and information in achieving any of these goals is paramount.

LEGACY is pursuing an active effort to develop and disseminate educational modules throughout the ecoregion. Current efforts include modules on principles of conservation biology; reserve design strategy; the role of large carnivores in the ecosystem; the hydrologic cycle and the fog drip connection; salmon as ecological, economic, and spiritual keystone species; watershed protection and the citizens role in such; and the role of hardwoods in the forest ecosystem.

Discussion

The bioreserve strategy for conserving biodiversity is fraught with difficulties but also shows great promise. The key to understanding the potential of such an approach is to recognize that it is a long-term strategy, not a quick fix to immediate problems. Given enough time, it may be possible to resolve the most intractable problems of land allocation.

Notwithstanding the serious limitations of the bioreserve strategy, it remains one of the most promising approaches to conserving biodiversity. The reason for this assertion is philosophical. This optimism is based on the belief that a strategy that tries to rely on natural ecosystem processes and components—developed through thousands of years of coevolution of plants, animals, and landscapes—is more likely to be successful in the long term than one that assumes that humans can do a better job of "managing" nature.

Bioreserves are not intended to serve as a stand-alone strategy for conserving biodiversity. Even though the bioreserve strategy allocates matrix lands as areas where human development and activities can be emphasized, these lands must also be treated with a certain amount of care. If, for example, agricultural practices within the matrix result in accelerated soil erosion, it will be detrimental to the entire regional ecosystem. Similarly, a bioreserve strategy must be complemented by many of the existing programs for protecting clean water, clean air, and so on.

The case history of the North Coastal Basin and the program of the nongovernment organization, LEGACY, to implement a bioreserve strategy in this region illustrates the complexity of trying to implement such an ambitious and long-term strategy. Education will be the key to successful implementation; only if people understand both the need for biodiversity conservation and the potential of bioreserves to protect it will progress be made.

This approach to implementation relies heavily on outreach to the resident citizenry and to various government and nongovernment organizations. David Johns and Michael Soule (1996) have described the approach as follows: "Think of the process as creating an ever expanding circle of people who understand and support wildness and biodiversity: networks of people defending networks of land and water." This philosophy is central to the approach being taken by LEGACY in the North Coastal Basin.

Finally, in the case history described, the proponents of the bioreserve strategy rely heavily on a bottom-up approach to ultimate resolution of land allocation problems. This contrasts somewhat from reports from other regions that appear to rely more on a top-down approach. The latter may be more feasible where much of the landscape remains in public ownership. We believe that the integrated approach described here, with an emphasis on watershed-level resolution of issues, is the appropriate one for this ecoregion. As Jim Walters (1996) has written regarding implementation of bioreserve strategies: ". . . to achieve even a measure of political possibility will require . . . acceptance and integration in the context of local struggles. We can never win on a larger stage what we can't win in our own homes and neighborhoods."

Literature Cited

Burkhardt, H. 1994. *Maximizing Forest Productivity—Resource Depletion and a Strategy to Resolve the Crisis.* Mendocino Environmental Center, Ukiah, California, 140 pp.

Carlson, S. A., L. Fox III, and R. L. Garrett. 1995. "Virtual GIS and Ecosystem Assessment in the Klamath Province, California–Oregon." Proceedings of the LIS–GIS Meeting.

Clark, T. W., and D. Zaumbrecher. 1987. "The Greater Yellowstone Ecosystem: The Ecosystem Concept in Natural Resource Policy and Management." *Renewable Resources Journal* 5(3):8–16.

Cooperrider, A. Y., L. Fox III, R. Garrett, and T. Hobbs. 1997. "Data Collection, Management, and Inventory." Paper presented at Ecological Stewardship Workshop, Tucson, Arizona, December 1995.

Crumpacker, D. W., S. W. Hodge, D. F. Friedley, and W. P. Gregg, Jr. 1988. "A Preliminary Assessment of the Status of Major Terrestrial and Wetland Ecosystems on Federal and Indian Lands in the United States." *Conservation Biology* 2(1):103–105.

Dawson, T. E. 1996. "The Use of Fog Precipitation by Plants in Coastal Redwood Forests." Pages 90–93 In Proceedings of the Conference on Coast Redwood Forest Ecology and Management, ed. John L. LeBlanc.

June 18–20, 1996. Humboldt State University, Arcata, CA, 170 pp.

Hunter, M. L. Jr., 1996. *Fundamentals of Conservation Biology.* Blackwell Science, Cambridge, MA, 482 pp.

Johns, D., and M. Soule. 1996. "Getting from Here to There." *Wild Earth* 5(4):32–36.

Keystone Center, The. 1991. "Final Consensus Report of the Keystone Policy Dialogue on Biological Diversity on Federal Lands." Keystone, CO: The Keystone Center.

Meffe, G. K., and C. R. Carroll. 1994. *Principles of Conservation Biology.* Sunderland, MA: Sinauer Associates, Inc. 600 pp.

Moyle, P. B. 1994. "The Decline of Anadromous Fishes in California." *Conservation Biology* 8(3):869–870.

Newmark, W. D. 1985. "Legal and Biotic Boundaries of Western North American National Parks: A Problem of Congruence." *Biological Conservation* 33:197–208.

Noss, R. F. 1983. "A Regional Landscape Approach to Maintain Diversity." *BioScience* 33:700–706.

Noss, R. F. 1993. "A Bioregional Conservation Plan for the Oregon Coast Range." *Natural Areas Journal* 13:276–290.

Noss, R. F. 1992. "The Wildlands Project: Land Conservation Strategy." *Wild Earth* (Special Issue): 10–25.

Noss, R. F., and A. Y. Cooperrider. 1994. *Saving Nature's Legacy—Protecting and Restoring Biodiversity.* Covelo, CA: Island Press, 417 pp.

Noss, R. F., and L. D. Harris. 1986. "Nodes, Networks, and MUMs: Preserving Diversity at All Scales." *Environmental Management* 10:299–309.

Pace, F. 1991. "The Klamath Corridors: Preserving Biodiversity in the Klamath National Forest." Pages 105–116 In *Landscape Linkage and Biodiversity,* ed. Wendy E. Hudson. Covelo, CA: Island Press. 196 pp.

Primack, R. B. 1993. *Essentials of Conservation Biology.* Sunderland, MA: Sinauer Associates Inc. 564 pp.

Primack, R. B. 1995. *A Primer of Conservation Biology.* Sunderland, MA: Sinauer Associates Inc. 277 pp.

Trombulak, S., R. Noss, and J. Strittholt. 1996. "Obstacles to Implementing the Wildlands Project Vision." *Wild Earth* 5(4):84–89.

U.S. Fish and Wildlife Service. 1997. "Klamath/Central Pacific Coast Ecoregion Restoration Strategy, 4 Volumes." Klamath Basin Ecosystem Restoration Office, Klamath Falls, Oregon.

Vance–Borland, R. Noss, J. Strittholt, P. Frost, C. Carroll, and R. Nawa. 1996. "A Biodiversity Conservation Plan for the Klamath/Siskiyou Region." *Wild Earth* 5(4):52–59.

Walters, J. 1996. "A Reality Check for TWP." *Wild Earth* 5(4):16.

Biodiversity Associates

Jeff Kessler

The lands administered by the Bureau of Land Management in the western United States hold many special natural values and important historical and cultural resources. In a few locations, several of these values come together in a relatively small area. The Jack Morrow Hills (JMH) study area located in the Red Desert of Wyoming is just such a place. In the JMH, one finds an incredible array of wildlife, rare plants, wildland recreational opportunities, wonderful scenery, archaeological resources, historic sites, and areas important to Native Americans.

While there is a relatively large amount of information available about the Jack Morrow Hills and its remarkable natural and human history, this information is generally scattered in dozens of government documents and hundreds of other articles, books, and reports. The draft environmental impact statement (DEIS) prepared by BLM for the Jack Morrow Hills Coordinated Activity Plan (CAP) collects a great deal of the most important information about the JMH in a single document. Unfortunately, this information is dispersed throughout more than seven hundred pages. There are many maps in the DEIS, but most of the maps display only a single group of features. Thus it is difficult for the reader to get a feel for the overall picture. And, more important, some key information about native species and other natural values simply have not been included in the DEIS.

Right now, the Jack Morrow Hills are under great industrial development pressure, primarily for oil and natural gas. The very features that make the JMH special are threatened by the push toward industrial development. Other threats exist as well. Through the JMH CAP, BLM will decide what level of protection is offered to the JMH and what level of development will be allowed. In a very real sense, the BLM's decision on the CAP could seal the fate of this remarkable place. Our goal in creating this document is to present the essential information, demonstrating the many special values of the JMH—and the threats to these values—in a more concise and easy-to-grasp format. Our belief is that only through full consideration of these values will they be protected.

Most maps show several different pieces of information, but each map is based on a particular unifying theme. For example, all rare plant and plant community information is shown on a single map and likewise for all big game information. Where there was simply too much to show on a single map, we attempted to break up the maps into logical subthemes.

Although we had visited the Jack Morrow Hills area numerous times in the past and had some familiarity with the literature describing many of the natural features, we were nonetheless quite surprised to observe the number of different and remarkable values found to occur in this one region of Wyoming. Whether one considers biological values like big game and rare plants or cultural and historical values like petroglyphs and historical sites, the confluence of so many special values makes the Jack Morrow Hills a truly exceptional place. Equally surprising was the fact that virtually no part of the Jack Morrow Hills area is lacking in special value.

The accompanying map (this page, top) shows the location of important big game habitats such as crucial elk, mule deer, and antelope ranges; birthing areas for elk and mule deer; severe winter relief for mule deer; and the important elk migration/connection zone. We obtained digital map coverages from the

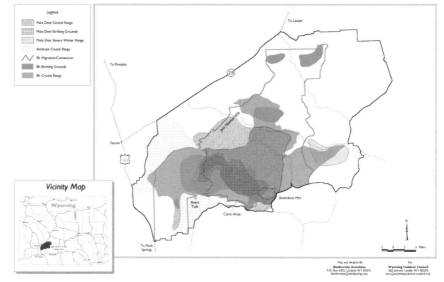

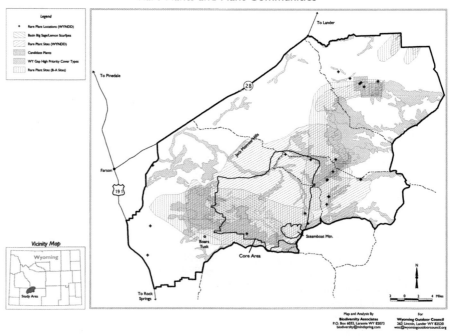

Wyoming Game and Fish Department and digitized the migration corridor/connectivity area from Map 59 in the DEIS. Habitat not considered "crucial" and severe winter relief areas are not shown on the map. Thus spring/summer/fall, winter/yearling, and noncrucial winter habitats are not shown on the map.

Big game, particularly elk, need protection from development activities. Otherwise, construction of roads, facilities, and other industrial activities can have severe consequences. "Maintaining connectivity between important habitats (crucial winter ranges, severe winter relief areas, calving/fawning habitats, migration corridors, topographic relief areas, mountain shrub communities, forest-type habitats) within the planning area is paramount to sustaining viable big game herds and other wildlife. Fragmentation of these crucial habitats will not sustain big game population objectives . . ." DEIS, page 235.

The accompanying map (see previous page, bottom) shows many of the known locations of rare plants and plant communities. The B1 "macrosite" is in the south-central portion of the Jack Morrow Hills area, shown on the map with SW–NE diagonal hatching. Specific areas of the basin big sagebrush/lemon scurfpea association are shown with NW–SE diagonal hatching. The latter were digitized by Biodiversity Associates from paper maps in Fertig, et al. 1998.

Predicted Vertebrate Species Richness

Although each type of habitat is important for some species, there are occasions when a particular kind of habitat is used by an unusually large number of different species. Protection of areas of unusually high species richness—biological "hot spots" if you will—is critically important for the conservation of biological diversity across the landscape.

The map below shows predicted vertebrate species richness for the Jack Morrow Hills planning area. The map was produced using the Wyoming Gap Analysis digital data on predicted total vertebrate species richness. The map uses different shades of red to show different levels of predicted vertebrate species richness—the darker the red, the greater the number of different species predicted to inhabit the area.

The habitats around Jack Morrow Creek, Pacific Creek, and Alkali Draw appear to be particularly rich in species. Interestingly, while the eastern portion of the Jack Morrow Hills area has generally been subject to less development, many of the areas of higher predicted richness are located in the western part of the area where there are fewer restrictions on development. Thus it appears that some of the most biologically diverse places are the least protected. The Action Alternatives considered in the draft EIS would allow further impacts to many biologically rich areas. These findings suggest a gap in biodiversity protection. The BLM must fill this gap by developing measures to protect the areas of unusually high vertebrate species richness.

This is not to suggest that areas with low predicted species richness are not important for biodiversity conservation or should not be protected. Indeed, a quick comparison of the "Species Richness" map with other distribution maps—"Rare Plants and Plant Communities" and "Animals of Special Concern"—reveals that many of the key habitats for rare and sensitive species do not occur in the areas of high predicted species richness. This may be due to the fact that species are of concern precisely because their habitats are uncommon or specialized, and therefore not likely to be used by many other species. Likewise, the key habitats shown on the maps of "Important Big Game Habitat," "Sage Grouse and Mountain Plover" habitats, and "Raptor Habitat" do not always occur in areas of higher predicted species

Predicted Species Richness

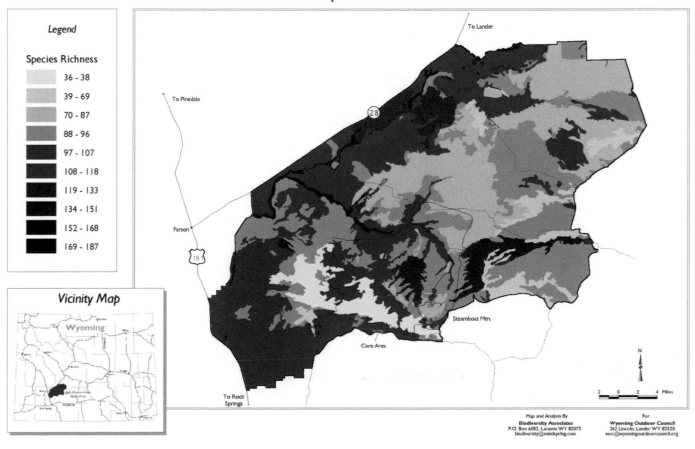

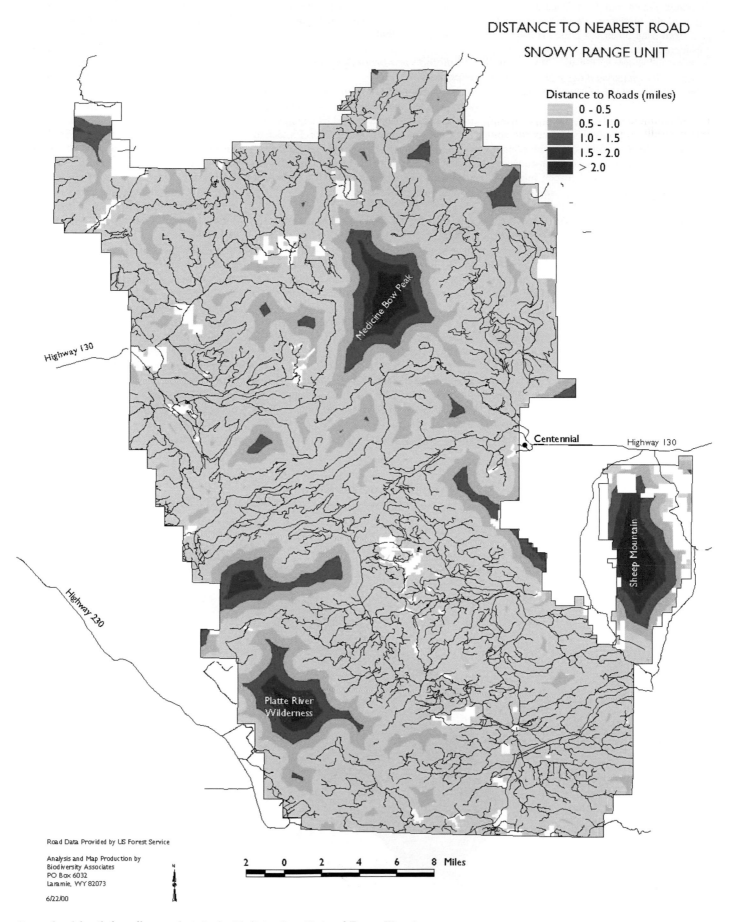

DISTANCE TO NEAREST ROAD

SNOWY RANGE UNIT

Distance to Roads (miles)

0 - 0.5
0.5 - 1.0
1.0 - 1.5
1.5 - 2.0
> 2.0

Medicine Bow Peak

Sheep Mountain

Centennial

Highway 130

Highway 130

Highway 230

Platte River
Wilderness

Road Data Provided by US Forest Service

Analysis and Map Production by
Biodiversity Associates
PO Box 6032
Laramie, WY 82073

6/22/00

2 0 2 4 6 8 Miles

Example of detailed roadless analysis in the Medicine Bow National Forest, Wyoming.

richness. Yet those key habitats are undoubtedly important for the species in question.

There are seven wilderness study areas (WSAs) within the Jack Morrow Hills boundary, totaling 117,000 acres. Wyoming BLM recommended 70,371 acres of these areas for wilderness designation, but Congress has yet to act on this recommendation. The WSAs are therefore managed under BLM's interim management policy for lands under wilderness review.

However, as in other parts of the country, the BLM inventory process overlooked many exceptional areas worthy of WSA status and wilderness designation.

Currently, citizens groups are conducting a new wildlands inventory for the Jack Morrow Hills study area using methods and criteria adopted by conservation groups in Utah (see map, previous page). Preliminary results indicate that both BLM and Wilderness at

Risk inventories overlooked some wild areas that possess wilderness qualities and deserve protection. The groups are also doing field investigations to determine the level of illegal damage (e.g., ORV use) to WSAs and other wildlands in the Jack Morrow Hills area.

Cultural and Historical Resources

The Red Desert is rich in cultural and historical resources (see map below). These include the famous South Pass National Historic Landscape, several Native American respected places, the White Mountain petroglyphs, and about 1,000 "cultural resource localities." The actual number is likely quite higher, however, because only about 2 percent of the region has been formally inventoried. Thus "a much larger number of resources should be expected in the area" DEIS, page 205. The Killpecker Dunes have been occupied by Native American cultures for at least 11,000 years (McGrew, et al. 1974). On the map (below), and in discussions below, we highlighted a few notable examples of cultural and historic resources. These and many other areas need protection or they will be irreparably damaged by mineral development, road construction, ORV abuse, or vandalism.

Cultural and Historical Resources - Jack Morrow Hills/Red Desert

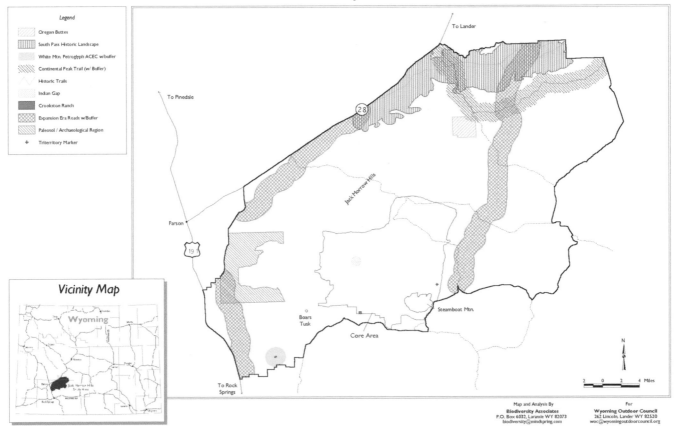

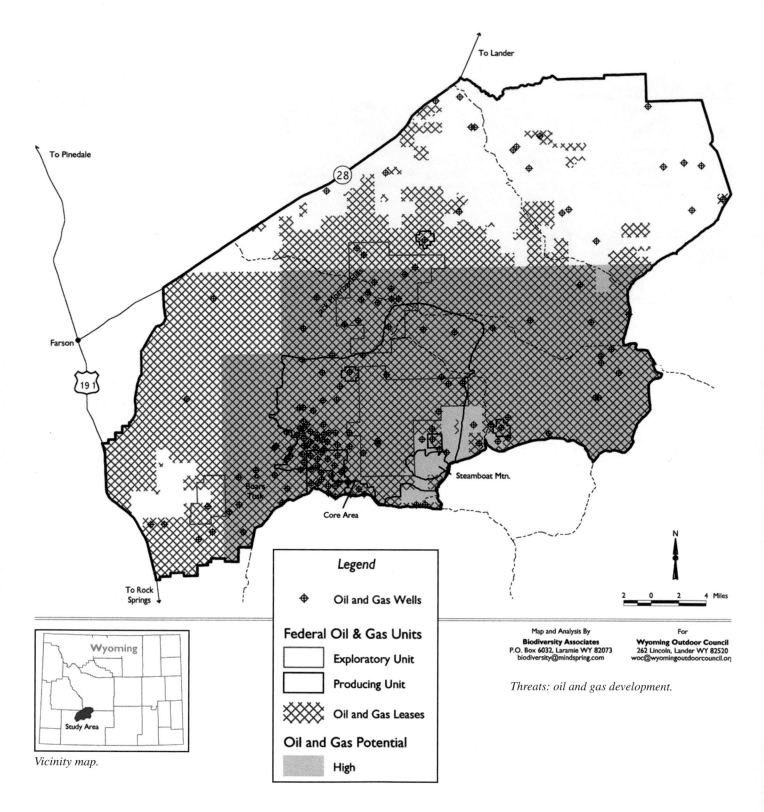

Legend

⊕ Oil and Gas Wells

Federal Oil & Gas Units

Exploratory Unit

Producing Unit

Oil and Gas Leases

Oil and Gas Potential

High

Map and Analysis By
Biodiversity Associates
P.O. Box 6032, Laramie WY 82073
biodiversity@mindspring.com

For
Wyoming Outdoor Council
262 Lincoln, Lander WY 82520
woc@wyomingoutdoorcouncil.org

Threats: oil and gas development.

Vicinity map.

Please note that many Native American sites and other sensitive cultural sites are not shown on the map. All of the descriptions below are paraphrased or quoted from the DEIS, pages 205–209, and *Green River Resource Management Plan,* page 361.

The above Threats map presents information on only a few of the many threats to the area such as oil and gas development, other mineral development, and range improvements and motorized vehicle use. Yet even considering

this sample, there were simply too many different kinds of threats to be meaningfully depicted on a single map. Moreover, the maps do not convey the true extent of impacts these kinds of developments pose to the special values.

For instance, although the maps show locations where oil and gas development may occur, they do not show how associated oil and gas development would impact the scenic quality of the landscape. Drill platforms, pump jacks, and other oil and gas

developments can be seen miles away from the locations where they occur. The same can be said of fences shown on the "Rangeland Developments and Motorized Vehicle Use" map. Fences can be seen from considerable distances. Fences also fragment wildlife habitat by creating barriers to wildlife migration and dispersal. The DEIS (page 236) recognizes this problem, but it is a problem that cannot be conveyed on a simple map.

Landscape Conservation

Appalachian Mountain Club's GIS Program

David A. Publicover, Senior Staff Scientist   dpublicover@amcinfo.org
Larry D. Garland, Staff Cartographer   lgarland@amcinfo.org

Founded in 1876, the 90,000-member Appalachian Mountain Club is the country's oldest conservation and recreation organization. Our mission is to promote the protection, enjoyment, and wise use of the mountains, rivers, and trails of the Northeast. The diverse programs of the AMC include recreational facilities management (more than fifty facilities ranging from backcountry shelters to full-service camps and mountain huts), trail maintenance, search and rescue, environmental education (natural history and outdoor skills workshops, guided hikes, school programs, and urban youth programs), publications, conservation advocacy, and scientific research. We believe that developing a love for the natural world enhances the desire and ability to be effective advocates for conservation.

Throughout its history the AMC has maintained a strong tradition of conservation advocacy focused on open space protection and sustainable management of natural resources. We were instrumental in the passage of the Weeks Act of 1911 that established the eastern National Forest system. We have been a leader in the Northern Forest Alliance (a coalition of more than forty organizations working on conservation issues in northern New England and New York), the Highlands Coalition (working on open space protection in the New York/New Jersey region), and Americans for our Heritage and Recreation (working to secure permanent funding for land conservation through the federal Land and Water Conservation Fund).

The professionally staffed Research Department supports these efforts by supplying technical and scientific information, knowledge, and analyses to the AMC, its conservation partners, and government agencies. The department's primary focus areas are land conservation, sustainable forestry, biodiversity conservation, alpine ecology and mountain stewardship, river shoreland protection and in-stream flow policies, and air quality. This work is done in cooperation with a wide range of partners including academic institutions (e.g., Dartmouth College, University of New Hampshire, and Harvard School of Public Health), public agencies (e.g., U.S. Forest Service, U.S. Fish and Wildlife Service, Massachusetts Department of Environmental Management, and Maine Bureau of Parks and Lands), and conservation alliances (e.g., Northern Forest Alliance, Clean Air Task Force, and Hydropower Reform Coalition). Research staff serve on numerous public policy committees and technical working groups.

The department's decade-old GIS program is a critical part of our work, especially in the area of land conservation. We have developed data layers not available from other sources including roadless area maps and historical clear-cutting patterns derived from satellite imagery. Our maps and analyses are a primary source of information for the Northern Forest region, and are widely used by other conservation groups, land trusts (e.g., Trust for Public Lands, Open Space Institute, The Nature Conservancy, and Forest Society of Maine), public agencies, and grassroots organizations in the development of and advocacy for large-scale land conservation projects. Over the past two years we have established a similar program for our mid-Atlantic region and are helping to inform conservation efforts from southern New England to northern Virginia. Our new initiatives include a major effort to develop a map-based Web site and support of an emerging broader regional conservation effort (the Appalachian Partnership for Eastern Forests).

New England Study

Previous donations from ESRI's Conservation Technology Support Program have been instrumental in helping the Appalachian Mountain Club (AMC) Research Department build our GIS capability. We are now a regional leader in providing landscape-level resource information, conservation planning and analysis, and trail mapping services to other conservation

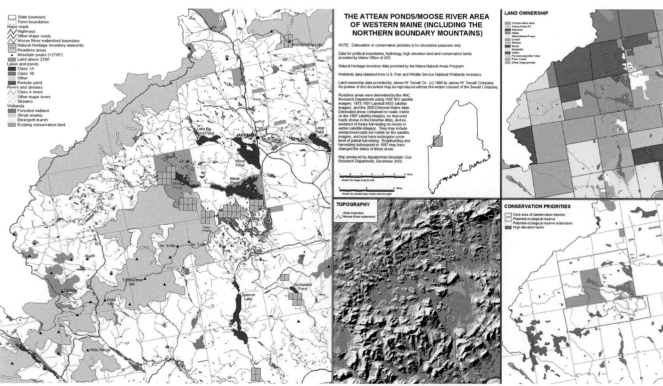

Example of a detailed conservation proposal map.

organizations, grassroots groups, and state and federal agencies in New England.

ESRI's recent donation to our New Hampshire-based Research Department of ArcView GIS updates, that is, additional basic GIS platform software (PC ARC/INFO® and ArcView GIS) and the ArcView Image Analysis and ArcView Spatial Analyst extensions, have enhanced our ability to provide regional resource information during a time of unprecedented conservation opportunity in northern New England (see map below). Extensive land sales by the forest industry, new landowners with greater interest in selling conservation easements of land in-fee, the availability of significant funding for land conservation, and a great increase in public support for land conservation have opened a window of opportunity in the region. AMC GIS capability is proving of great value in shaping potential conservation actions across the region.

Since April 1999 we have:

- Hired a full-time GIS research assistant giving us the ability to undertake more detailed projects as well as respond to short-term analytical needs in a more timely fashion.

- Completed a roadless area evaluation for northern New England using recent (1996–1999) satellite imagery and published road atlases. This data layer shows those portions of the region that have had relatively little impact from road building or heavy timber harvesting over the past few decades. This information has been widely circulated and is of high value in identifying lands with high potential for conservation, especially lands that may be suitable for ecological reserve status.

- Begun an update of our earlier analysis of historical clear-cutting patterns across the region that mapped forest clearing from 1973–1991. The current analysis will update this to the late 1990s using our recent satellite imagery.

- Provided satellite imagery and roadless area data to the Maine Chapter of The Nature Conservancy (TNC) for their conservation planning efforts in the St. John River watershed. This is one of TNCs most ambitious current projects; the upper St. John is the wildest and least developed major watershed in the eastern United States. Our work with the TNC included extending our roadless area analysis to the Canadian portion of the watershed.

- Provided maps, resource information, and satellite imagery to the Forest Society of Maine, which is developing a comprehensive conservation plan for 660,000 acres of land owned by the MacDonald Investment Company. The land contains most of the watershed of the West Branch of the Penobscot River, as well as the headwaters of the St. John River. The information we provided included our own conservation proposal developed with groups in the Northern Forest Alliance.

- Worked with outreach staff of the Northern Forest Alliance and local grassroots groups to develop detailed conservation proposals for a number of areas in Maine where landowners have expressed a willingness to consider such proposals (see map, previous page). AMC provides maps, data, analysis of high value parts of the landscape, and preliminary conservation proposals to these groups, which then suggest revisions based on their local knowledge and desires. These proposals are eventually presented to the State officials or land trusts responsible for actually negotiating conservation deals. Projects are currently underway with groups in Greenville (concerned with 900,000 acres around Moosehead Lake owned by Plum Creek Timber Company), Grand Lake Stream (450,000 acres formerly owned by Georgia–Pacific in east Maine), and Weld (former United Timberland lands in the Mt. Blue/Tumbledown Mountain region), as well as a statewide group

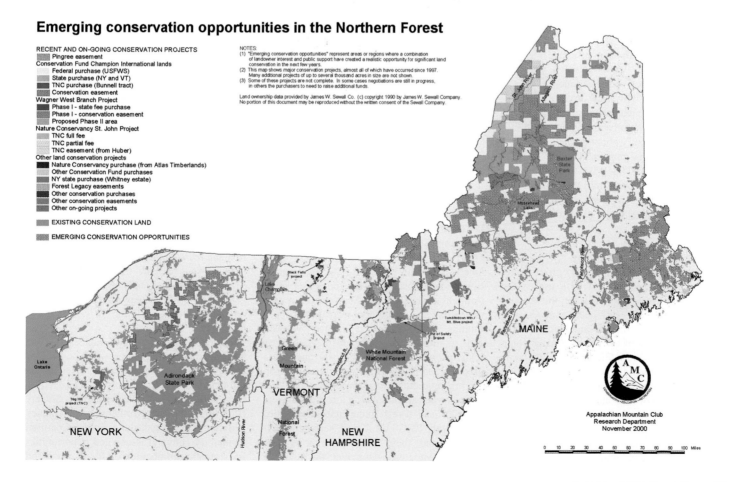

Emerging conservation opportunities in the Northern Forest

RECENT AND ON-GOING CONSERVATION PROJECTS
Pingree easement
Conservation Fund Champion International lands
    Federal purchase (USFWS)
    State purchase (NY and VT)
    TNC purchase (Bunnell tract)
    Conservation easement
Wagner West Branch Project
    Phase I - state fee purchase
    Phase I - conservation easement
    Proposed Phase II area
Nature Conservancy St. John Project
    TNC full fee
    TNC partial fee
    TNC easement (from Huber)
Other land conservation projects
    Nature Conservancy purchase (from Atlas Timberlands)
    Other Conservation Fund purchases
    NY state purchase (Whitney estate)
    Forest Legacy easements
    Other conservation purchases
    Other conservation easements
    Other on-going projects

EXISTING CONSERVATION LAND

EMERGING CONSERVATION OPPORTUNITIES

NOTES:
(1) "Emerging conservation opportunities" represent areas or regions where a combination of landowner interest and public support have created a realistic opportunity for significant land conservation in the next few years.
(2) This map shows major conservation projects, almost all of which have occurred since 1997. Many additional projects of up to several thousand acres in size are not shown.
(3) Some of these projects are not complete. In some cases negotiations are still in progress, in others the purchasers need to raise additional funds.

Land ownership data provided by James W. Sewall Co. (c) copyright 1990 by James W. Sewall Company. No portion of this document may be reproduced without the written consent of the Sewall Company.

Appalachian Mountain Club
Research Department
November 2000

0 10 20 30 40 50 60 70 80 90 100 Miles

working to enhance existing conservation lands around the Allagash Wilderness Waterway.

- Worked with The Nature Conservancy and the Trust for Public Lands to develop a conservation proposal for the nearly 400,000 acres of land still owned by Great Northern Paper. These critically important lands lie adjacent to the south and west borders of Baxter State Park (Maine's only significant large wilderness area).

- Conducted an analysis of roadless areas on the White Mountain National Forest in preparation for the revision of the forest's 10-year management plan. The work was released in a report entitled Mountain Treasures, developed cooperatively with the Wilderness Society and the Conservation Law Foundation. The report included maps showing proposed changes in land allocation on the forest that would protect these areas and increase the amount of the forest managed as natural area. Though controversial, the report has received widespread publicity and helped bring the debate over roadless area protection into the public's eye.

In the coming years we will continue these efforts. The forest products industry is in a state of transition, and additional conservation opportunities will undoubtedly present themselves. We continue to add data that will assist us in identifying those areas most appropriate for conservation action. For example, we recently obtained the detailed forest-type map for Maine developed by the U.S. Fish and Wildlife Service's Gap Analysis Program that allows much more detailed examination of the nature and distribution of wildlife habitats across the state.

Central Appalachian Study

In keeping with our mission to promote the protection, enjoyment, and wise use of the mountains, rivers, and trails of the Appalachian region, AMC has been conducting a study of the key natural and back-country recreational areas of the Central Appalachian region. The ArcView GIS and ArcInfo® software received through the ESRI Conservation Technology Support Program has enabled us to bring the powerful capabilities of GIS to bear in this endeavor. We have gathered GIS information from such diverse sources as the U.S. Geological Survey, state agencies, and a host of nongovernmental organizations into a

single database for the Central Appalachian region. Not only does this information meet our research needs, but it also serves as a resource for other conservation organizations looking for regional information for this area. The process of assembling and maintaining a GIS database brought us into contact with numerous organizations (such as Appalachian Regional Commission, USGS Biological Resources Division, Natural Lands Trust, etc.) with whom we can now share information. By sharing information, unnecessary duplication of work is minimized, increasing the overall efficiency and effectiveness of the entire conservation community. Regional landscape analysis is quickly becoming recognized as an

important way of considering environmental problems, and AMC's newly created Central Appalachian GIS program is poised to supply the information required to conduct studies operating at this scale.

One of our main goals is to use the information to assess priority areas for conservation throughout the Central Appalachian region, taking into account outdoor recreational opportunities as well as natural features. We were able to identify key large open space areas (chiefly by examining land cover and demographics information) throughout the region. We are proceeding to incorporate more data, such as water and air quality information from the EPA and forest fragmentation data provided by the USGS Biological Resources Division, to rank these areas in terms of their relative ecological value and the level of

threats they are facing. When complete, this assessment will be used to identify priority areas for AMC involvement and to guide our conservation, advocacy, education, and recreation management efforts in the region.

One of the initial priority areas that has emerged from the study is the Central Appalachian Highlands—the easternmost ridgeline of the Appalachians that forms a contiguous green corridor extending from the Berkshires in Massachusetts down to Harper's Ferry in West Virginia. In examining our initial maps of land cover in this region, this area was clearly identifiable and remarkable due to its proximity to the heavily urbanized and expanding areas of this region. As a member of the Highlands Coalition, a group of organizations working to protect a portion of this area (predominantly focused on New York and New Jersey), we have been utilizing these maps to demonstrate the potential extent of the highlands region and to engage conservationists in neighboring states in this effort. Our long-term goal is to create a Vision Map of priority areas for land protection in the highlands that will serve as the basis for the coalition's land conservation and advocacy work in the region.

During the course of our work, it became abundantly clear that there is a severe lack of spatial data in regard to recreational use of the landscape. We were able to use ESRI software (specifically PC ARC/INFO) to construct digital maps of trail systems in the New York/New Jersey area, and canoeable waters within the study area (including rapid class). We foresee the need to create more accurate and extensive trail information layers than currently exist, as well as the development of layers pertaining to outdoor recreational activities other than hiking (e.g., camping or climbing) to be used in supporting conservation and recreation management efforts in the region. We anticipate that the Appalachian Mountain Club and our mid-Atlantic GIS program will be able to play a vital role in developing this information base.

The establishment of a GIS database for the Central Appalachian region will also be valuable to the emerging effort to protect forests throughout the east. AMC is part of an exciting new partnership with the Wilderness Society, Biodiversity Project, Northern Forest Alliance, and Southern Appalachian Forest Coalition to identify common issues and opportunities for strategic coordination on forest protection issues. Our Central Appalachian GIS program will provide data as well as mapping and analytical capabilities for the central portion of this region. In conjunction with our Northern Forest GIS program and our partners to the south, we can begin to map and analyze the forests along the entire Appalachian chain in order to raise public awareness and identify priorities for land protection. AMC has received funding for our role in this larger effort, and has several grant proposals pending. Our Central Appalachian and Northern Forest GIS programs will be critical to undertaking this effort. The software grants we have received from ESRI have helped to lay the foundation for what we believe is a very important and overdue campaign to ensure that eastern forests receive the same level of attention and protection as their counterparts in the west.

Katahdin Alpine Plant Communities

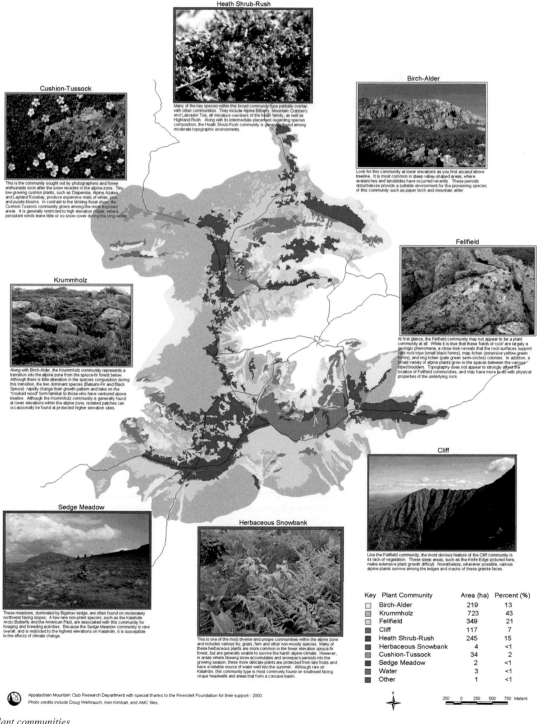

Katahdin alpine plant communities.

The Conservation Fund

Will Allen, GIS Analyst, The Conservation Fund

A Regional Approach to Conservation Planning and Hazard Mitigation

Why a Regional Approach?

The New River watershed encompasses more than 4.4 million acres of land in West Virginia, Virginia, and North Carolina and includes vast natural and engineered resources that must be protected against potential hazards to the fullest extent possible. Assessing hazard mitigation priorities for such a large geographic area requires both the local-scale, community-based approaches of hazard mitigation planning as well as the landscape-scale, regional approaches of land and water conservation planning. Understanding fully a local community's existing and potential hazards requires a regional analysis of a community's geographic location in the watershed, the characteristics of the surrounding watershed, and its physical proximity to potential risk factors. Based on this regional geographic analysis, some land conservation and restoration techniques may be helpful in reducing the hazard risk of communities and can complement traditional, community-based hazard mitigation approaches.

Why Conservation?

Selected land conservation and water quality protection/restoration techniques may be helpful in making communities disaster resistant through (a) retention of windbreaks that may reduce impacts from high winds and severe winter storms; (b) protection of high-quality water supplies; (c) retention of buffering functions of open space or forestlands; (d) reduction of potential impervious surface area coverage in the watershed; (e) reduction of storm water runoff, siltation, and sedimentation that exacerbate flooding events; and/or (f) detention of floodwaters and mitigation of hazardous material spills through creation or maintenance of riparian buffers. Existing state and federal natural areas provide some of these mitigation benefits; voluntary conservation techniques may include planting of vegetative buffers, sale of undeveloped lands to conservation organizations or agencies, or protection of working landscapes through the use of conservation easements.

Why Is Geography Important?

To gauge the vulnerability of the New River watershed communities to natural and man-made hazards, the communities were asked to respond to a hazard assessment questionnaire. Not surprisingly, the survey responses varied significantly based on the geographic differences between communities. Lowland communities more frequently identified flooding as a serious hazard faced in their community.

Upland communities often identified landslides or other geological hazards as a more serious threat than flooding. Generally, a higher percentage of communities with larger populations and more staffing resources responded to the survey than did less-populated communities, highlighting the need for additional outreach, education, technical assistance, and mentoring efforts. Although the hazard assessment questionnaire provided critical information about potential risks in communities that completed the survey, significant gaps remained in understanding the New River watershed's hazard mitigation priorities and the regional geographic context of the watershed's communities.

In addition to soliciting community input to defining hazard threats in the New River watershed, existing geographic data sources for the entire 4.4 million acre watershed were collected to inventory existing and potential hazards as well as significant natural resources. This data was integrated into a GIS to perform a watershed-based analysis that highlights the potential for coordinating hazard mitigation efforts with land and water protection strategies. The results of the analysis identify first-cut, community-based priority areas where land and water conservation and hazard mitigation techniques could work hand in hand to create more disaster-resistant communities while protecting resources with significant conservation values.

GIS Mapping Methodology

Mapping Goals

The goals of the regional GIS analysis were to (1) provide a regional geographic context for hazard mitigation activities in the watershed; (2) assess local communities' geographic location in the watershed, the characteristics of the surrounding watershed, and its physical proximity to potential risk factors; and (3) identify locations where land conservation and restoration techniques can work in tandem with hazard mitigation to create more disaster-resistant communities while protecting resources with significant conservation values.

To achieve these goals, we reviewed the inventory of existing GIS data; assessed the quality of the data and its fitness for use for this application; and selected, analyzed, and integrated GIS data that would assist in achieving the above goals. We were limited by the lack of available FEMA Q3 data and detailed hydrography data for the entire watershed, so we developed alternative indicators to assess flooding and landslide risks. In addition, we did not

have detailed data on hazardous material transportation (e.g., hazardous material manifests and railroad incidents), so we developed alternative indicators to assess abandoned mines and hazardous material risk. A review of the core data elements selected for these indicators is included below.

New River Corridor and Buffer

Conservation planning often focuses on the creation and maintenance of riparian buffers along stream and river corridors as an option to maintain water quality, protect riparian ecosystems, and reduce impervious surface areas. Establishment of riparian buffers through the use of conservation easements in the New River watershed provides a potentially cost-effective solution to reducing storm water runoff, siltation, and sedimentation that can exacerbate flooding and landslide events. Creation of upstream riparian buffers also has the potential of reducing hazard mitigation costs in some cases by reducing the cost of engineered solutions or relocations.

A river or stream corridor should be wide enough to effectively perform the function of controlling water and nutrient flows from upland to the stream. To accomplish this objective, the corridor should include the floodplain, both banks, and an area of upland on both sides that is sufficient to protect water quality and riparian ecosystems. Based on this recommendation and the resolution of the GIS data available, we buffered the New River corridor by 900 meters and designated it as a regional conservation priority in our analysis. We selected 900 meters to provide an effective visual representation on maps at a watershed scale and to ensure that we incorporated all of the existing wetland and riparian buffer areas. While this is appropriate for a regional conservation planning analysis, future local-scale efforts should analyze the hydrology and topography to determine an appropriate buffer width for particular subwatersheds of the river.

Without detailed hydrologic data for the entire watershed, we focused our attention on the primary New River corridor by virtue of its historic and cultural importance. As a result, the conservation priority analysis is inherently biased toward communities and natural features in close proximity to the primary river corridor. While many of the hazard questionnaire respondents with high flood risks were in close proximity to the New River, a few communities along the Bluestone River, a tributary to the New River, were clearly high-priority locations for hazard mitigation based on an overlay of flood risk, land cover types, and steep slopes. Future local scale efforts can better assess opportunities for integrating conservation and hazard mitigation activities in river and stream corridors other than the major New River corridor in the watershed.

Steep Slopes/Digital Elevation Model

In association with land-use disturbances, such as road construction, agriculture, and development, steep slopes increase erosion and storm water runoff velocity that exacerbates flooding and increases the risk of landslides. The Natural Resources Conservation Service (NRCS) has developed slope guidelines that identify high-risk lands. For agricultural applications, special precautions should be taken when slope inclines range between 8 and 15 percent. Agricultural activities on slopes with a gradient above 15 percent are at high risk for erosion and runoff problems. For developed land, special precautions should be taken when slope grades range from 12 to 20 percent. Development in areas where slopes exceed 20 percent grade are at high risk for erosion and runoff problems.

As a result of these guidelines, we used the Digital Elevation Model (DEM) topography data set to identify all areas in the watershed with a slope greater than 15 percent and designated those areas as conservation priorities. We also overlaid the steep slopes' data with the watershed's human settlements to identify communities with greater potential risks based on their physical proximity to steep-sloped areas. The Conservation Priorities Map (not shown) overlays human settlement/disturbed land cover, the EPA sensitive areas data layers, and a conservation priority model. The model is a simple additive overlay of slopes greater than 15 percent, managed areas plus a 300-meter buffer, and the New River corridor plus a 900-meter buffer. Lands were identified as high priority if they met all three of those geographic conditions, medium priority if they met two conditions, and low if they met one condition. Areas not meeting any of the three conditions were not considered a conservation priority at this regional scale. Note that most of the sensitive area features fall within the conservation priorities model, which both validates the model and the accuracy of the managed areas data layer that we utilized.

A large percentage of the area included in the conservation priority map is already protected as national or state parks or designated as a Wild and Scenic River corridor. Of the remaining areas, national and state forests are managed for multiple use and therefore may or may not have sufficient protection to achieve conservation or hazard mitigation goals. Land management changes, rather than land acquisition or conservation easements, would be an appropriate tool in these locations to reduce hazard risks. Coordination with local land managers in these locations may reduce flood and landslide risks downstream. The remainder of the conservation priorities are privately held lands that are either (1) adjacent to existing managed areas, (2) inside the New River buffer area, or (3) have slopes greater than 15 percent. When local-scale conservation and hazard mitigation activities are undertaken for a priority watershed community, lands that meet these conditions within a community's watershed should be precisely identified to determine if easements or other conservation planning techniques are an appropriate tool to complement local hazard mitigation activities.

Conclusion

This regional approach to conservation and hazard mitigation has provided the critical geographic context for identifying priorities and has exemplified where these two often separate planning processes that complement one another create more disaster-resistant communities and protect natural, cultural, and historic resources of significant value.

New River Watershed

The Conservation Fund

Will Allen, GIS Director
The Conservation Fund
North Carolina Office
P.O. Box 271
Chapel Hill, NC 27514
919-967-2230
WAllen@TCF.arcane.com

Map Prepared by The Conservation Fund
UTM Zone 17, NAD27 Datum, Meters

Scale 1:400,000

Legend

Cover Type

- Open Water
- Wetlands
- Forest
- Agriculture
- Human Activity

Pacific Biodiversity Institute

Peter Morrison, Executive Director   peter@pacificbio.org
Jason Karl, GIS Analyst   jason@pacificbio.org
http://www.pacificbio.org

Watson Falls in the proposed Mount Bailey National Monument.

The Pacific Biodiversity Institute (PBI) conducts research and provides scientific and technical support in ecology, conservation biology, and natural resource management. Our activities focus on conservation of biodiversity and maintenance of ecological integrity in the Pacific region. We develop advanced analytical tools and information that aid the conservation of biodiversity. We provide access to these tools and information to public agencies, educational institutions, and nonprofit conservation organizations. PBI also produces publications related to conservation and natural resource management, sponsors conferences and public forums, and develops educational materials and programs. We have provided extensive GIS support for hundreds of conservation efforts in North America.

Pacific Biodiversity Institute's mission is to serve as a source of credible information and technical support on critical conservation issues. We serve as a public resource center, providing critical conservation research, conservation support, information, and data related to conservation and biodiversity issues. PBI functions as a GIS service center, particularly for conservation groups in the Pacific region. We have provided technical assistance to more than eighty different groups, agencies, and institutions and have gained a reputation as one of the premier conservation GIS service providers in the Pacific Northwest. We have worked on more than a hundred critical conservation projects. Our work on many of these projects has led to tremendous conservation victories that have protected thousands of hectares of pristine habitat and aided the survival of many endangered species.

One of our goals has been to build the capacity of conservation organizations to use advanced technology such as GIS. We have developed conservation GIS training materials and offer conservation GIS training classes that have

enabled many individuals and groups to use this technology to help protect nature. We helped form and have participated in CTSP and the Society for Conservation GIS (SCGIS).

Pacific Biodiversity Institute makes extensive use of GIS- and Internet-based technologies to provide technical support and conservation information to aid many conservation efforts. One of our flagship projects is the Endangered Species Information Network (ESIN)

found at www.pacificbio.org/ESIN/Infopages/intropage.html. This Internet application provides extensive information on endangered species and other species of conservation concern in the Pacific Northwest. It provides maps of species ranges and occurrences, information about population status, natural history, threats, and so forth. One of its goals is to help in developing a deeper public understanding of the nature and seriousness of the current biodiversity crisis. Through the

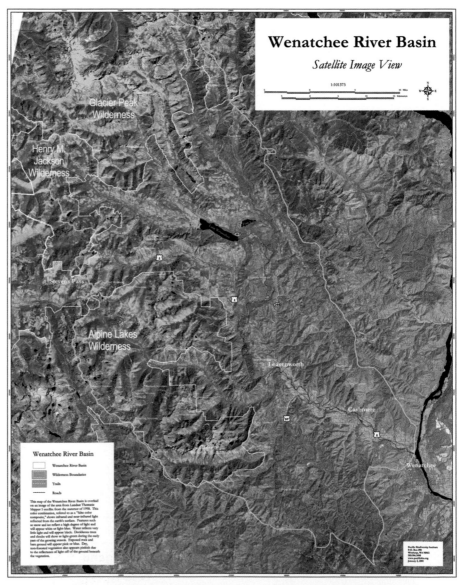

This map of the Wenatchee River Basin in Washington is overlaid on an image of the area from Landsat Thematic Mapper 5 satellite from the summer of 1998. This color combination, refered to as a "false color composite," shows infrared and near-infrared light reflected from the earth's surface. Features such as snow and ice reflect a high degree of light and will appear white or light-blue. Water reflects very little light and will appear black. Deciduous trees and shrubs will show as light-green during the early part of the growing season. Exposed rock and bare ground will appear pink or blue. Dry, nonforested vegetation will also appear pinkish due to the reflection of light off the ground beneath the vegetation.

ESIN we are helping develop a better public understanding for why species are becoming extinct, how the threats occur, and most importantly, what can be done to reverse the current trend. Over the next two years PBI is greatly expanding the information provided via the ESIN.

We are continuing to develop a second flagship project—the Wildland Information Network (WildInfoNet). A prototype is already operational; see www.pacificbio.org/wildinfonet/wildinfonet.htm. This Internet information network will provide information on size, location, management status, vegetation, fish and wildlife habitat, presence of endangered and threatened plants and animals, wetlands, erosion potential, geology, mineral potential and mining claims, past logging, insects and diseases, climate, topography, and other important attributes of every wild area in the Pacific Northwest. We also are making available satellite imagery and aerial photography covering many wild areas. We will be putting this information on an Internet map and data server with interactive database capability. This will enable all conservationists in the region to have full access to this extensive informa-

tion. The WildInfoNet allows access to vast conservation information without requiring individuals to acquire new computer hardware, new software, or to learn new computer skills.

Undisturbed wildlife habitat in Washington is disappearing at a disturbing rate. We estimate that an average of 250,000 acres per year has been lost over the last 100 years. Regional biodiversity depends on these wildlands. To achieve protection and intelligent management of wild areas in the state, there is a need for much better information on the characteristics of and conditions found within these areas—and to make this information available to conservation activists, scientists, agency personnel, decision makers, and the public at large. PBI has accumulated an immense library of information on Washington state's wildlands, and we receive regular requests for pieces of this information library. But there is a need to make all of this information available in a form that is readily accessible, easy to access, and easy to use.

Our goal is to have a greatly enhanced version of the WildInfoNet up and running by August 2001 and to continue to enhance its capability

over the next two years. Once the new WildInfoNet is functional, PBI will assist the Wild Washington Campaign, a coalition of more than twenty conservation groups working for the protection of Washington's remaining wild areas, in assessment of the characteristics and

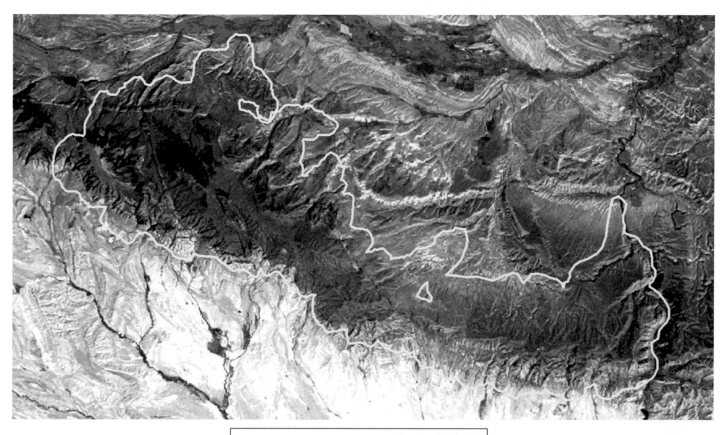

This satellite image from 1992 shows the landscape characteristics of the Kate's Basin fire area. The white, blue, and light green colors indicate sparse vegetation or grasslands. The bright red areas are lush vegetation (e.g., irrigated agriculture, riparian). The gray-red color indicates juniper woodlands. The very dark red patch in the western portion of the burn is coniferous forest.

Kate's Basin fire perimeter with satellite image.

Kate's Basin Fire Perimeter (8/19/00)

Data Sources:
Fire Perimeter: PNW National Interagency
    Incident Management Team 3
    www.wildlandfires.com\pnw_team3
Satellite Image: Landsat Multi-Spectral Scanner
    edcwww.cr.usgs.gov

Map produced by:
Pacific Biodiversity Institute
www.pacificbio.org
August 25, 2000

5   0   5 Miles

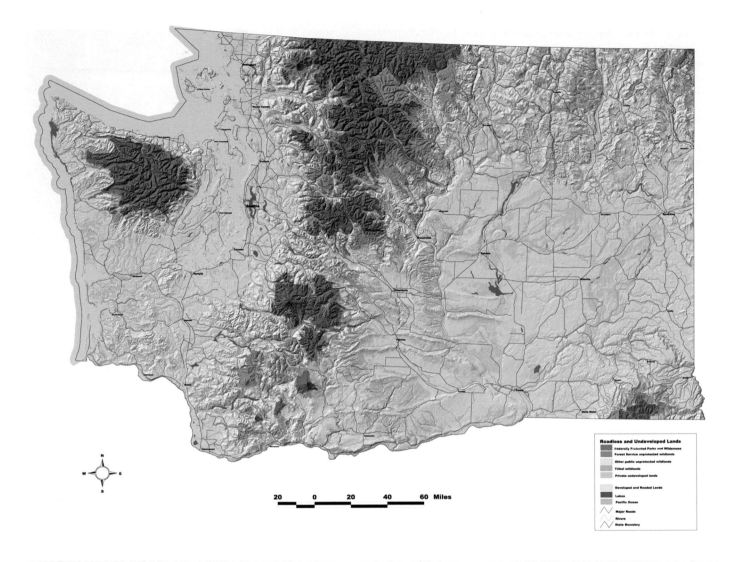

20   0   20   40   60 Miles

Wild and Roadless Lands in Washington State
Pacific Biodiversity Institute http://www.pacificbio.org

This map portrays a comprehensive analysis of all roadless and undeveloped regions in Washington state, completed by the Pacific Biodiversity Institute(PBI). This work is a continuation of an independent evaluation of wild landscapes in eastern Oregon and Washington that Sierra Biodiversity Institute (SBI) began in 1993 as part of the Eastside Forests Scientific Society Panel Study (Henjum et al. 1994). It represents part of our effort to map the wildlands of the western United States. This analysis is the first to look at roadless/undeveloped regions across all ownerships (federal, other public, tribal and private).

The methods to delineate roadless/undeveloped regions are similar to those used by the Eastside Forests Scientific Society Panel Study (Henjum et al. 1994), with a number of enhancements. All undeveloped lands (i.e., not urban, agricultural, or otherwise permanently developed) more than 100 meters from a road and over 400 hectares (1,000 acres) in size are designated as roadless/undeveloped regions. Appendages and narrow connections less than 200 meters in width are eliminated from the roadless/undeveloped regions because these areas are dominated by edge effects. Small, irregular roadless and undeveloped regions that approach the minimum size but have area/perimeter ratios of less than 80 percent of that of a circle are also eliminated because of the strong influence of edge habitat. Complex GIS analysis was used to identify landscapes that meet these criteria resulting in precise, repeatable, and consistent coverage across large regional landscapes.

The source data layers used in this analysis are:

Description scale date Land use / Land cover data from the USGS 1:250000, 1980s TIGER road data from the US Dept. of Commerce 1:100000, 1992 Colville National Forest road data 1:24000, 1995 Gifford Pinchot National Forest road data 1:24000, 1995 Mt. Baker–Snoqualmine National Forest road data 1:24000, 1995 Okanogan National Forest road data 1:24000, 1996 Olympic National Forest road data 1:24000, 1995 Umatilla National Forest road data 1:24000, 1995 Wenatchee National Forest road data 1:24000.

The accuracy of the roadless/undeveloped region mapping is dependent on the accuracy of the source data layers. We used the most reliable data that was available for each area of the state. The mapping is more reliable on National Forest lands than other ownerships, due to the recent, large scale source layers. Although no accuracy assessment has been made of this map, our reviews suggest that the accuracy of mapping on National Forest lands exceeds 90 percent. The accuracy on other ownerships appears to exceed 70 percent.

Federally protected wildlands (Wilderness Areas and undeveloped portions of National Parks) amount to about 4.4 million acres—only 10 percent of the land area of the state. There are currently about 3.7 million acres of unprotected wildlands (over 1000 acres in size) in the state's National Forests. Some of these lands have partial administrative protection. In other public ownerships (federal, state, and local) there are over 2.3 million acres of additional wildlands. Most of this land has no formal protection. There are over 1.5 million acres of undeveloped tribal land and over 4.4 million acres of undeveloped private land in the state. Overall, undeveloped wild lands still remain on about 38.5 percent of the state's land base.

This work was made possible through the generosity of the following corporations and foundations: Environmental Systems Research Institute, The Hewlett Packard Co., The Flintridge Foundation, Northwest Fund for the Environment, The Hewlett Foundation, The Bullitt Foundation.

condition of Washington's wild areas. The WildInfoNet will be critical to this work, allowing PBI to quickly post updated information, the activists to produce maps to meet their needs, and the ability to submit information on wild areas online. While this work will continue for a long time, the majority of it will occur within the next eighteen to twenty-four months.

PBI has also initiated work to further develop our ESIN, an invaluable public educational resource available on our Web site. We have found that one of the challenges in achieving effective conservation policy is to develop a deeper public understanding of the nature and seriousness of the current biodiversity crisis. That is, to get beyond the lists of endangered species and to develop an understanding of what lies behind the lists: why species are becoming extinct, how the threats occur, and most importantly, what can be done to reverse the current trend. By raising the level of understanding of species endangerment, we can increase the effectiveness of policy responses to it. This is the goal of the ESIN. PBI is in the process of developing the ESIN beyond its current skeleton form into an engaging, comprehensive, and up-to-date information source on the biodiversity crisis that can be easily maintained. Our project will:

- Provide the latest information on endangered species success stories. This is a critical element of building public support for biodiversity and the Endangered Species Act; informing people about how the act can be effective (and what would have happened without it).

- Expand the current ESIN into a larger bioregion-based network. This can serve as a model for other regions of the country and ultimately would be interconnected with those regions.

- Provide up-to-date lists of all organisms that have special state or federal status.

- Provide the latest information: photographs, legal status, population trends, historic and current distribution, habitat, natural history, threats, what is being done to restore/protect them, research and conservation references, and Web links.

- Provide the latest information on species extinctions—case studies with pictures of species that have recently gone extinct—to document the losses and to show people what it is we are losing and the reasons for the loss.

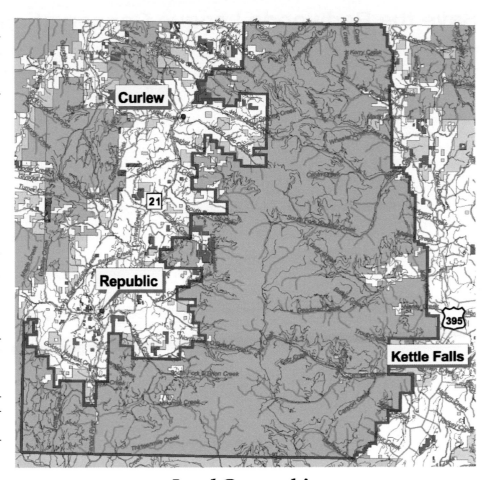

Land Ownership
Columbia Mountains National Monument (proposed)

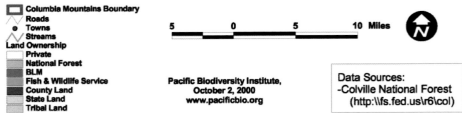

Pacific Biodiversity Institute,
October 2, 2000
www.pacificbio.org

Data Sources:
-Colville National Forest
(http:\\fs.fed.us\r6\col)

- Provide general information on endangered species; explain the various levels of endangerment and the ESA.

- Provide a mechanism for gathering information and maintaining currency.

For all of this to happen, the ESIN needs to be restructured from its current format and implemented on advanced Internet serving technology. The initial restructuring of the ESIN will be completed by September 2001. The expansion of the ESIN's scope and information resources will be completed by spring of 2003 with the help of PBI interns and students from Sonoma State University in California.

ESRI is proud to announce that PBI is the winner of the 2001 "Special Achievement in GIS" award for Conservation

Save the Redwoods League Conservation GIS Program

Ruskin Hartley
redwoods@savetheredwoods.org
www.savetheredwoods.org

Summary of GIS Activities

The GIS program of the League began in summer 1997 through the support of the Conservation Technology Support Program (CTSP). The original grant of an HP computer and ESRI's ArcView GIS software has since been augmented through the generosity of the ESRI Conservation Program (ECP). Through this program, we have expanded our capabilities with the addition of the following:

- ArcInfo

- ArcView 3D Analyst™

- ArcView ArcPress™ extension

This has greatly increased the range and quality of the maps we have been able to produce. GIS maps are now an integral part of all land acquisition decisions, presentations, donor communications, and grant requests.

Land Acquisition Program Presentation to State Parks

The California Department of Parks and Recreation is developing a process to prioritize land acquisitions made possible by the passage of Proposition 12 (the "Park Bond") in 2000. The League has been working with the department to prioritize redwood-related land purchases. In December 2000 we produced an internal report, "A Conservation Strategy for the Coast Redwood," for presentation to state parks. This identified priority projects at three scales:

- Landscape-scale projects

- Critical watershed projects

- Land currently held by the League pending transfer to state parks

ArcInfo and ArcView GIS were used to compile the database on which this analysis was conducted. Report maps were produced using ArcView GIS (see example right).

Corridor from the Redwoods to the Sea Project

To date, the League has acquired more than 5,000 acres of land for permanent protection in the Gilham Butte Conservation Area including Douglas Fir forests and critical upper watershed lands of the Mattole Valley. Much of the funding for this project has come through the Wildlife Conservation Board (WCB), which approved a Conceptual Area Plan for the project. We are currently in the process of transferring parcels to the BLM and state parks for long-term management and are negotiating conservation easements with local ranchers.

ArcInfo and ArcView GIS were used to compile a database and develop maps supporting the successful application to WCB. We continue to update these maps as new properties are considered for addition to the project. A recent map for a League Board of Directors meeting is shown on the next page.

Mill Creek Project

The League is working on a major project to permanently protect 25,000 acres of forestland at the northern end of the redwood range. The project will permanently protect the entire Mill Creek and Rock Creek watersheds, tributaries to the Wild and Scenic Smith River, and home to some of the state's best salmonid runs. It also greatly strengthens landscape connections between the Redwood National and State Park and the national forests of the Klamath–Siskiyou ecoregion.

We are currently negotiating a purchase option and working to secure major funding. A land acquisition evaluation, highly ranked by the Lands Committee of the Wildlife Conservation Board, included GIS mapping and analysis. In addition, a briefing book has been prepared that included vegetation transects from the property generated using ArcView Spatial Analyst and Adobe® Illustrator®.

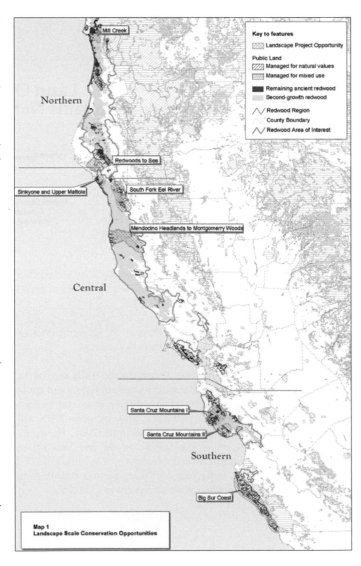

Map 1
Landscape Scale Conservation Opportunities

Master Plan for the Redwoods

The master plan is being developed to refocus the League's conservation action toward protecting the ancient redwood forest landscape. It is a scientifically based conservation plan that uses GIS tools to focus, assess, and present conservation action. The land acquisition projects presented above embody the concepts developed in the master plan. In addition, the following master plan projects made use of GIS tools and techniques.

GeoWorld Article

The February 2001 cover article for *GeoWorld* featured the league's master plan for the redwoods titled, "Map the Redwoods, Save the Redwoods." GIS is playing a critical role in the continuing development of the master plan. This article placed in context this work and the central role of GIS in developing and presenting the plan. For complete text of the article see www.geoplace.com/gw/2001/0201/0201emg.asp.

North Coast Stewardship Group

In partnership with the Bureau of Land Management, the League convened a group of twenty-five federal and state agencies, and national and local nonprofit organizations to discuss stewardship strategies for the California North Coast region. The goal of the group is to "identify the key nonregulatory actions to be taken over the short to medium term (up to five to ten years) that contribute materially to the goal of maintaining and restoring a fully functioning (natural and healthy) ecosystem across the region, being rooted in the low elevation ancient forest, riparian/aquatic systems, and coastal habitats."

The group, led by the League, is developing a briefing book to describe the region's key resources, identify ongoing conservation action, and set the context for future conservation actions. GIS maps have been used as group facilitation tools and will be used extensively in the briefing book.

Future Development

GIS will continue to play a central role in our land acquisition program and in the master plan. We expect to widely distribute the master plan in a document making extensive use of maps, color photos, and illustrations. We continue to work with GreenInfo Networks to supplement staff resources in support of our GIS program.

We are in the process of developing a redwood visitor center at the Hartsook Inn, Humboldt County ("Hartsook Inn Redwood Center"). We are working with a team of consultants to create a site plan, at the heart of which is a world-class redwood visitor center. We envisage implementation of the plan to take a number of years. To meet the needs of the coming summer season we plan to develop a simple visitor center focused on our work (including the master plan) and future plans for the inn. The maps created using GIS will be at the heart of this center, creating striking visual images of the current state of the redwood forest. We expect these to reach a wide audience of summer travelers.

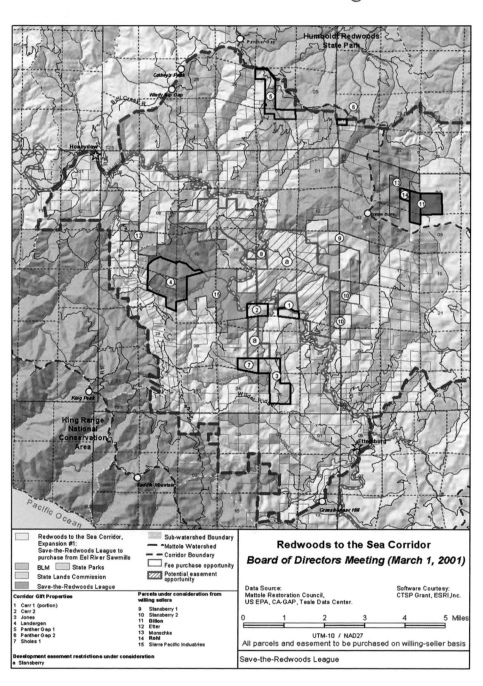

Southern Appalachian Forest Coalition: GIS Program Progress Report

Hugh Irwin, Conservation Planner   hugh@safc.org

The mission of the Southern Appalachian Forest Coalition is to protect and restore the wildlands, waters, and native forests and ecosystems of the Southern Appalachian landscape. We seek to ensure protection of the existing public land base and add substantially to that base through voluntary means in order to provide benefits for low-impact recreation, fisheries and wildlife, scenic beauty, and sustainable, healthy communities. Through advocacy and education we will build a supportive constituency, foster conservation leadership, and enhance the regional land ethic to create a political reality that enables protection and restoration of all that we cherish in the Southern Appalachians.

The Southern Appalachian region is one of the most biologically diverse regions in North America and has unique qualities that make it globally significant. The Southern Appalachian Forest Coalition seeks to protect and recover the biological diversity of the region through conservation planning and public lands management. The region is swiftly changing from a rurally based economy to an urban economy. This is a critical time to plan for the biological health of the region because development, population, and resource pressures are increasing dramatically. The potential exists to protect the region's tremendous biological values and even recover species that are extirpated or have declined. This protection and recovery depends to a great extent on the region's public lands. The public lands, if managed for biological diversity, could form the cores of bioreserves with river riparian areas, greenways, and conservation easements unifying these core areas into a network of conservation areas.

Besides high technology development, tourism forms a basic building block for the economy of the region. In addition, the natural beauty of the region is an attraction for business as well as individuals. Quality of life in uncongested communities close to nature is an appeal to many people. Nature-based tourism, an appreciation of quality of life, and high-value conservation areas form a set of supports that could create sustainable communities living in conjunction with high-quality natural areas. However, this potential must overcome unbridled growth, an aversion to planning, and pressure to destroy the remaining wildlands.

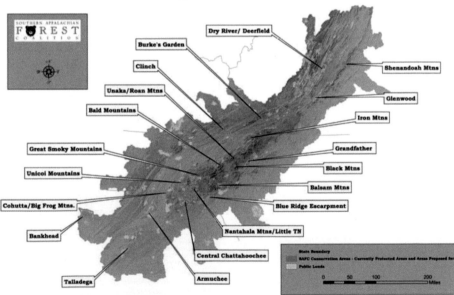

One of SAFC's goals is work aimed to create conservation plans to answer the question of what must be done to protect and recover the region's biological diversity. However, SAFC's mission goes beyond this to attempt to create the right economic conditions and community support to make protection of the region's natural heritage a regional priority. Accomplishing this mission will take a scientifically defensible plan that is presented in a persuasive manner. GIS is an essential tool to develop and present the vision of sustainable communities and naturally functioning ecosystems.

Southern Appalachian Landscape Conservation Areas

SAFC GIS and Conservation Planning Program

SAFC's GIS and Conservation Planning Program is managed and coordinated by Hugh Irwin, who started the program with a basic Conservation Technology Support Program (CTSP) grant in 1995. In addition, a GIS technician, Glen Locascio, works full-time for SAFC. The program maintains two PCs, one provided by a 1998 grant through CTSP and provided by a 1998 grant through CTSP and one purchased by SAFC. Both are running ArcView GIS, one copy provided through a 1995 CTSP grant and upgraded by SAFC, and the other obtained through a 1998 CTSP grant.

SAFC's Regional Conservation Plan

A major thrust of SAFC's work over the past five years has been the development of a plan for a regional conservation network built around landscape-scale conservation areas centered on public lands. SAFC's GIS program has spearheaded this effort and plans to publish a book promoting this vision. The executive summary of the book, *A Southern Appalachian Conservation Network: Reclaiming an Ancient Forest Refuge,* appears on the next page. This presentation has already been shown in several venues including the 2000 Southern Appalachian Man and the Biosphere Conference. A Web version of this presentation will be posted to our Web site.

This conservation vision and plan will serve as the foundation for a campaign effort to gain protection of Southern Appalachian public lands and to acquire additional public lands. We expect to develop and promote legislation over the next two years for protection and acquisition of regional public lands. We will also continue to work with cooperating groups to place key lands in conservation easements. Materials already developed and in the planning stage will play a key role in our legislative and outreach campaign.

In addition to the legislative campaign, SAFC continues to play a crucial role in USFS plan revisions and agency policies and reform (as both chief participant and organizer for other groups' participation). SAFC has been the major conservation group involved in plan revision efforts. SAFC has also played a pivotal role in negotiating the roadless policy through to administration approval. SAFC preparation of maps and analyses for these processes has been significant and decisive.

SAFC's GIS Support to Southern Appalachian Groups

SAFC has continued its GIS support of groups within the region. In some cases this assistance has been in the form of GIS maps produced at SAFC. In other cases this assistance takes the form of providing data for groups that have access to ArcView GIS. SAFC has developed or modified many data sets that are of broad use in the region. SAFC has developed public lands coverages for the Southern Blue Ridge and the Southern Ridge and Valley ecoregions. This data is much more comprehensive and accurate than any other source. Besides using this data with The Nature Conservancy and state heritage programs for biological hot spot analysis, this data has been shared with several groups for their own work. SAFC has also developed coverages for specific parts of the region. For example, SAFC developed coverages for an acquisition priority along the Little Tennessee River. These

Black Mountains - Conservation Easements

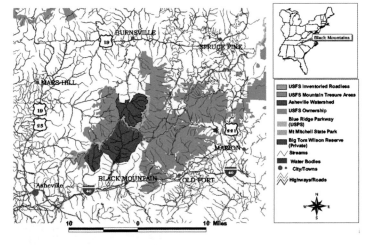

coverages have been shared with the Little Tennessee Land Conservancy, The Nature Conservancy, the Southwest Commission of North Carolina, and the North Carolina Wildlife Commission. SAFC has also provided data, assistance, and tutoring to groups that have access to ArcView GIS and ArcExplorer. These groups include the Harvey Broome Group of the Sierra Club (ArcView GIS) and Cherokee Forest Voices (ArcExplorer™).

In summary, GIS capability continues to play a pivotal role in SAFC's work. We are very grateful to ESRI and CTSP for the support they have provided in the past to help establish and build SAFC's GIS program and for the ongoing support that allows us to continue to protect and restore the lands, species, and ecosystems of the Southern Appalachians.

A Southern Appalachian Conservation Network: Reclaiming an Ancient Forest Refuge

Executive Summary

The 20th century dawned across a Southern Appalachian region widely cut, burned, and barren after just a few decades of rapacious timbering. Only in the wake of devastating floods fed by denuded mountainsides did the population at large begin to question the real cost of such industry. From that concern grew a movement, out of Asheville, North Carolina, that eventually led to the establishment of national forests in the east.

It is these forests, many of which grew back around the stumps and the rubble, that promise so much for the 21st century and beyond. But while much of the green has returned to the slopes, peaks, and valleys of the Southern Appalachians, that is only one component of an incredibly complex picture. In many ways that very canopy masks problems and concerns every bit as real and immediate as those a hundred years ago.

Our region is home to a breadth of biological diversity found in few places on earth. Yet that tapestry is frayed and in danger of unraveling as roads, timbering, urban development, and other encroachments, such as acid rain, take their toll. The threats are not restricted to wildlife and habitat. Humans, too, rely on our national forests as sources of clean drinking water, filters for clean air, and avenues for recreation and renewal.

The Southern Appalachian Forest Coalition believes the dawning of the new century sheds light on an environmental, economic, and social crossroads.

Proposals

The Southern Appalachian Forest Coalition envisions a future where the landscape serves and sustains the interests and viability of all communities, from those on the forest floor to those in our largest urban centers. Such a landscape would ensure healthy populations of all native species, perform a critical role in the generation of clean water and clean air, and satisfy increasing demand for backcountry recreation.

Toward this vision, SAFC proposes a network of conservation areas across the Southern Appalachian region. These areas would:

- Be of sufficient size and connectivity to provide habitat and sustain the ecological processes necessary for all native species, plants, and animals.

- Stand permanently protected from development and uses that conflict with the interests of wildlife, habitat, and the natural forces that play out over time.

- Provide a landscape suitable for the reintroduction of species lost to the region.

- Exist within buffer zones that provide gradual transitions from natural to urban lands.

Why?

This goal of mutual sustainability and the effort necessary for its realization are critical if we really treasure the interests of future generations and honor those past. The rich and rugged terrain of the Southern Appalachians shaped the regional character and heritage we celebrate today. As populations and cities grow, and more and more open lands are paved over, the value of our public forests and all that they provide, intact, grows exponentially.

How?

The building blocks of this regional network of conservation areas already exist in public ownership. Yet individually, none of these areas is large enough to perform as a microcosm of the original native forest. The task then is to maximize protection of these land blocks while linking them across the landscape so that they realize a capacity far greater than the sum of the component parts. Land acquisitions from willing sellers, conservation easements and other complementary private management options, would act in concert with local, state, and federal regulations to deliver such an outcome.

New Mexico Wilderness Alliance

Kurt A. Menke, GIS Coordinator

The New Mexico Wilderness Alliance is wholly dedicated to the conservation of wilderness and wildlands in New Mexico. We use a biology conservation methodology in our wilderness proposals and wildlands planning to ensure connectivity throughout the landscape to protect entire ecosystems and not just individual areas.

Carson National Forest
Tres Piedras Ranger District

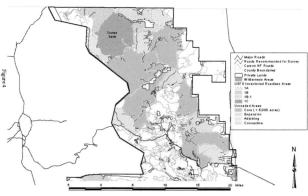

Created as part of the comments on the USFS Roadless Proposal by the New Mexico Wilderness Alliance, this map is an example of how GIS was used to locate unroaded areas not identified by the Forest Service.

Today, road building, resource extraction, and off-road vehicle abuse endanger many of our remaining wildlands. These are lands that have been used by local communities for hunting, fishing, and spiritual renewal for centuries. They are places on which New Mexico and Arizona's incredible biological diversity depends. They are key watersheds for rural communities and historical reminders of the thousands of years of indigenous culture in the region.

There are well over 2.5 million acres of potential wilderness in New Mexico; only an aggressive wilderness advocacy campaign will create the public demand to permanently protect these areas as wilderness.

We have inventoried more than 1.8 million acres of potential wilderness. The information fills more than two full filing cabinets, and we still have more than six million acres of land to go. We plan to utilize GIS technology to analyze a wide array of information available about our public lands to put together the most comprehensive wilderness proposal we can. With stronger justifications grounded in ecological, biological, and on-the-ground information that are only available through GIS technology and field work, we can make the strongest case possible for wilderness protection of these areas.

The New Mexico Wilderness Alliance is implementing an aggressive grassroots organizing and media outreach strategy to invigorate traditional supporters of wildlands protection while motivating a new generation of wilderness advocates. At the same time, we are working to provide interim protection for these wild places so that they are not ruined through the benign neglect and apathy of our land managing agencies.

We are advancing a legislative strategy to protect wilderness areas. While we obviously must have the grassroots power to incite change, we must also be able to assemble and utilize the most up-to-date information about the public lands of New Mexico to change the way they are managed.

The New Mexico Wilderness Alliance is currently undertaking a comprehensive survey of New Mexico's public lands to assess their potential to be included in the New Mexico Citizens' Wilderness Proposal. For this endeavor, we comprehensively mapped and photo documented all human intrusions and impacts on our public lands throughout the state.

We plan to utilize mapping information to produce reports and analyses of wild areas that will aid in advocacy through public education, media outreach, and administrative action by the following means:

- Produce visual analyses of on-the-ground impacts of road building, off-road vehicle abuse, oil and gas development, and overgrazing (see map above).

- Produce visual representations of critical habitat needs of important species and ecosystems.

- Utilize statewide information to aid in the production of a conservation area design for the "New Mexico Link" region between the Sky Islands and the Southern Rockies.

Following the protection of core wilderness areas, the New Mexico Wilderness Alliance will continue to focus on the protection of compatible use buffer zones and corridors to provide ecological connectivity across the landscape through the NM Link Conservation Area Design Project and the Sky Island Wildlands Network Conservation Area Design Project.

Our GIS outreach strategy has three parts:

1 Produce easy-to-understand maps that contain relevant information to educate targeted audiences in the general public.

These would include a hard-copy map of the Citizens' Wilderness Proposal. This map will be an E-sized map and will be mass-produced. This will be sold to gather revenue, generate interest in the program, and educate the public on the content and extent of the proposal.

In addition, we would produce 1:100,000-scale maps of citizens' proposed wilderness areas. Roughly fifty E-sized maps will be needed to cover the entire statewide inventory. The maps will correspond to USGS 1:100,000 quads. The inventoried areas will be shown on color-shaded relief with roads and towns. These will aid future field work efforts.

2 Produce sophisticated maps to advocate for land protection among key decision makers.

- Produce letter-sized basemaps of each citizen's proposed wilderness area. There will be roughly one hundred areas mapped for this series. Background layers will include roads, towns, BLM areas of critical environmental concern, BLM wilderness

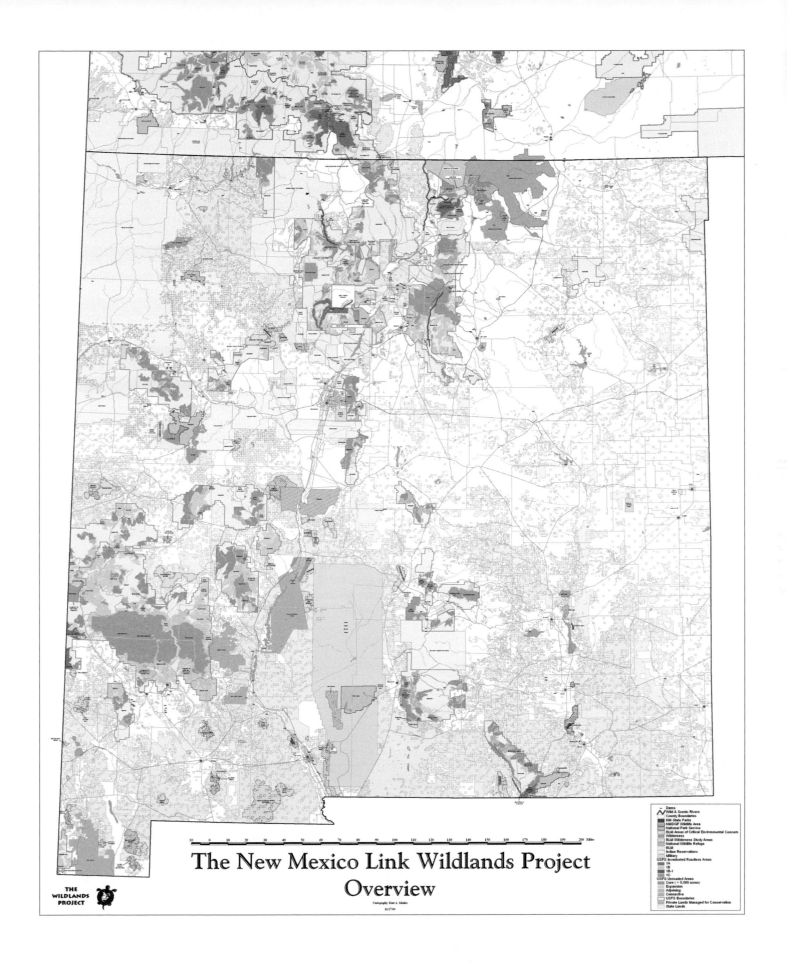

The New Mexico Link Wildlands Project
Overview

Cartography Kurt A. Menke

11/27/00

study areas, and wilderness areas. These maps will help put each area into context within the landscape to illustrate connectivity between wild areas and proximity to roads and towns. These will help to characterize each specific area and as such will help to advocate for land protection among key policy makers.

3 Utilize mapping information to produce reports and analyses of wild areas that will aid in advocacy through public education, media outreach, and administrative action.

a Produce visual analyses of on-the-ground impacts of road building, off-road vehicle abuse, oil and gas development, and overgrazing.

• Analyze acreage, ownership information, cattle grazing activity, other resource activity, and ecological conditions/character of each proposed wilderness area. This will be accomplished by clipping spatial layers to individual wild areas and gathering statistics. Data will include land ownership, GAP vegetation data, BLM grazing allotment data, and field inventory data. The analysis will be presented in an Microsoft Excel spreadsheet.

• Analyze off-road vehicle impacts on New Mexico's wildlands. The off-road vehicle impact data will come from the field inventory, and the impacts will be determined with GAP vegetation data, wildlife data from the New Mexico Department of Game and Fish, endangered species data, and the Biota Information System of New Mexico (BISON-M). This will be accomplished by buffering selected roads segments and determining what habitats are being adversely impacted. This will allow NMWA to recommend road closures necessary to protect critical habitat and endangered species. This analysis will be mapped and printed as hardcopy plots printed at a scale of 1:100,000. This will amount to 50 E-sized plots.

• Produce an off-road vehicle impact ArcView GIS application from field and office data to highlight negative impacts of off-road vehicles on wilderness values and "edge effects" on habitat, wildlife, hydrology, and native vegetation. This will incorporate some of the analyses above into a customized ArcView GIS application (programmed in Avenue) that can be circulated to the federal agencies responsible for enforcing road closures on public lands. This will also incorporate images from a digital camera hyperlinked to point locations of impact.

b Produce visual representations of "wilderness values" across the state to show high wilderness quality of proposed areas.

• Develop a "remoteness index" for proposed wilderness areas incorporating usage statistics (if available), distance from urban centers, road densities, and census data. Population data will be obtained from Census 2000. Road densities will be accomplished by computing the number of miles of road per square mile of a region.

• Analyze the wilderness characteristics and value aspect of the wilderness inventory and develop a standardized index for evaluating "wildness" and "naturalness" in proposed wilderness areas. This will be a combination of remoteness, off-road vehicle impacts, species diversity, size, and proximity to other wild areas.

• Compare wilderness values in proposed areas to those in areas not proposed for wilderness designation, and produce "Wilderness Character Map" to show overlap of high wilderness value areas with proposed wilderness areas. This will be an E-sized plot and will be mass-produced. It will be sold to gather revenue, to promote the program, and to educate the public and the federal agencies on the state of wilderness in New Mexico. Initially, thirty copies of this map will be produced.

c Produce visual representation of NMWA membership to demonstrate broad support for wilderness protection in rural areas (we already have a diverse and spread-out membership).

• Conduct a demographic analysis and begin mapping regions of support for wilderness designation within the state. This will help us determine needs for greater outreach efforts. Our membership will be geocoded, and data from other local conservation organizations will be synthesized.

d Utilize statewide information to aid in the production of a conservation area design for the "New Mexico Link" region between the Sky Islands and the Southern Rockies.

• Develop a conservation area design for the "New Mexico Link Wildlands Project" (see map on previous page). This will include those areas in central New Mexico not already included in other emerging conservation area designs, and include potential core areas, buffer zones, and corridor linkages for more holistic landscapewide habitat protection. The wildlands project will have several dozen experts from multiple backgrounds, including many NMWA staff members, develop the reserve network. GIS support for the New Mexico Link will be provided by NMWA's GIS shop.

• Produce final map of New Mexico Link Conservation Area Design. Map product is already 75 percent complete. A hard-copy E-sized map of the New Mexico Link Wildlands Network will be produced on a color-shaded relief image. As the analyses in year two are completed, they will be incorporated into future iterations of this map.

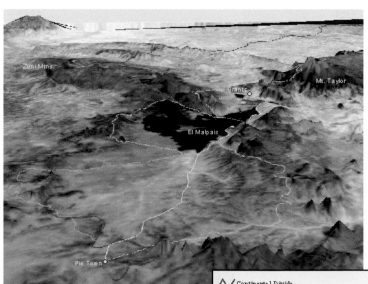

The four southern routes and a Landsat Thematic Mapper satellite image draped on a DEM view looking northeast from Pie Town.

Southern Rockies Ecosystem Project

Bill Martin, GIS Manager   wwmartin@indra.com
http://csf.Colorado.EDU/srep/

The Southern Rockies Ecosystem Project (SREP) employs science, technology, and advocacy to protect and restore ecological integrity throughout the Southern Rocky Mountains of southern Wyoming, Colorado, and northern New Mexico. SREP uses science and GIS technology to identify and map areas critical to the preservation of native biodiversity and ecosystem health in the Southern Rockies. SREP works closely with numerous other conservation organizations to influence important land management decisions, protect the region's wildlands, and serve as the main GIS resource for the Southern Rockies Forest Network (SRFN), which includes twenty grassroots, regional, and national groups.

The Southern Rockies' remaining wildlands provide incredible opportunities to protect biodiversity, yet these areas are increasingly threatened by a multitude of destructive land uses including urban sprawl, logging, road-building, and irresponsible motorized recreation use. SREP seeks to maintain and restore ecological integrity in the Southern Rockies by identifying and protecting a network of biologically critical wildlands. To accomplish this, SREP has three main programs: (1) designing and implementing the Southern Rockies Wildlands Network (SRWN) conservation plan, (2) providing GIS mapping services to local environmental organizations, and (3) conservation education, outreach, and advocacy.

- The SRWN design process utilizes conservation biology and GIS technology to identify and protect wildlands and other habitat critical to native biodiversity. In 2001, SREP will complete a mapped wildlands network

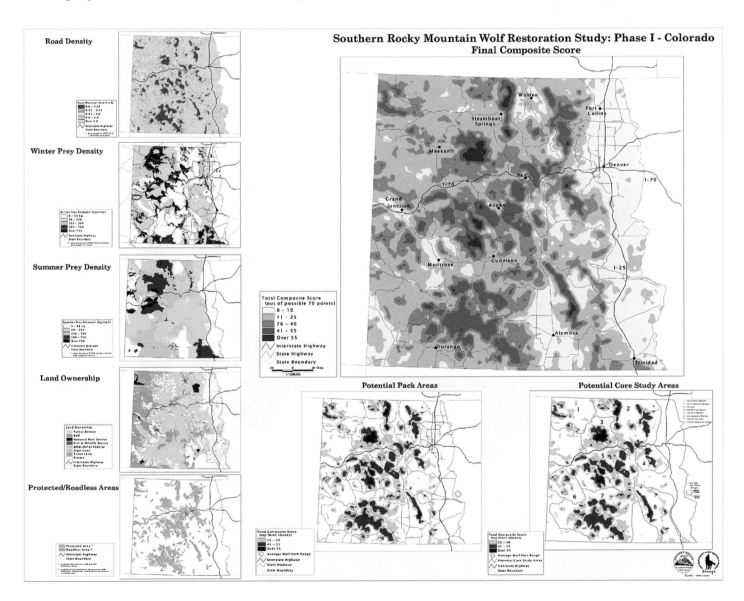

The Greater Southern Rockies Ecoregion and Interregional Connectivity

design for the Southern Rockies ecoregion that will provide a regional-scale conservation plan to protect native species and maintain sustainable human activities (see example on previous page).

• Our GIS mapping and analysis services to other conservation organizations in the region are tightly linked to the SRWN design. By providing these services to conservation organizations for public outreach efforts, litigation, and general advocacy purposes we are ensuring that the very places

SREP identifies as biologically critical are actually being protected.

• SREP works closely with conservation partners, works to forge new partnerships, educates the general public, and works with land management agencies in order to advocate for the protection of biodiversity and wildlands. These are all necessary components of implementing an SRWN conservation plan.

As a component of our GIS mapping service, SREP provides GIS services to the SRFN. This informal alliance of twenty conservation groups is dedicated to protecting and restoring the region's wildlands and native biological diversity through collaborative efforts, by using its member groups' diverse areas of expertise, and by leveraging its joint power for effective action.

SREP uses its science and GIS expertise to help SRFN succeed in its Roadless Area Protection Campaign. This campaign aims to protect remaining wild, roadless areas of

public lands through responsible management of motorized recreation, protection and restoration of habitat and native biodiversity in the region, and promotion of wilderness designation. SREP's efforts with SRFN's campaigns complement the SRWN design and conservation plan as roadless area protection is a key element in the design process.

Roadless Area Analysis

SREP has been working on roadless area issues since its inception in 1992. Utilizing data collected by citizen survey teams along with data from federal agencies (primarily the Forest Service and Bureau of Land Management), SREP has recently undertaken a multi-year process to remap and analyze the region's remaining roadless wildlands. Part of this process will involve building a comprehensive database for groups to use. SREP will also continue to produce reports, analyses, and maps about the ecological role that roadless areas play. Hopefully, this will lead to more informed decisions about how to manage these remaining wildlands. Based on the threat by

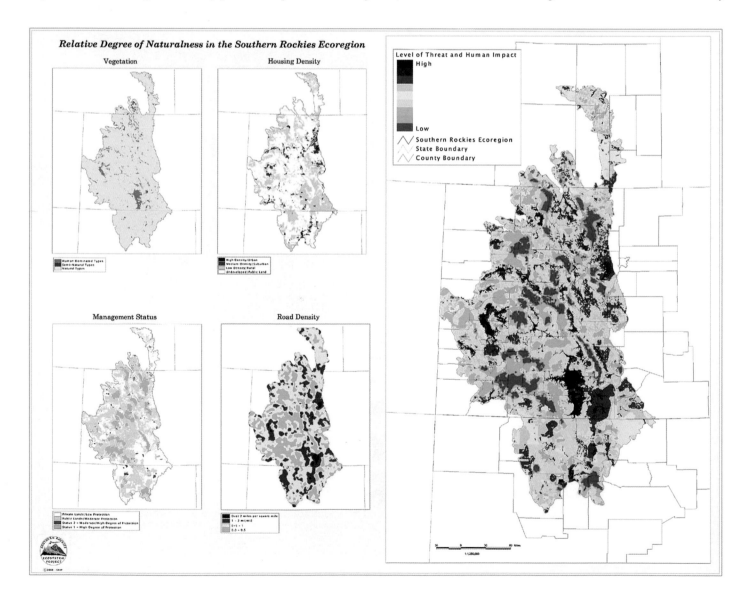

Relative Degree of Naturalness in the Southern Rockies Ecoregion

Vegetation

Housing Density

Management Status

Road Density

Level of Threat and Human Impact
High
Low
Southern Rockies Ecoregion
State Boundary
County Boundary

the new administration to overturn federal protection for roadless areas, our work is needed now more than ever.

SREP's roadless area work utilizes the power of GIS in three primary ways. First, by examining all of the recent mapping efforts to date (both citizen and federal surveys), we seek to reconcile the differences in boundaries that have evolved by differences in mapping techniques. Second, by surveying all existing and potential roadless areas and building a comprehensive database around them, we can provide data to a larger audience. And third, we will integrate all of the data into both near-term protection schemes and long-range conservation planning efforts (see map right). The reconciliation process involves overlaying all of the various survey data to establish a baseline of what has been examined and where areas have been overlooked or not considered for permanent protection. After this initial effort is done, field crews will go out and investigate all of the possible remaining roadless areas and consolidate their boundaries. This will involve both GIS and field mapping techniques to identify the "missing pieces." It will also involve the "cleaning up" of overlapping areas that have been surveyed by multiple parties. Last, data will be integrated into a comprehensive database for use by citizen activists fighting for roadless area protection as well as long-term conservation planning efforts in the region.

The most important aspect of this project is the need to expose people to why these areas are important. Utilizing GIS for education and outreach to the public, as well as helping to inform decision makers, is critical. GIS analysis has been effective in the past, relaying spatial information such as how few remote areas remain in the Southern Rockies, as well as analyzing trends, such as the loss over time of roadless areas, which help people see which regions have been most impacted. Mapping can also reinforce or dispel popular misconceptions about issues that are in the national spotlight such as where leased lands for oil and gas drilling may occur. By examining inventoried roadless areas that have active oil and gas leases on them (and hence, access constraints), GIS helped to display how few and far between these leases actually were in Colorado. Mapping efforts like these are all helping SREP and its affiliates move forward with their mission to protect and enhance the biodiversity of the Southern Rockies.

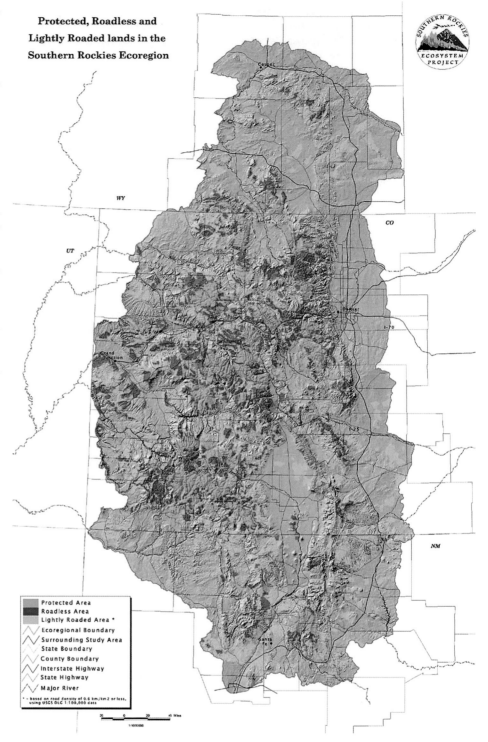

Protected, Roadless and Lightly Roaded lands in the Southern Rockies Ecoregion

Protected Area
Roadless Area
Lightly Roaded Area *
Ecoregional Boundary
Surrounding Study Area
State Boundary
County Boundary
Interstate Highway
State Highway
Major River

* - based on road density of 0.6 km/km2 or less, using USGS DLC 1:100,000 data

The Southern Rockies have more than nine million acres of magnificent wildlands that provide critical habitat for thousands of species. It is one of the few regions in the United States with potential to protect and restore all of its native species, even wide-ranging carnivores like the gray wolf. However, we must work quickly and credibly to identify and protect these areas, as most roadless lands and other important natural areas do not have permanent protection. Many are threatened by rampant development, resource extraction, road building, and recreational activities.

Yet, land use decisions are typically made without the benefit of regional-scale assessments of conservation opportunities, resulting in further loss and fragmentation of wildlife habitat. Thus, conservation efforts would benefit from a concerted effort to map these last remaining wildlands and to integrate these areas into a comprehensive conservation plan for the entire Southern Rockies, one that emphasizes protecting and restoring biologically critical areas in an interconnected system of wildlands.

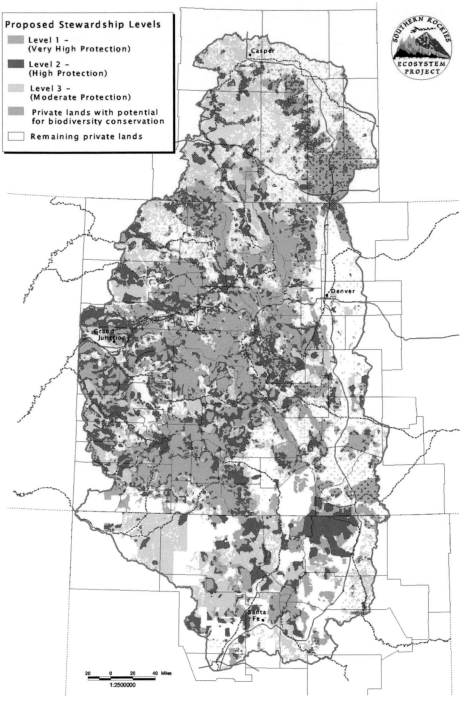

Western Slope Environmental Resource Council

Eli Lindsey–Wolcott, Mapping and GIS Coordinator   <Eli@wolcott.to>

Western Slope Environmental Resource Council (WSERC) is a 250-member community-based citizens group in western Colorado dedicated to protecting the natural environment and enhancing the quality of life for residents of Delta County. WSERC formed during the 1977 energy crisis with the goal of limiting and containing the environmental and social impacts resulting from the exponential increase in coal production in Delta County's North Fork Valley. WSERC has grown and its mission has become much broader, but its focus on finding solutions instead of perpetuating conflict has remained unchanged.

Within the past ten years, a new issue—all-terrain vehicle (ATV) use—has eclipsed the impacts of timber harvesting on public lands. Approximately 15,000 logging trucks travel through our national forests each day, the same number as in 1950. However, in this same 50-year period, the number of recreational vehicles (4-wheel drive and ATV) on forest roads has increased tenfold. An average of more than 1,700,000 recreational vehicles use national forest roads each day. This exponential increase has resulted in the creation of "ghost roads" that are not included on U.S. Forest Service maps and the reopening of formerly closed, and in some cases, bermed, logging roads.

Traditionally, this issue has been fought out as a wilderness/roadless battle, wherein the mostly urban Front Range Colorado environmental community has campaigned to fix the problem via wilderness designation. Meanwhile, the mostly rural West Slope governments, communities, and elected officials have opposed the wilderness designation as locking up public land and keeping out all local users.

But while these great political wars were raging, WSERC heard voices on the West Slope start to ask new questions. Long-time residents in small communities like ours—ranchers, farmers, outfitters, irrigators, and loggers—still oppose the wilderness designation, yet they are becoming increasingly alarmed at the damage caused by ATVs.

WSERC has been monitoring and mapping this increase of user-created roads. We have continued to document site-specific examples of problems West Slope communities are now experiencing due to the ATV boom. Perhaps the most important aspect of this research is that they are local examples and ring true

with personal experience of most area residents. The community's growing frustration with ATV onslaught has brought us nontraditional, nonenvironmental allies agreeing and supporting us. Our field mapping is working to verify the problems people are seeing on their own and gaining us respect.

WSERC Responsible Motorized Recreation Campaign

ATV Campaign

Since 1996, WSERC has worked to build a campaign documenting the impacts that new roadbuilding and the boom in recreational motorized traffic have had on forest health, wildlife habitat, and the degradation of roadless and wilderness areas. WSERC utilizes a two-pronged strategy: (a) to use mapping, photography, wildlife research, and other site-specific data to win support from other forest

stakeholders to stop the ATV onslaught, and (b) to use this same data to influence Forest Service decision making on all applicable projects, from timber sales to travel plan revisions. The following narrative report highlights our campaign activities over the past year.

Background

The Grand Mesa, Uncompahgre, and Gunnison (GMUG) national forests, located in the heart of the Colorado Rocky Mountains together encompass more than three million acres, one of the largest national forest administrative units in the United States. The GMUG forests feature spectacular mountain scenery, 14,000-foot peaks, abundant wildlife, five wilderness areas, and the headwaters of several major tributaries of the Colorado River. These forests are also home to old-growth timber, unique roadless areas, and a rich array of plant and animal species.

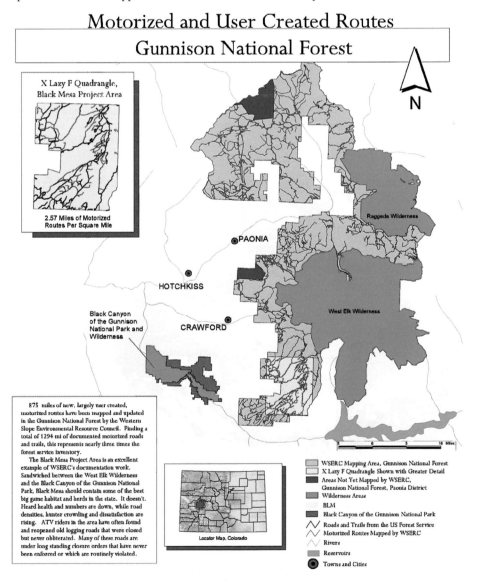

Over the last four years (1997–2000) WSERC, in collaboration with neighboring groups, has conducted detailed field surveys, documenting the devastating impacts unrestricted ATV use is causing throughout the region. With this documentation it has been possible to develop and organize support for forest planning among forest stakeholders and diverse sectors of the community.

WSERC has worked in 1999–2000 to continue with our detailed research of ATV-caused damage to wildlife habitat in local forests, compiling the data into a GIS database developed with our sister group, the High Country Citizens Alliance in Crested Butte. We have combined our documentation with the Colorado Division of Wildlife big game data and other agency information to create specialized GIS maps of wildlife impacts and road densities, "ugly" photo albums exposing the ATV problem, and a touring slide show.

Our ATV Hunting Brochure was picked up by the Colorado Wildlife Federation and Great Outdoors Colorado for statewide publication.

We have also used our GIS data and photodocumentation to advocate for better Forest Service decisions on timber sales and as the basis of WSERC's portion (HABCAP analysis) of the environmental coalition's recent appeal, the Uncompahgre Forest Travel Plan.

Site-Specific Documentation:
Field Inventories and GIS Capabilities
Field teams, including staff, several board members, volunteers, and interns, have now mapped approximately two-thirds of the Paonia Ranger District of the Gunnison National Forest (about 300 square miles), and plan to complete the remaining third in the coming months. During time in the field, mapping groups have had numerous opportunities to share viewpoints and information with local stakeholders and

forest users. This work has documented powerful site-specific examples of unacceptable damage to wildlife habitat and the forest. These damages are occurring in areas that diverse members of the community have known and used throughout their lives.

The Black Mesa Project Area is an excellent example of our documentation work. Sandwiched between the West Elk Wilderness and the Black Canyon of the Gunnison National Park, Black Mesa should contain some of the best big game habitat and herds in the state. It does not. Herd health and numbers are down, while road densities, hunter crowding, and dissatisfaction are rising. In the summer of 1999, WSERC's mapping project discovered far more roads than existed in the Forest Service inventory. Overall, we found 253 new road/trail segments and updated 71 existing nonmotorized trails segments that now see motorized use, for a combined 93.7 miles of new road. This averages out to 2.13 miles per square mile, and 2.57 miles per square mile on the X-lazyF Quadrangle. ATV riders in the area are often finding and reopening old logging roads that were closed but never obliterated. Many of these roads are under long-standing closure orders that have never been enforced or which are routinely violated.

The fieldwork completed to date has been digitized and added to the USFS GIS inventories. Over the last year our GIS capabilities and resources have grown to the point that with continued support we will soon be capable of providing GIS services to diverse regional groups and federal agencies. We now support a seasonal full-time GIS mapping coordinator who works closely with High Country Citizens Alliance (WSERC's sister group advocating similar issues on the Taylor–Cebolla District of the Gunnison National Forest), and other regional and State-wide groups to ensure consistency and collaboration on a wider scale.

We are now able to query a comprehensive roads coverage for the Gunnison National Forest. This enables us to calculate road mileage and densities, buffers, wildlife security areas, impacts and incursions to Rare II and other areas of special interest, motorized route conditions and linked photos, and much more. Other compatible coverages include hydrological and watershed information; agency jurisdictional boundaries; land ownership and parcels that illuminate development on private land adjacent to the forest; county-wide aerial photographs; digital topographic maps; RARE II areas; wilderness study areas; elevation models; DOW summer, winter, critical range, and potential suitable habitat for elk, deer, lynx, bear, goshawk, and so forth; cities; State highways; county roads; reservoirs; and more. This sophisticated GIS capacity will allow us quick verification, unique insights, and

Rare II areas in the Grand Mesa, Uncompahgre and Gunnison National Forests

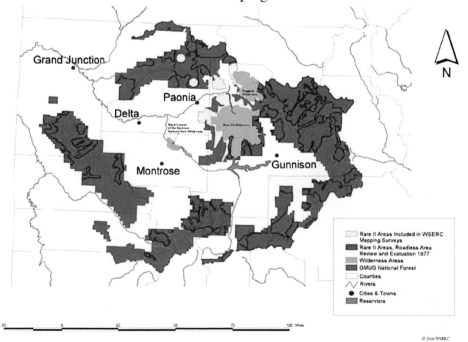

Working Area of the
Western Slope Environmental Resource Council

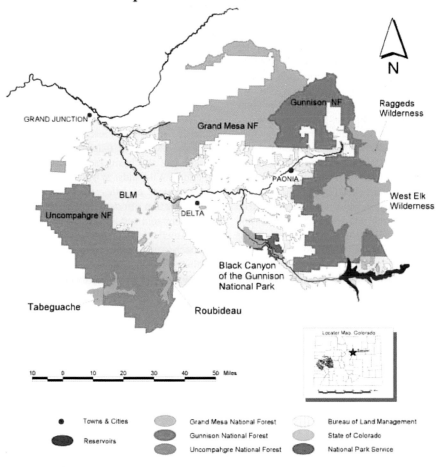

Towns & Cities ·   Grand Mesa National Forest ·   Bureau of Land Management
Reservoirs ·   Gunnison National Forest ·   State of Colorado
·   Uncompahgre National Forest ·   National Park Service

We worked with the agency to fix these problems; however, due to staffing shortages they were unable to rerun HABCAP and instead asked us to do it for them. The results found that under current conditions, 78 percent of wildlife units fail Forest Plan standards for the minimum acceptable HABCAP levels. During hunting season results drop to an 87 percent failure rate.

The Uncompahgre Travel Management Final EIS was released on April 14, 2000, but the preferred alternative still failed to bring the forest up to even a 50 percent compliance rate. It instead turned hundreds of illegal user-created two-tracks into official Forest Service system roads and trails, in our view. Motorcycles would have even been allowed to ride through a proposed wilderness area.

WSERC, along with neighboring and regional groups, filed a forest plan appeal. In the appeal, we charged that the Forest Service was violating its own regulations, documented the damage done by off-road vehicles, and drew attention to Forest Service data that recommends closure of these areas. The Uncompahgre Plateau once had the largest mule deer population in the country. It is now at one-tenth of its historical number. Elk populations are also showing signs of decline. The appeal also claims that the Forest Service's own data proves that its plan would be devastating to elk and deer populations.

Forest Service Decision Making: Motorized Restrictions

Our documentation, publications, and presentations have helped organize local, nonenvironmental constituency support for regulating ATV use on public lands. This pressure, by us and others, has led to a Forest Service interim plan to restrict motorized travel on the Gunnison National Forest. The Forest Service plans to publish an environmental assessment and implement this plan prior to the start of the next hunting season. This action by the Forest Service will restrict all motorized travel to existing routes in the national forest. Following this, over the next several years, Travel Management Revision for the GMUG national forests will focus on detailed route-by-route decisions.

WSERC's campaign will, in the coming year, move forward with efforts at effective research, presentation, and publication to improve mainstream community and agency support for habitat protection during route-by-route negotiations on travel management revisions in the Grand Mesa, Gunnison, and Uncompahgre national forests.

powerful analytical capabilities during route-by-route decision making in travel plan revisions in the GMUG national forests.

Community Outreach: Brochures, Community Dialog, and Slide Presentations

In 1999, we created a brochure about ATV use aimed to protect the region's hunting heritage that has been universally applauded and has opened the door to nonconfrontational and solution-oriented discussions with ranchers, inholders, outfitters, ATV enthusiasts, irrigators, loggers, and the thousands of hunters that flow into our valley each fall. This brochure, with an ATV slide show developed with support from the New-Land Foundation in 1999–2000, uses graphic local examples aimed at local audiences and diverse, nonenvironmental constituencies that share many of the same values about the surrounding mountains and wildlife, but that do not realize the scale of ATV impacts. In the last year, we have presented the slide show and other materials to our WSERC membership, local forest users and Forest Service officials, a local ATV club, and the Delta–Montrose Public Lands Partnership (a regional coalition of county

and city governments, economic development coalitions, business leaders, environmentalists, ranchers, loggers, motorized recreation groups, and federal land managers). We have also presented it in Aspen, Colorado, and to the Colorado Wildlife Federation Board of Directors in Denver.

Forest Service Decision Making: Appeals

As part of the environmental community's response to the draft Uncompahgre Travel Plan revisions, WSERC was asked to use its expertise to evaluate road impacts on big game. To do this, we replicated the Forest Service's HABCAP (mathematical) model, which is used to test environmental consequences to various wildlife species. We tested their data for elk, as this is the only species in the model that is sensitive to road densities. Our results were astounding.

First of all, charts in the draft environmental impact statement figures resulted in conclusions we found to be mistaken.

Second, we discovered that the Forest Service GIS had accidentally inflated road mileage by a third (800 miles), again resulting in bad data.

American Wildlands

Lauren Oechsli, GIS specialist   loechsli@wildlands.org
Linda Bowers Phillips, GIS lab manager

American Wildlands (AWL) is a science-based nonprofit conservation organization with a 22-year history of wilderness legislation and natural resource advocacy in the American West. American Wildlands' mission is to promote, protect, and restore biodiversity and advocate sustainable management of the West's wildlands, watersheds, and wildlife with special attention to the Northern Rocky Mountain region.

American Wildlands has been working on natural resource issues of the American West for more than twenty years and has developed a long list of impressive accomplishments. AWL was instrumental in securing legal protection for 100 million acres of Alaska's wildlands as well as millions of acres of wilderness and miles of river areas in the West. In 1981, AWL initiated one of the first wildland resource research programs in the country focusing on the economic values and benefits of wildland resources, and in 1985 hosted the first policy dialog ever held between conservationists and hydro-permitting authorities about the protection of free-flowing rivers. AWL founded the first successful coalition effort to address management problems of Glen Canyon Dam and its operation, which damages resources of the Grand Canyon. AWL was also one of the first organizations in the country to combine wilderness experiences with conservation policy— "ecotourism"—through its former American Wilderness Adventure Program. AWL staff and board members have functioned in similar capacities for many other coalitions, served on scores of citizen advisory commissions and committees, and been advisors to numerous members of Congress and three U.S. Presidents. Currently, AWL is actively involved in both the scientific and activist efforts of the Yellowstone-to-Yukon initiative; is a leader in the efforts to protect and restore the aquatic indicator species, the westslope cutthroat trout (WCT); and is distributing the results of the best available science on the corridors of the Northern Rockies.

Conservation biology informs us that to protect many of the region's widely distributed or wide-ranging species, such as grizzlies, wolverines, WCT, or bull trout, our thinking must encompass vast landscapes. Conservation strategies must, out of necessity, be regional in scale. However, they must be implemented at the local level. And that is exactly the way we conduct the business of conservation at American Wildlands. Whether on land, with our Corridors of Life project and roadless area protection efforts, or by water, with the WCT campaign and watershed protection program, American Wildlands is taking a regional approach to ensure the ecological health of the Northern Rockies. At the same time, we collaborate with individuals, local citizens, and grassroots activists to effectively implement real change.

Bozeman Pass Wildlife Corridor

The Bozeman Pass area between Bozeman and Livingston is key linkage habitat for animals traveling between the Bridger/Bangtail and the Absaroka and Gallatin Mountains. The Pass contains Numerous obstacles to wildlife movement, including Interstate 90, a frontage road, and the Burlington Northern railroad. The maps on the poster illustrate the current conditions, threats and opportunities for wildlife linkage in the key area.

Inset map of the *Corridors of Life* Model

American Wildlands
AWL GIS Lab
40 E. Main Suite 2
Bozeman, MT 59715
www.wildlands.org

Poster created by Linda Phillips

Land ownership

Land Management
- Bureau of Land Management
- US Forest Service
- State of Montana
- Big Sky Lumber
- Conservation Easement
- Designated Roadless Areas

3-Dimensional view of elevation

Vegetation Cover
- Herbaceous vegetation
- Shrub
- Hardwood forest
- Conifer forest
- Riparian vegetation
- Cropland
- Conservation easements
- Designated roadless areas

Vegetation Classified by a Landsat Thematic Mapper Satellite image

Species Geography

Skulls of some of the giant lemur species that have gone extinct in Madagascar since the relatively recent arrival of human settlement.

Photograph courtesy of Frans Lanting. © 2001 Frans Lanting

You too can Join the California Academy of Sciences!

When you become a member of the California Academy of Sciences, you open a door to the natural world. The Academy is the oldest scientific institution in the West and one of the ten largest natural history museums in the world. Academy members support the important work of our scientists, who travel the world to discover and research animals, plants, and artifacts. The Academy's Natural History Museum, Morrison Planetarium, and Steinhart Aquarium bring the Academy's research to the public with educational exhibits and programs. Through these efforts the Academy hopes to inspire people of all ages to preserve and protect this planet we share. In addition, you'll receive the following:

- Unlimited admission to the Steinhart Aquarium and Natural History Museum

- Eight Morrison Planetarium tickets—a $20 value

- Four guest passes—a $28 value

- A subscription to California Wild, our full-color quarterly magazine—a $13 value

- Bimonthly Academy newsletter

- A lecture or event every month

- 10 percent discount at the store and café

- Invitation to Open House—where you can tour behind the scenes

- Discounts on adult classes, field trips, and special lectures

- Membership in the junior academy for young people 6 through 16—a $10 value

- Many travel opportunities

- Admission to more than 250 science museums worldwide

- Borrowing privileges from the Academy library's Community Lending Collection

Ways to Join the California Academy of Sciences

Call us   (415) 750-7111 or 1-800-794-7576

E-mail us   membership@calacademy.org

Fax us   (415) 750-7346

Or write to   Membership Office
California Academy of Sciences
Golden Gate Park
San Francisco, CA 94118-4599

On-site   membership desk at the main entrance

California Academy of Sciences
Membership Categories

Family $60
Admission for 2 adults and their children under 18—two cards

Grandparent $60
Admission for two grandparents and their grandchildren under 18—two cards

Individual $45
Admission for one adult and his/her children under 18—one card

Dual Senior $40
65 or over—two cards

Senior $35
65 or over—one card

Out-of-Towner $35
One card—one individual; for those who live more than 200 miles from San Francisco

Curator's Circle $125–$249
($70 tax deductible)
All basic benefits plus the following:
- *Invitation to curator's reception and special programs*
- *Annual donor listing in Academy newsletter*

Director's Circle $250–$499
 ($170 tax deductible)
All Curator's Circle benefits plus:
- *Invitation to an exclusive exhibit showing*
- *Four additional guest passes*

President's Circle $500–$999
($410 tax deductible)
All Director's Circle benefits plus:
- *Exclusive trustee's reception with program*
- *Donor listing in Academy Annual Report*

Friend of the Academy $1,000 and above

Educators Please Note!
The Academy is committed to giving you special assistance in teaching your students about the natural world. Educator benefits are available with any category of membership.

Many companies gladly match employee contributions to nonprofit organizations. To see if your employer has a matching gift program, check with your human resources office.

The Role of Geographic Tools at Natural History Museums

Nina Jablonski, California Academy of Sciences   njablonski@calacademy.org

Natural history museums are repositories of vast stores of biological specimens and the data associated with those specimens. They are also homes to systematic biologists, whose charge is to describe the earth's species and analyze the relationships of species to one another. Thanks to major philosophical and methodological advances in the 1970s and 1980s, the science of systematic biology has undergone a renaissance. The field now is seen as a vigorous scientific discipline that is central to the understanding of biological diversity and the evolutionary patterns and processes by which that diversity arose. Further, systematics now interfaces more effectively than ever before with other areas of biology, from molecular biology and pharmacology through ecology.

Systematics has a critical role to play in the science of conservation biology and the practice of environmental stewardship because of its essential contributions to the understanding of biological diversity. The significance of biological collections and the value of the information they contain have gone largely unnoticed outside the natural science professions. Until very recently, there has been little recognition that collections provide critical information that allows us to understand global change, evolution, and biodiversity. Systematic collections constitute the definitive scientific evidence of biodiversity, and the data derived from them are reproducible and verifiable. In this regard, collections provide evidence superior to that provided by observations of species occurrences in the wild (e.g., bird sightings), which depend on the quality of the observation and the experience of the observer. Biological collections, as historical entities, also provide unmatched time depth for the study of spatiotemporal trends in organismal distributions. Although this time depth often extends only to the age of the first formal collections (typically 100–150 years for American museums), this period marked the beginning of a heightened intensity of human influence on North American ecosystems and is, thus, of great value. These realizations, coupled with a new surge of interest into the study of biodiversity, have led to enhanced public recognition of the importance of biological collections and systematics.

To capitalize on broader scientific and public interest, however, systematics must strengthen its connections to ecology through more coordinated multidisciplinary research. It must also enhance its abilities to communicate its findings to a broader public.

The collection of biological data may be undertaken as an end in itself, but more often today there is an increasing demand for information about wildlife, and biological records are valuable to a wide range of organizations and individuals, inside and outside of the world of museums. Users of biological records such as scientists, planners, land management, and conservation workers have overlapping needs and interests, but share a common need for accurate, usable, and readily available data. Enhancing the accessibility of collections data is of critical importance to ecologists, conservation biologists, and other more practically oriented users interested in monitoring the impacts of humans on the environment.

The specimen collections of all major natural history museums suffer from documentation problems (i.e., from a high level of discipline specificity in database field descriptors, and irregularities in descriptions of collecting localities). GIS permits the development of standard geocoding and georeferencing tools. This is of tremendous practical importance insofar as it will allow collection management personnel to use a software product designed and tested specifically for use on the documentation of natural history collections. It will also facilitate the use of collection-based data by natural heritage programs (e.g., The Nature Conservancy) because data on species distributions can be supplied in a standardized and highly usable format. The scientific benefits deriving from the implementation of standard geocoding tools and the adoption of strict georeferencing standards across museums are, perhaps, even greater. As a result of such standardization, one can envision with little difficulty the electronic unification of specimen databases between institutions. The vision of such a megadatabase has long been shared by CAS and ESRI. ESRI's Geography Network[SM] is now laying the foundation for such distributed global databases. This type of electronic union will, in turn, make possible the interrogation of not one, but many, specimen databases representing many natural history museum collections. Such a resource would make possible an enormous range of biodiversity surveys and analyses, and will be critical to basic and applied science and environmental decision making in the 21st century.

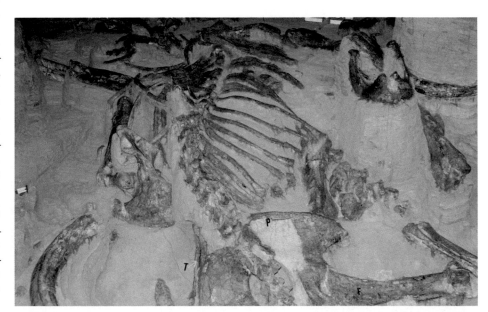

New York Botanical Garden GIS Program

Dr. Charles M. Peters and Kate E. Tode, Curator of Botany
Berry Brosi, MESc Institute of Economic Botany

The New York Botanical Garden (NYBG) is an advocate for the plant kingdom. The garden pursues its mission through the wide-ranging research programs of the International Plant Science Center; through its role as a museum of living plant collections arranged in gardens and landscapes across its National Historic Landmark site; and through its comprehensive education programs in horticulture and plant science.

With more than 6.5 million plant specimens, the New York Botanical Garden has the largest herbarium in the western hemisphere and the fourth largest herbarium in the world. Research activities at the garden are divided between two institutions. The Institute of Systematic Botany (ISB) focuses on nomenclature, taxonomic relationships, and the development of monographs and floras. It is no exaggeration to say that all biodiversity and conservation studies in the neotropics ultimately depend on output from the ISB to identify their plant specimens. The Institute of Economic Botany

(IEB) is concerned with the interrelationships between people and plants. IEB scientists study ethnobotany, indigenous resource management and land use planning, and sustainable plant use. IEB projects and field activities are currently underway in eighteen countries.

Much of the information generated by these two research institutions is spatial in nature. Herbarium specimens are georeferenced to the point of collection; forest types, harvest areas, and the densities of different plant resources collected in ethnobotanical surveys can be mapped to depict spatial distributions and use patterns. With assistance from the ESRI Conservation Program, GIS is becoming an increasingly important tool in botanical science at the garden. Of particular note are recent ISB projects to map species distributions and collection intensity using data from herbarium labels and new efforts by the IEB to develop sustainable management plans for community forestry projects.

Oaxaca–Bursera fieldwork: Mexican collaborators Mirna Ambrosio Montoyo and Dr. Silvia Purata work with Berry Brosi (left-right) to measure a Bursera glabrifolia tree, used in woodcarvings by Oaxacan communities, in San Juan Bautista Jayacatlan, Oaxaca.

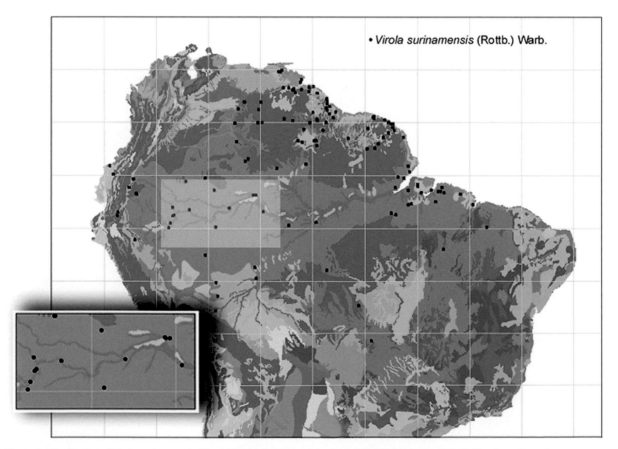

Figure 1. Distribution of Virola surinamensis *(Rottb.) Warb. Soil information from UNESCO/FAO World Soils Database. Inset shows association of* Virola *and alluvial soils near the conjunction of the Amazon and Japura Rivers.*

Systematic Botany and GIS

Every herbarium label contains information about where the plant was collected. With the advent of inexpensive GPS receivers, these locality records have become quite accurate. By compiling and mapping the label data for selected collections in the NYBG Herbarium, preliminary estimates of the distribution and habitat affinity of different species can be obtained. An example of this type of analysis is presented in figure 1, which shows the distribution of *Virola surinamensis* (Rottb.) Warb in the Amazon Basin. By adding a soils map or precipitation data to the project, as shown in figure 1 (previous page), the distribution of a species in relation to certain biophysical parameters can be assessed.

A major objective of the GIS work at the NYBG is to develop digital maps of selected plant collection data and to put these maps on our Web site (http://www.nybg.org) to be used interactively by other users. In our first foray into the world of ArcView IMS®, we used all of the georeferenced specimen data from the hepatic (liverwort) collections contained in the

American Bryophyte Catalog. As can be appreciated at http://www.nybg.org/bsci/hcol/bryo/bryomap.html, our efforts have been reasonably successful, and visitors to the "North American Hepatic Distribution from NY Collection Map" link can select hepatic family names from a list and then view their North American distribution.

A "type" specimen is an herbarium specimen that is used to represent a species when it is described for the first time. These are invaluable reference points for botanists when they are trying to reconstruct taxonomic relationships or determine the correct application of a plant name. The NYBG Herbarium contains more than 125,000 type specimens from all over the world. A map showing the number and country

Local field assistants involved in inventory work, community ecoforestry project, Kikori River Delta, Papua New Guinea.

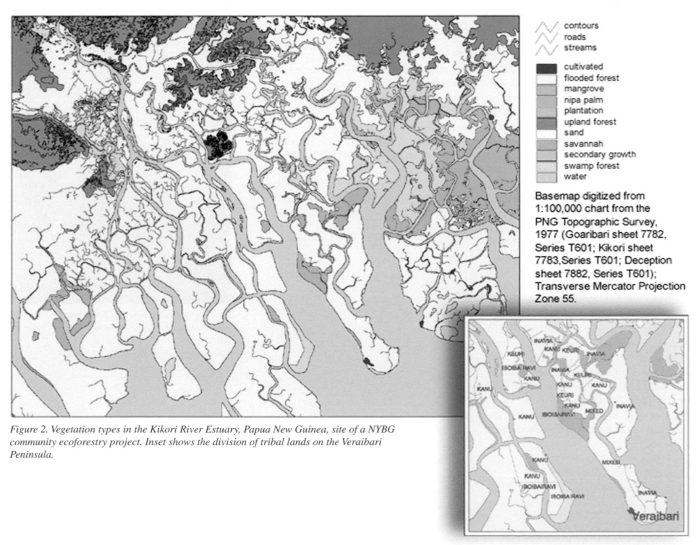

contours
roads
streams

cultivated
flooded forest
mangrove
nipa palm
plantation
upland forest
sand
savannah
secondary growth
swamp forest
water

Basemap digitized from 1:100,000 chart from the PNG Topographic Survey, 1977 (Goaribari sheet 7782, Series T601; Kikori sheet 7783, Series T601; Deception sheet 7882, Series T601); Transverse Mercator Projection Zone 55.

Figure 2. Vegetation types in the Kikori River Estuary, Papua New Guinea, site of a NYBG community ecoforestry project. Inset shows the division of tribal lands on the Veraibari Peninsula.

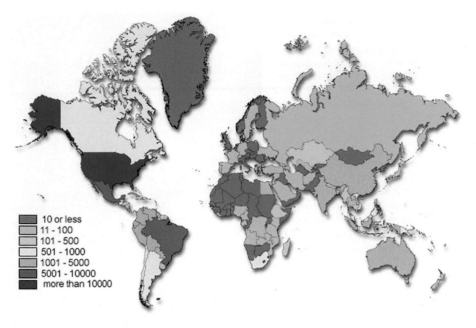

■	10 or less
■	11 - 100
■	101 - 500
□	501 - 1000
■	1001 - 5000
■	5001 - 10000
■	more than 10000

Figure 3. Number of vascular plant type specimens by country in the New York Botanical Garden Herbarium.

of origin of the type specimens curated in the NYBG Herbarium is shown in figure 3 (above).

Economic Botany and Resource Management

IEB scientists study the ecology, use, and management of important plant resources throughout the world. Much of this research is done in collaboration with local community groups and is focused on learning more about the ecology of different species to develop management strategies for its sustainable use. Two pieces of information fundamental to any management plan are the distribution of the resource and its abundance in different habitats. Compiling this information in a usable format involves building a basemap and then conducting systematic forest inventories and forest type mapping to quantify resource abundance and define local habitats. One such project, in the dry forests of the Central Valley of Oaxaca, put together a management plan for Bursera glabrifolia, an important carving wood for the local handicraft market. A 10 percent sample of selected forest areas was conducted, and total tree volumes were calculated using allometric equations derived for the species. The lack of adequate regeneration in some of the high-volume plots (e.g., plots 41–50) was detected with ArcView Spatial Analyst using the neighborhood analysis functions.

A broader mapping of vegetation types and community lands was required by an ecoforestry project currently underway in the Gulf Province of Papua New Guinea. The clans and tribes of this region own their land and can be receptive forest stewards provided with the appropriate motivation and technical assistance. The first step toward achieving this objective was to map the forest types and kinship group boundaries within the overall management area together with individual landowners. The results from this fieldwork are shown in figure 2 (previous page). Larger copies of this map are prominently displayed on the walls of the village leaders' homes in many of the participating communities.

Dr. Charles Peters (left) and field assistants install a dendrometer band to measure tree growth, Kikori River Delta, Papua New Guinea.

African Elephant Monitoring Project
North Carolina Zoological Society and Park, World Wildlife Fund, Camaroon Government Collaborate on Study

Mark MacAllister, Coordinator, Online Learning Projects

The team at work with one of Waza's elephants. To the left are Dr. Mike Loomis of the North Carolina Zoological Park (back to camera) and Dr. Martin Tchamba of World Wildlife Fund–Cameroon (green cap). Other team members to the right are securing a VHF/UHF transmitter collar on the animal.

transmitted to NOAA weather satellites several times daily. That stream is in turn relayed to an Argos facility in France, which processes it and sends it via e-mail to members of the research team. The data is analyzed and used in the preparation of GIS maps that illustrate the movement patterns of various herds. Analysis of these movements over time, and in the context of other data, helps researchers to better understand elephants' land-use patterns and to monitor and anticipate interactions between elephant herds and human settlements. The ultimate goal of the project is to discern a nationwide elephant management plan that will limit harmful human/elephant interactions.

GIS Methodologies

Currently, data from four elephants—three in the northern savannah region and one in the southeast—are sent to NCZP chief veterinarian Dr. Mike Loomis; Mark MacAllister, North Carolina Zoo Society's Coordinator of Online Learning Projects; and WWF. The

Project Overview

As part of its commitment to conservation research, the North Carolina Zoological Society has for four years supported a collaborative effort among the North Carolina Zoological Park (NCZP), Cameroon's Ministry of Environment and Forests (MINEF), and the World Wildlife Fund (WWF)–Cameroon Programme Office to study elephant migration patterns in the northern savannah region of Cameroon, Africa. The joint study has recently extended its activities into the heavily forested southeast region of that country.

The data gathering process begins when a team composed of veterinarians, field biologists, and trackers locates a high-ranking female elephant whose movements will effectively represent those of a herd. The elephant is then darted and anesthetized. While under anesthesia, the animal is provided oxygen, and her medical condition is closely monitored. A collar that holds both VHF radio and UHF satellite transmitters is fitted onto the elephant. The anesthetic is then reversed and in less than an hour after being darted, the elephant returns to her herd.

A data stream that includes latitude, longitude, time of day, and other information is

Dr. Mike Loomis, North Carolina Zoological Park's chief veterinarian, with a pulse oxymeter and respiration equipment. All elephants darted in this project are supported with oxygen while under anesthetic.

data is transferred from e-mail to spreadsheet format via a Java applet developed specifically for this project. The key data fields, including each animal's ID, transmission date and time, latitude/longitude, and location accuracy index, are then reformatted in a variety of GIS tables. Data in the GIS tables is added as event themes to various views. Some of the data sets for these additional layers are provided by World Resource Institute's Africa Data Sampler (1995, http://www.wri.org/wri/data/ads-home.html). Commonly used layers include national borders, rivers and lakes, roads, villages, elevation contours, and national parks.

Elephant location data is also used in developing animated tracking maps using ESRI's ArcView Tracking Analyst software. These tracking maps are particularly valuable for comparing the migratory paths of two or more herds over long periods of time (see map below). For example, one Tracking Analyst map demonstrates the way in which two distinct herds effectively share Waza National Park in northern Cameroon. One herd occupies the dry western section of Waza throughout much of the country's rainy season. At the onset of the dry season, though, that herd migrates north out of Waza, while a second herd from the south migrates north into the wetter western section of the Park.

Tracking maps will become even more valuable as location data spanning multiple years becomes available to the project, and researchers are able to compare annual migratory paths of specific herds. Also, the NCZP and North Carolina Zoo Society are intending to more fully utilize various other tools for analyzing animal movement including the Animal Movement Analysis Extension (AMAE) available from the United States Geological Survey. Some home range maps have already been produced, and additional maps will be generated as data becomes available. What is lacking at this point is access to the computing resources necessary to interactively display GIS data via the World Wide Web. The North Carolina Zoo Society is seeking a partner willing to provide the hardware, software, and technical support necessary to do so.

Dissemination/Activism

To share information about the research project with the widest audience possible, the North Carolina Zoo Society supports The Elephants of Cameroon (http://www.nczooeletrack.org). The Web site, designed for use in K–12 classrooms, offers information about elephants, the nation of Cameroon, and the daily activities of the field team. Interactive tools that allow users to communicate with field researchers via the Web are also available. GIS maps and associated instructional activities are featured, and recent location data is downloadable for those who wish to build their own GIS projects. Classrooms from all fifty U.S. states and more than a hundred nations worldwide have

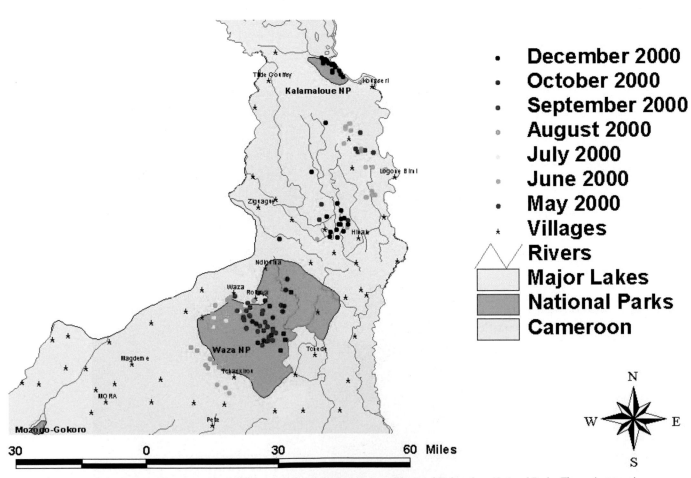

This map focuses on the movement of just one animal, Saleh, whose herd migrates between Waza and Kalamaloue National Parks. The project team's use of GIS mapping to place location data in the context of other information enhances its ability to develop protective strategies for this and other elephant herds in Cameroon.

used the site. Classrooms in North Carolina and other states, as well as university classrooms in Scotland, have utilized the site's data for GIS-related instruction.

The North Carolina Zoo Society also operates a GIS computer in its Wolf Bay Traders Gift Shop on the grounds of the NCZP. As part of a larger display regarding the NCZP/WWF program, the GIS machine loops a Tracking Analyst map that traces the migratory routes of the Waza elephants described above.

"Our access to GIS supports our attempts to better understand where, when, and why Cameroon's elephants migrate," explains Loomis.

"And, by sharing what we've learned with zoo visitors in North Carolina and with Web site users worldwide, we're better able to advocate on behalf of their protection."

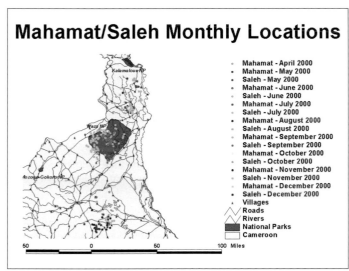

This map shows the migration routes of Saleh and Mahamat, two females collared in the northern savannah region of Cameroon, for the period of April–December 2000. By monitoring monthly movements with maps like this, researchers were able to discern that two distinct herds were sharing Waza National Park on a seasonal basis. Data for the period beginning January 2001 will be added to the maps to determine if this phenomenon continues over time.

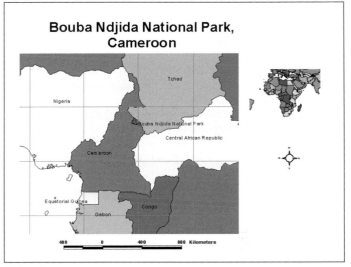

Orientation map.

Elephant research in the grasslands of northern Cameroon. This animal's collar is in place, and the team is preparing to reverse the anesthetic and release her. Satellite data generated by this collar will provide location data and other information to the research team for approximately two years.

Predator Conservation Alliance

David Gaillard, Program Associate

Predator Conservation Alliance (PCA), founded in 1991, is dedicated to conserving, protecting, and restoring native predators and their habitats in the Northern Rockies and High Plains. In short, we are saving a place for America's predators. This "place" is on the ground where we seek remedies to threats facing predators and their habitats through our advocacy work. This "place" is also in the human heart and mind, where we improve the public's understanding of and appreciation for the ecological role of predators through our education, outreach, and organizing work. We advocate on behalf of fourteen species, many of whom receive little other attention: black bear, black-footed ferret, burrowing owl, coyote, ferruginous hawk, fisher, grizzly bear, lynx, marten, mountain lion, northern goshawk, swift fox, wolf, and wolverine. Of these, nine are imperiled.

Predator Conservation Alliance focuses our advocacy and public education, outreach, and organizing work in the Northern Rockies and High Plains because these two regions provide the best opportunity to conserve and restore native forest and grassland predators.

The Northern Rockies is the only region in the United States where all of the native forest predators still exist. The High Plains were once the most vast and diverse grassland ecosystem on the continent. During the next five years, PCA will focus on protecting and restoring the Northern Plains—eastern Montana, eastern Wyoming, and North and South Dakota—because most of the grassland predators still exist there, and the largest proportion of public lands also exists in the Northern High Plains, as potential refugia for these species.

Through our programs, we seek to conserve wildlife predators by changing perceptions and behaviors of opinion leaders, agency decision makers, elected officials, and the public. We also seek to improve how state and federal wildlife and land agencies manage predators and their habitats in the Northern Rockies and High Plains grasslands. We provide people who live and work in predator habitats with the information and tools needed to reduce human/predator conflicts and to resolve them nonlethally. We promote a vision to shift the public discussion from simply protecting predator species from extinction to fully restoring them in numbers and distribution so they can play out their natural function across the region.

Predator Conservation Alliance fills an important, unique niche in the conservation community because it:

- Is the only organization that focuses on all predator species (not just wolves and bears) in the Northern Rockies and High Plains regions.

- Collects and uses the best scientific information (including our own field-based inventories) to determine each predator species' ecological needs and condition, the threats they face, and where to focus our monitoring and advocacy efforts.

- Shares this information with other conservation groups and public agencies to help guide and prioritize their efforts.

- Pursues all administrative, legal, and public pressure avenues to secure adequate protections for predators and their habitats.

- Is building an informed and active constituency that supports protecting predators, their habitats, and PCA's work.

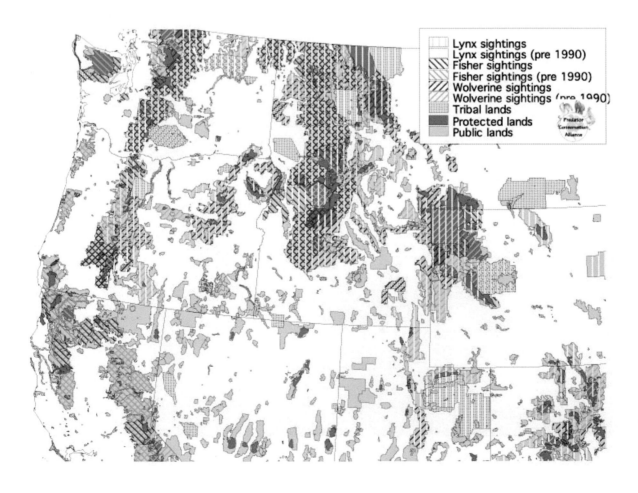

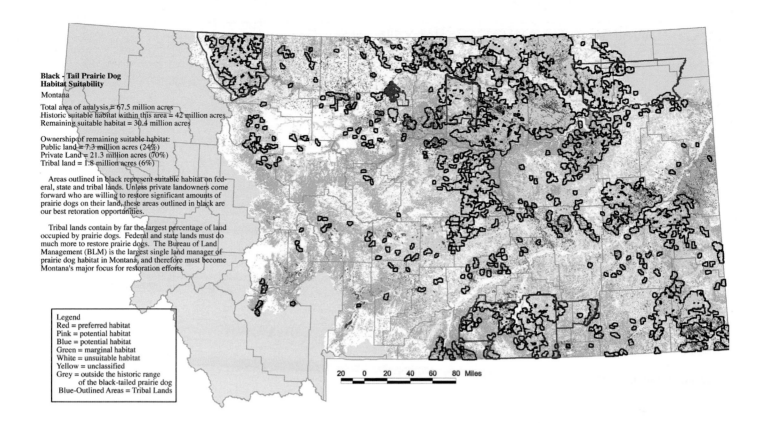

Black - Tail Prairie Dog Habitat Suitability

Montana

Total area of analysis = 67.5 million acres
Historic suitable habitat within this area = 42 million acres
Remaining suitable habitat = 30.4 million acres

Ownership of remaining suitable habitat:
Public land = 7.3 million acres (24%)
Private Land = 21.3 million acres (70%)
Tribal land = 1.8 million acres (6%)

Areas outlined in black represent suitable habitat on federal, state and tribal lands. Unless private landowners come forward who are willing to restore significant amounts of prairie dogs on their land, these areas outlined in black are our best retoration opportunities.

Tribal lands contain by far the largest percentage of land occupied by prairie dogs. Federal and state lands must do much more to restore prairie dogs. The Bureau of Land Management (BLM) is the largest single land manager of prairie dog habitat in Montana, and therefore must become Montana's major focus for restoration efforts.

Legend
Red = preferred habitat
Pink = potential habitat
Blue = potential habitat
Green = marginal habitat
White = unsuitable habitat
Yellow = unclassified
Grey = outside the historic range
 of the black-tailed prairie dog
Blue-Outlined Areas = Tribal Lands

20 0 20 40 60 80 Miles

Our GIS program is focused around the following three areas:

1 Forest Predator Ecosystem Program (FPEP)–Identifying priority habitat for forest predators in the U.S. Northern Rockies and Northwest. PCA plans to achieve the following objectives within FPEP with the aid of GIS.

A Identify and secure increased protection for priority habitat areas on public and private lands in the Northern Rockies region for the protection of forest predators and their habitats.

i Identify, through the use of the best available science, expert consultation, and GIS mapping, priority habitat areas for the wolf, grizzly bear, lynx, wolverine, fisher, marten, goshawk, black bear, and mountain lion.

ii Prioritize, based on level of threat and opportunities for protection and restoration, these priority habitat areas for each species or suite of species.

Initial progress has already been made toward these efforts by compiling presence/absence data from state heritage programs for most of the species listed above. Flagging areas of public lands with occupancy of these species will generate an important foundation map of where the animals still survive and need to be maintained. We can identify other areas important for restoration efforts based on historical occupancy and fragmentation/isolation of currently occupied habitat. Finally, we can use

GIS to overlap ranges of all forest predator species to refine which areas are priority habitats for which species (see map on previous page). The existing data must be refined, and we need to add distribution data for wolves, grizzly bears, black bears, mountain lions, and goshawks. Much of this has already been digitized in mortality databases.

B Secure the restoration of priority habitat areas by identifying and advocating for (1) the effective closure or obliteration of motorized routes to reach an average density of no more than one mile of motorized route per square mile in developed priority habitat areas, and (2) preventing establishment of new routes in undeveloped priority habitat areas.

i Develop a comprehensive list of key motorized routes to be effectively closed or obliterated within our priority habitat areas. (Note: First, this will be largely focused on national forest lands. Second, this will involve continuing to conduct field-based inventories of the numbers, condition, types, and levels of use of roads and trails open to off-road motorized vehicles, as needed. Until this listing is completed, we will use existing Roads Scholar Project field data to identify those routes we will work to effectively close or obliterate.)

ii Make significant progress toward getting the U.S. Forest Service to effectively close or obliterate excessive or inappropriate motorized routes within our priority habitat areas.

PCA has already made much progress toward identifying important roads for closure and obliteration in priority areas for predators. What remains is to compile this information into a GIS database for some additional analysis and to publicize these priority areas to land managers, the media, and the general public. Much initial GIS work has been done by working with the Ecology Center in Missoula, and while we will continue to collaborate with them on this project, the limitations of having our data elsewhere have become increasingly clear. We have found that without the staff support that we have here in Bozeman—for data input, maintenance, and ongoing analysis—we have not been able to make the most effective use of our field monitoring data. Another critical challenge is that under this program, we are collecting original data (versus synthesizing and analyzing data sets already available), and have been recording our findings on paper maps as a temporary stop-gap measure until our staff can travel to Missoula to input data. We need basic GIS capability in-house to make more significant, timely, and meaningful progress in this program area.

2 Grasslands Predator Ecosystem Protection Program–Mapping priority habitat for the restoration of the black-tailed prairie dog (see map above), black-footed ferret, and other dependent predators in the High Plains.

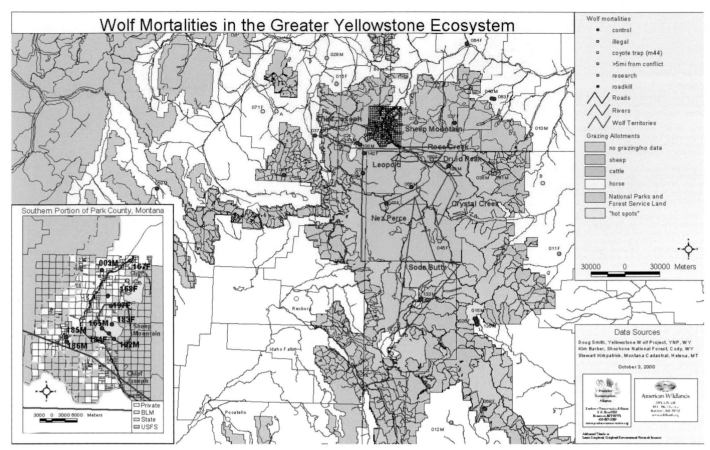

Wolf Mortalities in the Greater Yellowstone Ecosystem

An example of predator/people conflict analysis.

A Restore prairie dogs to a minimum of 10 percent of suitable habitat on all federal lands within the Northern High Plains (as a percentage of jurisdictional land base).

i Identify one secure core refugia area for prairie grassland predators on each of the agencies' jurisdictional units in the Northern High Plains by December 2001.

ii Produce a prairie dog ecosystem recovery and restoration plan based on the core refugia in item A,i (above).

PCA's GIS modeling of suitable restoration areas for the prairie dog is already well underway, thanks to the ongoing support of the Ecology Center (see map on previous page); these maps have already been purchased by the Ft. Belknap and Northern Cheyenne reservations and the Montana State Department of Fish, Wildlife, and Parks for use in developing their black-tailed prairie dog restoration plans. But, as mentioned above, it would be far more efficient for both PCA and the Ecology Center if PCA had GIS capabilities in its own office. Over the next two years, we will be undertaking more ambitious projects than we have to date by mapping suitable habitat for black-tailed prairie dogs and other critical grassland species in all eleven states of the historic range of the black-tailed prairie dog (as we have already done for Montana). In addition, several of the groups and individuals we are working

most closely with in developing a rangewide conservation and restoration plan are located in Bozeman, and the ability to generate, display, and print these maps at our office is crucial in helping us move forward quickly (none of these Bozeman groups/contacts have this capability in-house).

3 Living with Predators Program–Mapping and Analyzing "Hot Spots" of Predator/People Conflicts in the U.S. Northern Rockies.

A Establish and maintain PCA as a preeminent source for information and referrals on resolving human/predator conflicts nonlethally within the Northern Rockies and High Plains.

i Identify human/predator conflicts that need to be addressed and strategies for resolving these conflicts nonlethally.

ii Develop and begin implementing a two-year outreach plan to promote Predator Conservation Alliance as a preeminent source for information and referrals on resolving human/ predator conflicts nonlethally within the Northern Rockies and High Plains.

With assistance from American Wildlands' GIS lab in Bozeman, PCA has made some initial progress toward these goals by obtaining the wolf mortality database from the National Park Service and plotting these in association

with land ownership, livestock grazing, and elk distribution in the Greater Yellowstone Ecosystem (see map above). PCA intends to expand on this foundation both in geographic scope by adding similar wolf data from Idaho and northwestern Montana and by adding grizzly bear mortality data from throughout the Northern Rockies region. Much of the remaining wolf data has been compiled by the U.S. Fish and Wildlife Service and Nez Perce Tribe (in Idaho); the grizzly bear mortality data is largely compiled by Montana Department of Fish, Wildlife and Parks and much of it has already been processed by the Ecology Center in Missoula. Ideally, we would like to supplement this database with point locations of nonlethal conflicts to identify problem areas whether or not they resulted in dead bears and wolves. We would also like to add conflict areas of mountain lions and black bears, which would require some digging into records kept by state wildlife officials. We believe this database will prove an invaluable resource to predator advocates, land and wildlife managers, and all members of the public interested in reducing and preventing conflicts between predators and people.

California Academy of Sciences Herpetology Project

Michelle Koo

Established in 1853, the California Academy of Sciences in San Francisco is a natural history museum dedicated to exploring the globe and to describing the diversity of plant and animal populations and their evolutionary histories. Approximately 14 million biological specimens are housed by the academy, representing a wide breadth of disciplines including anthropology, botany, entomology, herpetology, ichthyology, invertebrate zoology, ornithology, and mammalogy. This sampling of the earth's diversity has traditionally been the

realm of academic exploits but more and more has supplied the persuasive facts for the arguments that conservation organizations make on behalf of habitat and species preservation.

Fundamental to that goal is the concept of locality data, the basic information about where a specimen was found, which has always been critical to museum curation. However, only within this last decade, after more than a century and a half of collecting, has the perception of "good" locality data radically changed. Early scientific collectors often simply described localities as the nearest populated place resulting in such nonspecific locality descriptions as "near Los Angeles." The recent advent of GIS technology has signaled a new awareness of the pinpoint accuracy we can assign our specimens and the new ways we can access and analyze our locality information.

A relative newcomer to GIS technology, the Department of Herpetology began in 1996 to

retrospectively geocode the California amphibians and reptiles in its collection (over 43,000 specimens) and then map them on a simple statewide map. During this process, it became increasingly obvious that many areas of the state had been little sampled. This was a concern to us for several reasons including the widespread use of potentially misleading range maps to indicate where amphibian and reptile species occur.

Range Maps Versus Point Distribution Maps: The Importance of Evidence

Most field guides or statewide maps of herpetofaunal distributions display species occurrence with range maps of shaded areas usually based on habitat ranges. It is assumed that if a habitat is favorable for a species in one given area, then that is applicable to any other area of similar habitat. Essentially, these maps are an assumption or prediction of species occurrence.

For most generalized use, these assumptions are good enough; however, in a state that is developing as rapidly as California, and yet one that is concerned with maintaining its rich, native natural history, such assumptions are not sufficient, particularly if resource management decisions of a specific area are relying on such information.

The value of museum data lies in that museum specimens serve as verifiable (and thus indisputable) evidence of species occurrence. This is critical with respect to endangered and threatened frog species in California, often considered "indicator" species of an ecosystem's health, which are commonly misidentified in the field or from photographs. The need for such baseline data is clear. For example, invoking our survey results, a wildlife biologist in the Tahoe National Forest prevented mining on certain streams due to the documented presence of Foothill yellow-legged frogs (*Rana boylii*).

Dominant Vegetation Types and *Ambystoma macrodactylum* (Longtoed Salamander)

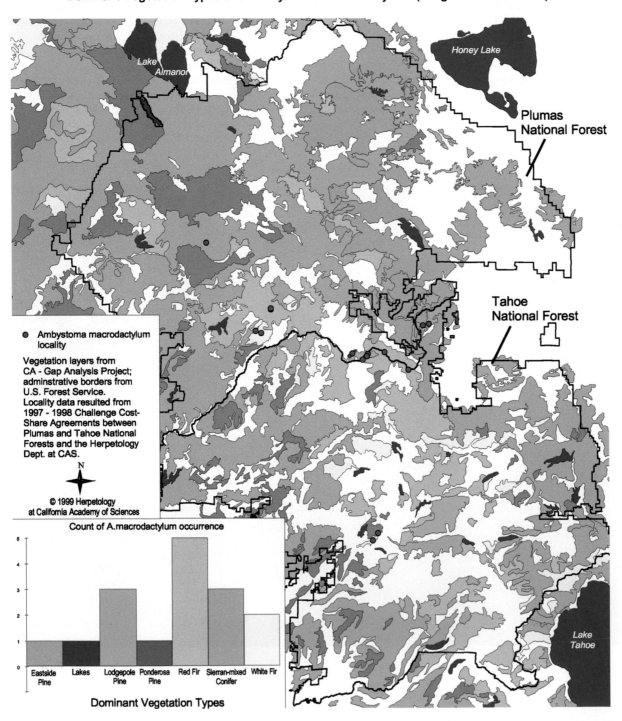

● Ambystoma macrodactylum locality

Vegetation layers from CA - Gap Analysis Project; adminstrative borders from U.S. Forest Service. Locality data resulted from 1997 - 1998 Challenge Cost-Share Agreements between Plumas and Tahoe National Forests and the Herpetology Dept. at CAS.

© 1999 Herpetology at California Academy of Sciences

Count of A.macrodactylum occurrence

Dominant Vegetation Types: Eastside Pine, Lakes, Lodgepole Pine, Ponderosa Pine, Red Fir, Sierran-mixed Conifer, White Fir

National Forest Surveys and GIS Use

The year 1997 marked the beginning of a cost-share agreement with the Tahoe National Forest, California, which has resulted in three seasons of amphibian and reptile surveys (see map at right). To begin, we mapped all known specimen localities for the Tahoe National Forest area since the mid-19th century from museums nationwide. Before 1997, the Tahoe National Forest had few known specimens from within its borders. Our subsequent surveys increased the number of documented specimens within the national forest 75 percent. Last year, we initiated fieldwork in the Plumas National Forest which shares an extensive boundary with the Tahoe National Forest, and have since increased the number of their known specimens by 150 percent. This year, we are continuing surveys in both the Tahoe and Plumas and initiated fieldwork in the Mendocino and Sierra national forests.

Gathering digitized data from many different sources, we have begun to make geospatial correlations from our data. For instance, using vegetation layers from the Forest Service and the California Gap Analysis Program (CA-GAP), we can now correlate certain dominant plant biomes with amphibian and reptile species to identify potential sensitive habitats.

Most immediately, we expect to use the locality data we have amassed from the national forest surveys to test predicted distributions of herpetofauna such as those produced by CA-GAP at the national forest level. Our hope is that the baseline data from our fieldwork will be used for new and improved predictive models.

The Impact of GIS and Herpetology

Certainly, the Department of Herpetology will never be the same again since the incorporation of GIS technology. It has made us reevaluate the nature and acquisition of collection data and has helped to focus our field efforts. It has allowed us to distribute and display our data much more effectively. More broadly, GIS has provided museum specimens and their associated data a new level of relevance to environmental and conservation efforts, at the same time lending scientific rigor and credibility to those efforts. As we begin to analyze our data in new geospatial terms, it will become increasingly valuable to conservation efforts and will increase our basic understanding about how this century has impacted amphibians and reptiles.

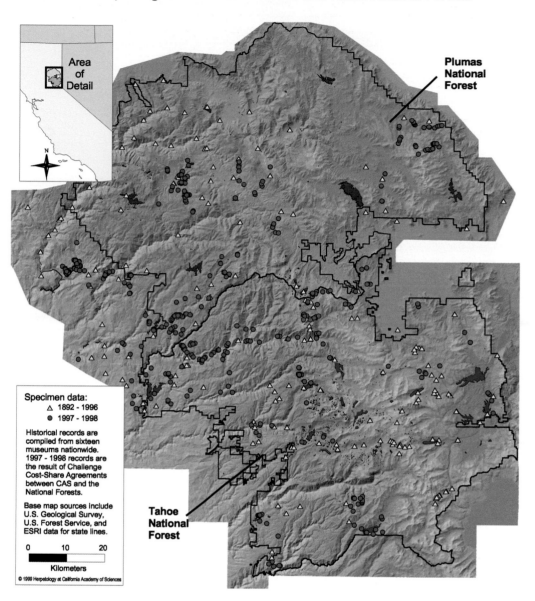

Herpetological Records of Plumas and Tahoe National Forests

Area of Detail

N

Plumas National Forest

Tahoe National Forest

Specimen data:
△ 1892 - 1996
● 1997 - 1998

Historical records are compiled from sixteen museums nationwide. 1997 - 1998 records are the result of Challenge Cost-Share Agreements between CAS and the National Forests.

Base map sources include U.S. Geological Survey, U.S. Forest Service, and ESRI data for state lines.

0   10   20
Kilometers
© 1999 Herpetology at California Academy of Sciences

Bird GIS

Platte River Whooping Crane Maintenance Trust

Robert J. Henszey, Wetland ecologist

Bob Henszey, Wetland ecologist for the trust, samples wetland plants to study relationships between plant species and hydrology.

The Platte River Whooping Crane Maintenance Trust (www.whoopingcrane.org) is a private nonprofit organization dedicated to the conservation of migratory bird habitat along Nebraska's Platte River. Our mission is to protect and maintain the physical, hydrological, and biological integrity of the Big Bend area of the Platte River in south-central Nebraska, so that it continues to function as a life support system for whooping cranes and other migratory bird species.

The central Platte River valley has hemispherical significance as a staging area for migratory water birds and offers critical habitat for a variety of migratory and nonmigratory birds. The region is best known for the 500,000 sandhill cranes and five million to seven million ducks and geese that migrate annually through the valley, but twenty-two endangered, threatened, or candidate plant and animal species are also found here. These species are largely dependent on the river and adjacent riparian habitat to meet their needs. During the past one-hundred years, however, the central Platte River valley has undergone a substantial transformation. Water development projects reduced a once wide and treeless river (up to one mile wide) to a number of narrow channels with extensive woody vegetation. Since 1865, channel widths in many areas have narrowed by 70 percent or more. In addition, agricultural policy and practice have led to extensive monocropping, habitat fragmentation, and loss of wetlands. Native grasslands, wetlands, and wet meadows, which provide important feeding and nesting habitat for migratory birds and other species, now exist primarily as remnants within a matrix of agricultural land.

To maintain and enhance this important migratory bird habitat, the trust acquires land and water rights; manages, protects, and restores habitat; and conducts research related to migratory birds and their habitat needs. Where compatible with the trust's mission, existing agricultural and other traditional land uses are used and promoted by the trust. The trust currently owns and manages about 10,000 acres of habitat along seventy miles of the Platte River. Most of this land is in native pasture, hayland, or other riparian habitats, with about 1,700 acres in row crop agriculture. The trust also has cooperative habitat protection programs with the National Audubon Society, The Nature Conservancy, and the U.S. Fish and Wildlife Service.

In 1982, the trust pioneered a GIS for the Platte River with its MOSS inventory of land use and land cover within three and a half miles of the river, but this system was cumbersome and is now obsolete. With a grant from ECP-CGISC, the trust purchased ArcView GIS and began utilizing the current features available for GIS. Examples from some of our recent projects follow.

Wet Meadow Restorations

The trust has an active program of converting marginal cropland back to wet meadows. Since 1981, the trust has restored 1,195 acres to mesic grasslands and wet meadows. Early restorations incorporated low-diversity seed mixtures on leveled crop fields. More recent restorations include locally collected high-diversity seed mixtures on partially recontoured land surfaces. Since recontouring is expensive, GIS has helped us to develop the best design for our budget while incorporating information such as soils and previous topography. Until detailed digital soil maps become available for our area, we have been using digital orthophotoquads as a basemap to indicate soil boundaries. Digital line graphs (DLGs) from USGS 1:24,000 topographic maps are also used

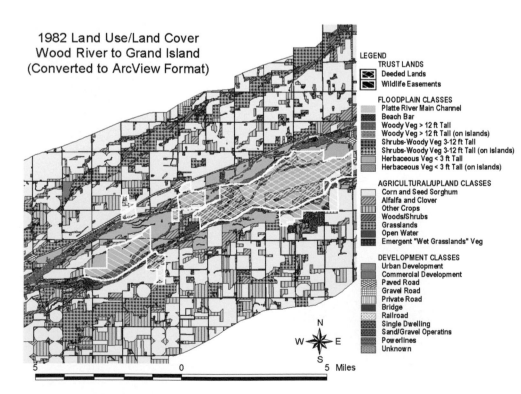

1982 Land Use/Land Cover
Wood River to Grand Island
(Converted to ArcView Format)

LEGEND

TRUST LANDS
- Deeded Lands
- Wildlife Easements

FLOODPLAIN CLASSES
- Platte River Main Channel
- Beach Bar
- Woody Veg > 12 ft Tall
- Woody Veg > 12 ft Tall (on islands)
- Shrubs-Woody Veg 3-12 ft Tall
- Shrubs-Woody Veg 3-12 ft Tall (on islands)
- Herbaceous Veg < 3 ft Tall
- Herbaceous Veg < 3 ft Tall (on islands)

AGRICULTURAL/UPLAND CLASSES
- Corn and Seed Sorghum
- Alfalfa and Clover
- Other Crops
- Woods/Shrubs
- Grasslands
- Open Water
- Emergent "Wet Grasslands" Veg

DEVELOPMENT CLASSES
- Urban Development
- Commercial Development
- Paved Road
- Gravel Road
- Private Road
- Bridge
- Railroad
- Single Dwelling
- Sand/Gravel Operatins
- Powerlines
- Unknown

5   0   5 Miles

Craig Davis, avian ecologist for the trust, bands a female northern oriole to study species movements and site fidelity.

Nature Center restoration after recontouring. Photo courtesy Nebraska Game and Parks Commision.

with ArcView GIS to suggest topography and drainage patterns for sites where the topographic map was compiled before conversion to cropland. The total area to be restored for each habitat type, computed with ArcView GIS, has been very useful in determining the amount of seed we need to collect by hand for each species and how much dirt work will be required. In the past, we marked the location to excavate topographic patterns on the ground "by eye," but we hope to improve upon our placements in the future with a recently purchased GPS.

Crane Roost Maintenance and Enhancement

At the peak of the spring migration in mid-March, up to 40,000 sandhill cranes per river mile return to the safety of the Platte to roost for the night after feeding in the surrounding cornfields and wet meadows during the day. The only remaining migratory flock of endangered whooping cranes also migrates through our area, and each year several birds stop to roost and feed along the river. Both crane species prefer a wide (>750 ft. for sandhill and >1,000 ft. for whooping cranes), shallow, unobstructed river to roost. Over the past hundred years, however, cottonwood and willow encroachment has narrowed the river, forcing cranes to roost in the few remaining suitably wide reaches of the Platte River.

Since 1982, the trust has been actively clearing selected areas on more than 20 miles of river channel to maintain and enhance crane roosting habitat. A combination of shredding and disking with heavy equipment is used to remove the vegetation. The focus of these efforts has been on river islands, but banks have also been cleared to achieve a minimum unobstructed channel width. If clearing is followed by high flows, some of the smaller islands and sandbars erode, creating wider channels. However, periodic maintenance is necessary on most sites every two to five years to maintain an unobstructed channel width.

For clearing on private lands, the trust must prepare a plan for the landowner and the U.S. Fish and Wildlife Service. Before using ArcView GIS, preparing these plans was difficult and time-consuming. We now use ArcView GIS

with a recent georeferenced aerial photograph, converted from MrSID™ to TIFF format, as a basemap. Section lines are overlaid on this basemap, and the area of potential riparian forest and willow/sandbar communities to be cleared is calculated. Photo points for long-term monitoring are also labeled on these maps.

Additional ArcView GIS Projects

Besides the projects highlighted above, the trust is using ArcView GIS for several other projects. We use ArcView GIS to monitor the spatial distribution and status of trust lands (e.g., facilities, rangelands, croplands, restorations, prescribed burns, and study areas) and to help with management planning, habitat assessments, and scientific studies of migratory birds and their habitat. With our new GPS, we plan to monitor the location and status of individual plants from a small population of the threatened western prairie fringed orchid (*Platanthera praeclara*), which occurs in one of our pastures. Whenever we install an observation well for studying wet meadow water table patterns, we use a map produced with ArcView GIS to register the well with the state of Nebraska. Finally, we use ArcView GIS to view and use spatial data from other organizations, such as the U.S. Fish and Wildlife Service's wetland inventory, and the spatial databases produced by the Platte River Endangered Species Partnership (www.usbr.gov/platte/).

Future ArcView GIS Projects

As a major participant in the Platte River Endangered Species Partnership, the trust will most likely use ArcView GIS in the future to accomplish the following:

1 Monitor spatial and temporal changes in Platte River channel habitat and determine how these changes may affect habitat use by endangered whooping cranes, least terns, and piping plovers and by other nonlisted bird species (e.g., sandhill cranes, waterfowl, and shorebirds).

2 Monitor spatial and temporal changes in Platte River grassland and woodland

communities, agricultural land uses, and rural and urban development, with special emphasis on how these changes may affect grassland and woodland birds.

3 Identify priority areas for channel, wet meadow, and woodland community protection, habitat enhancement, and restoration.

4 Inventory the annual status of Platte River channel management areas (e.g., clearing woody vegetation to enhance crane and waterfowl roost habitat and nesting habitat for terns and plovers).

Data and analyses from these efforts will help the trust educate decision makers, conservationists, natural resource professionals, and landowners about protecting and enhancing the Platte River ecosystem. In the past, these efforts were hampered by a lack of quality spatial data and computer-based analytical tools. With ArcView GIS, we have an invaluable tool to promote the wise use and management of the Platte River.

Protected Habitat - Central Platte River East Half

Bird Studies Canada GIS Program

Andrew Couturier, GIS Analyst

Bird Studies Canada (BSC) is recognized nationwide as a leading and respected non-profit conservation organization dedicated to advancing the understanding, appreciation, and conservation of wild birds and their habitats in Canada and elsewhere through studies that engage the skills, enthusiasm, and support of its members, volunteers, staff, and the interested public.

BSC conducts leading-edge ornithological research across Canada and elsewhere. GIS forms a major part of its analytical toolbox. ARC/INFO has been instrumental in helping us achieve this magnitude of success. Some of our achievements include:

- The development of land management plans that balance economic needs with biological diversity concerns.

- The coordination of a priority-setting system for Ontario's breeding birds. The system acts as a tool for the conservation of avian (and other) biodiversity and has become a standard reference for planners and ecologists across the province.

Our future plans include:

- The design and management of the Ontario Breeding Bird Atlas (see below). In addition to contributing to the development of sampling methods for the atlas, we will be managing the overall database and developing tools for viewing results interactively over the Web.

- The compilation of a polygon layer for Important Bird Areas across the nation. Once complete, the IBA polygons will be used for interactive data display and entry over the Web.

- Analysis of forest cover/breeding bird relationships in southern Ontario. The results of this study will be of special interest to both biologists and planners alike.

GIS, and specifically the tools afforded by ARC/INFO, are indispensable to the success of the aforementioned projects.

Bird Studies Canada has been dedicated to enhancing our understanding of bird ecology and bird populations in Canada for more than forty years. From the outset, BSC's programs have depended on participation by volunteer members of the public: "Citizen Scientists." BSC makes a special commitment to involve Citizen Scientists in our work. Citizens learn

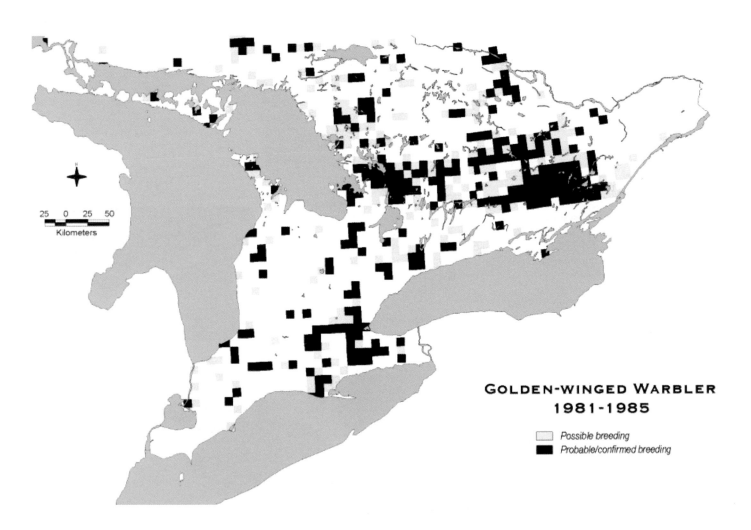

GOLDEN-WINGED WARBLER
1981-1985

☐ Possible breeding
■ Probable/confirmed breeding

by doing, contribute by doing, celebrate their achievements, and become engaged. Our volunteer Citizen Scientists are found throughout Ontario and increasingly throughout the rest of Canada. Overall, BSC projects engage nearly 25,000 participants throughout North America. Over the last forty years, BSC has developed many programs in partnership with many agencies. Since 1960, BSC's staff and volunteer Citizen Scientists have monitored population trends of birds on migration at Long Point, Ontario. This program continues today as one of the longest running bird monitoring initiatives in North America. Based on our success at Long Point, BSC and the Canadian Wildlife Service have recently developed the Canadian Migration Monitoring Network—a chain of fifteen migration monitoring stations across the country (five of which are based in Ontario). Over the years, thousands of volunteers have received intensive training in field research techniques at these research stations and provided data critical to bird conservation. The Marsh Monitoring Program (MMP) was launched across the Great Lakes basin in 1995, in partnership with Environment Canada and the U.S. Environmental Protection Agency, to assess and monitor the health of aquatic habitats. Each year, about five hundred volunteer Citizen Scientists participate in this program in Ontario and the adjacent Great Lakes states. The MMP is now serving as a model for the development of a continental monitoring effort focused on wetlands and wetland-dependent species. The Important Bird Areas Program is a worldwide initiative of BirdLife International to identify and protect the world's most important sites for birds. In Canada, the BirdLife partners are BSC and the Canadian Nature Federation. More than 750 sites across Canada (including about 100 in Ontario) have been identified as "Important Bird Areas." The Canadian Lakes Loon Survey (founded as the Ontario Lakes Loon Survey in 1987) enables more than 1,000 cottagers across the Province and across the country to contribute to conservation and enhances enjoyment of their natural world by keeping track of the reproductive success of loons on their lakes. Project FeederWatch (founded as the Ontario Bird Feeder Survey in 1976) encourages people with bird feeders to keep track of the numbers of each species visiting their feeder. This project attracts nearly 2,000 Citizen Scientists annually in Canada (more than 1,000 of whom reside in Ontario) and 12,000 more in the United States. BSC also coordinates several provincial programs including the Ontario Nocturnal Owl Survey, the Ontario Red-Shouldered Hawk and Spring Woodpecker Survey, and the Ontario Birds at Risk Program.

Through its varied programs (volunteer based or other), BSC is helping ordinary citizens become more connected with the natural environment and to make a difference in ecological conservation.

The organization is celebrating its 40th anniversary this year, while the GIS program began in 1997. GIS is still a one-person show in terms of core maintenance and analysis; this situation will remain for the foreseeable future. The organization has undergone significant change in the last five years, and this change is reflected in the evolution of the GIS program. The GIS program has expanded in scope, geographic extent, and in usage within the organization. The complexity of GIS projects undertaken currently is far greater than when the GIS program was first instituted. We are increasing our focus on national and international projects and are forming a network of GIS partners to match this scale of expansion. Desktop use of GIS for "spatially challenged" staff within the organization is increasing. ArcView GIS is now used by several scientific staff for basic data viewing and querying.

We are currently working on a wide variety of projects concerning GIS and conservation. The grant will be used to support our ongoing work with respect to integrated land management plans, priority-setting for Ontario bird conservation, and analyzing habitat/breeding presence relationships with Marsh Monitoring Program data, as well as providing basic mapping needs for BSC's other flagship programs.

While virtually all projects at BSC utilize GIS in some manner, the two listed below make extensive use of the advanced GIS analysis and modeling capabilities afforded by ARC/INFO. These two projects constitute major elements of our original ECP application four years ago. The initial stages of these two projects are now complete, and reports are available on our Web site at the addresses mentioned.

1 Conservation Priorities: This multiscale project involves prioritizing conservation needs for Canada's breeding birds. GIS is used to calculate the breeding area of all birds for a variety of spatial units—provinces, ecozones, counties, watersheds, and so on—so that species may be assigned a rank of "jurisdictional responsibility" for the spatial unit of interest. Finally, with the GRID module, distributions of high-ranking species are then overlaid together to produce maps showing areas of high species richness for species of conservation concern. From this analysis, we are able to advise municipalities and conservation authorities as to which species require special consideration in land-use planning and development activities. This work is directly contributing to the conservation of natural heritage values in Ontario, while a related project seeks to develop a similar conservation ranking system for ecoregions across North America. (www.bsc-eoc.org/conservation/conservmain.html)

2 South Walsingham Forest Management Project: The aim of this project was to provide forest management recommendations to public and private landowners within the South Walsingham Sand Ridges/Big Creek Floodplain Forest (see map next page), a provincially designated Area of Natural and Scientific Interest (ANSI). Through field surveys of birds and vegetation, the technical team designed a five-year plan that balances both wood production and biodiversity conservation within the forest. GIS modeling was used intensively in designing the management strategy. By using data gathered at transect points throughout the forest, we were able to create data layers (grids) depicting (i) the "breeding activity zone" for each bird species inhabiting the forest, (ii) patterns of forest density/maturity, and (iii) other sensitive features including riparian zones and steeply sloped areas.

These data layers were incorporated into the GIS model, and a framework was built to guide forestry operations for the next five years. Based on this information, we have provided forest management recommendations to the Ministry of Natural Resources and have coauthored a comprehensive forest management plan for the forest. The management plan will be implemented by the local conservation authority and private landowners. (www.bsc-eoc.org/swalsreport.html)

A Guide for Setting Conservation Priorities

Given the bewildering array of plants and animals inhabiting the landscapes of southern Ontario, accommodating the specific needs of each individual species in local planning and development activities is not practical. Conservation objectives will be most attainable if planning actions are targeted to critical habitats and to sites containing regionally important assemblages of species. The approach described in this report aims to help planning authorities prioritize conservation efforts by targeting bird species (and their associated habitats) that are significant within their region and not necessarily just rare. Specifically, this report advocates the use of prioritized lists of birds as tools that planning authorities might use when developing or amending official plans, including the identification of significant natural heritage features, and when

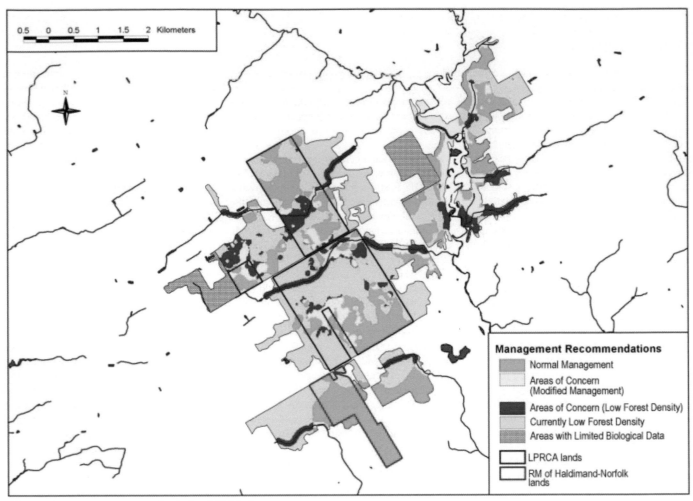

Five-year forest management map for the South Walsingham Forest.

Management Recommendations

- Normal Management
- Areas of Concern (Modified Management)
- Areas of Concern (Low Forest Density)
- Currently Low Forest Density
- Areas with Limited Biological Data
- LPRCA lands
- RM of Haldimand-Norfolk lands

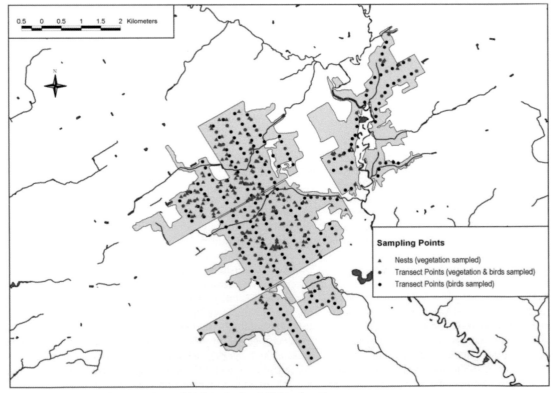

Sampling Points

- ▲ Nests (vegetation sampled)
- ● Transect Points (vegetation & birds sampled)
- • Transect Points (birds sampled)

Data collection points for vegetation and birds in the South Walsingham Forest.

evaluating development proposals. Lists of birds of conservation priority thus represent another tool within a municipality's toolbox that can be used to fulfill its obligations under the Planning Act with regard to the protection of natural heritage features. These lists can be combined with information on significant plants, mammals, and so forth, in a multi-faceted approach to identifying and protecting significant natural heritage features. The approach has the advantage of being standardized throughout southern Ontario, thus facilitating collaborative work among municipalities that may not have equal resources at their disposal to carry out detailed ecological studies on their own. While this approach focuses on birds, the same principles can be applied to other groups of wildlife.

Logic of the Prioritization Scheme: Components of the Conservation Priorities Approach

Determining conservation priorities for breeding birds at the municipal level involves the assessment of three criteria: Jurisdictional Responsibility (JR), a scale-dependent measure related to breeding distribution within a given spatial unit; Preservation Responsibility (PR), a scale-independent measure based on the biological characteristics of the species; and Area Sensitivity (AS), a scale-independent measure related to the habitat area requirements of the species. Each species breeding within a municipality is objectively assigned a score for each component, and these scores are then summed to provide a total score. The list of priority species for a municipality comprises species that score highly on at least one of these components and have a total score greater than 8.5 (of a total 15 possible points).

Component 1: The Concept

Jurisdictional Responsibility is based on the concept that some species have distributions that are concentrated in some jurisdictions more than others, regardless of the species' abundance or status. In such cases, because some jurisdictions have a proportionally greater share of a species' population, these jurisdictions also carry a proportionally greater responsibility for the conservation of the species in question. The breeding habitat of these species will need special consideration in land-use decisions, which are invariably made at the local jurisdictional level.

Component 2: The Concept

Preservation Responsibility focuses on species that are at risk at the provincial level. These species, whether because of rarity, very limited distribution, low reproductive output, or declining numbers, may warrant priority conservation throughout their range regardless of jurisdictional boundaries. Essentially, this component serves as a warning that the species may be in trouble and that extra care

may be necessary in local level planning and development activities

The Application

Species are assigned scores based on the following criteria: abundance—rare species score highest; breadth of breeding/winter range—restricted ranges score highest; reproductive output—smallest clutches, or most infrequent breeding, score highest; and population trend over time—greatest declines score highest. These individual scores are averaged together to form a composite Preservation Responsibility score. Species scoring highly on this component are identified as species of conservation priority wherever they occur within the Province, regardless of their Jurisdictional Responsibility score within any particular municipality.

Component 3: The Concept

The area sensitivity criterion identifies species whose presence or absence is closely related to the amount of breeding habitat area within a given spatial unit. For example, some species may not be sensitive to habitat amount at all (i.e., they are present almost everywhere—whether there is 5 percent suitable habitat cover in a region or 90 percent suitable habitat cover), while others may be very sensitive (i.e., they only regularly breed in areas containing larger amounts of habitat). Because of their sensitivity to changes in habitat amounts, area-sensitive species may require special consideration in planning and development activities.

The Application

We used breeding range information from the Ontario Breeding Bird Atlas (Cadman, et

al. 1987) and land cover information from OMNR's Landsat imagery to determine relationships between breeding presence/absence and habitat area. Since this project focuses on southern Ontario, the analysis of area sensitivity was restricted to breeding distribution and land cover data within this region. Species scoring highly on the Area Sensitivity component are flagged as species of conservation priority.

Limitations of the List

The approach described in this report is one of many tools a municipality can use in efforts to conserve biodiversity. Other approaches could be used in conjunction with the list of priority species to make informed decisions. The resulting list of priority species was developed with broad scale, static data (breeding presence/absence within 10×10 km squares). Since species ranges will likely have changed slightly since the atlas data was gathered (approximately fifteen years ago), the list for a given municipality will not contain species that have recently colonized the municipality in question. While the list cannot possibly account for the dynamic nature of bird communities, it represents the best information on species' ranges currently available. The new Breeding Bird Atlas, scheduled to commence in 2001, will be used to create revised lists that reflect these changes in breeding distribution, but this information will not be available until 2005 at the earliest.

Point Reyes Bird Observatory

Diana Stralberg, GIS Specialist
Point Reyes Bird Observatory   prbo@prbo.or

PRBO, founded in 1965 as Point Reyes Bird Observatory, operates the oldest bird observatory in the United States. Located at the threshold of the Point Reyes National Seashore, PRBO has become an internationally acclaimed authority dedicated to conserving birds, other wildlife, and their ecosystems through innovative scientific research and outreach. Our forty-five staff scientists and more than thirty-five seasonal biologists work throughout the Pacific West to:

- Research and protect common and endangered birds, marine mammals, and sharks.

- Monitor, enhance, and restore related habitats and ecosystems.

- Provide science-based recommendations to ecosystem managers, policy makers, and private interests.

- Educate others about birds and conservation.

PRBO's conservation partners include dozens of federal, state, and local agencies and other nonprofit organizations such as U.S. Fish and Wildlife Service, Bureau of Land Management, National Marine Fisheries Service, National Oceanic and Atmospheric Administration, National Park Service, California Department of Fish and Game, The Nature Conservancy, and National Audubon Society.

In the thirty-five years since its founding, PRBO has:

- Established the first permanent bird banding station in the United States to perform long-term songbird population monitoring.

- Advanced the establishment of three National Marine Sanctuaries off the northern and central California coast.

- Cofounded the Riparian Habitat Joint Venture with eighteen other federal, state, and nonprofit organizations and served on many other technical advisory panels.

- Published the *Riparian Bird Conservation Plan: A Strategy for Reversing the Decline of Birds and Associated Riparian Species in California,* a model for habitat conservation and restoration for other regions throughout the country.

- Published more than six hundred scientific publications in peer-reviewed literature (journals and books).

- Launched a permanent research station on the Farallon Islands, located twenty-seven miles west of San Francisco and host to the largest colony of breeding seabirds in the United States south of Alaska.

- Identified and restored key habitat for common, threatened, and endangered species of birds and other wildlife including the threatened snowy plover.

- Trained nearly five hundred post-baccalaureate students in advanced field biology skills.

- Educated thousands of students, teachers, and members of the general public annually, illustrating basic environmental principles through the wonder of birds.

While in the past we have used GIS primarily for creating simple maps of our study sites and plotting visual overlays of bird and habitat data, we are now expanding into several new areas of GIS analysis, as follows:

- Generating detailed habitat maps for specific study sites based on manual and automated aerial image classification (San Francisco Bay), see map below.

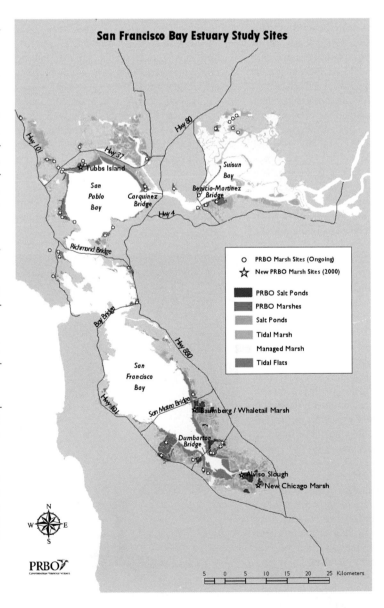

- Landscape-level analyses of bird habitat relationships and spatial patterns (Sacramento River riparian, San Francisco Bay).

- Spatially explicit modeling of bird habitat suitability (Point Reyes spotted owls, San Clemente Island shrikes, Sacramento River riparian songbirds, see map at right).

The San Francisco Bay and Sacramento River projects have been underway for five years or more and will likely continue through the next five years, depending on funding availability. The Point Reyes and San Clemente Island habitat modeling projects are one-year projects scheduled to start in 2001, with ongoing GIS mapping and analysis.

We also intend to make GIS an integral component of all research projects, from site selection to GPS data collection to data analysis and modeling. Issues of spatial scale and geographic location are starting to become an explicit component of all projects conducted over large geographic areas.

Finally, within the next three years, we plan to set up and administer an Internet map server as part of the California Partners in Flight conservation plans, allowing biologists and land managers to query a database of breeding bird data using a custom GIS interface. This would be an expansion of our current Web site, which presents static representations of species range maps and breeding status.

Point Reyes Bird Observatory Research Reserve on the Farallones Islands, California.

Photo courtesy Charles Convis

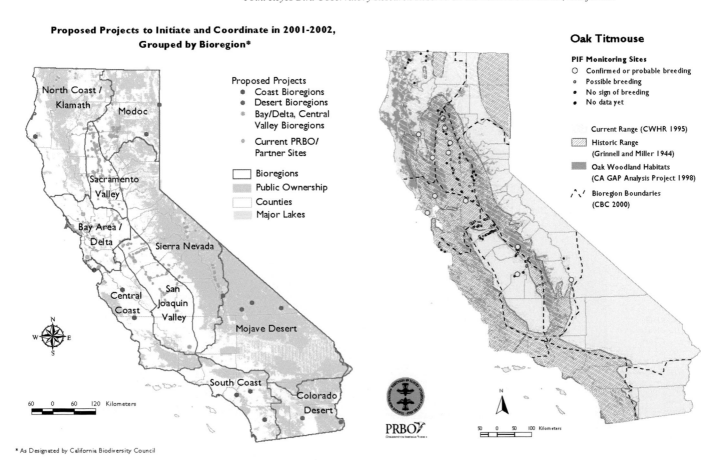

Proposed Projects to Initiate and Coordinate in 2001-2002, Grouped by Bioregion*

North Coast / Klamath
Modoc
Sacramento Valley
Bay Area / Delta
Sierra Nevada
Central Coast
San Joaquin Valley
Mojave Desert
South Coast
Colorado Desert

60   0   60   120 Kilometers

Proposed Projects
- Coast Bioregions
- Desert Bioregions
- Bay/Delta, Central Valley Bioregions
- Current PRBO/ Partner Sites

☐ Bioregions
☐ Public Ownership
☐ Counties
☐ Major Lakes

PRBO

Oak Titmouse

PIF Monitoring Sites
○ Confirmed or probable breeding
◌ Possible breeding
● No sign of breeding
• No data yet

☐ Current Range (CWHR 1995)
▨ Historic Range (Grinnell and Miller 1944)
▓ Oak Woodland Habitats (CA GAP Analysis Project 1998)
⌇ Bioregion Boundaries (CBC 2000)

50   0   50   100 Kilometers

* As Designated by California Biodiversity Council

The Vermont Institute of Natural Science

Kent McFarland
Andrew Toepfer, atoepfer@vinsweb.org

Kent McFarland (center with glasses) on Bicknell's thrush field work in the Dominican Republic.

The Vermont Institute of Natural Science (VINS) has made extensive use of ArcView GIS over the past two years, thanks to an initial grant from the ESRI Conservation Program. Most of our conservation biology research projects are now developed and implemented with GIS integrated as a critical component. The following provides an overview of how VINS has used each of the software components acquired through the ECP grant program, and the report concludes with a summary of a specific Conservation Biology Department project.

ArcView GIS

ArcView GIS is used on a daily basis in the Conservation Biology Department. Both GPS and radiotelemetry point data is converted into ArcView GIS shapefiles and usually linked to one or more project databases. From this data researchers are able to determine species movement, home range, territoriality, nesting patterns, habitat and elevation preferences, and so forth. We have also made use of the Animal Movement Analyst extension 2.04 (Hooge and Eichenlaub 1997).

In the Education Department, ArcView GIS is used routinely as a tool for the Community Mapping Program. VINS coordinates and directs this middle and high school place-based community and landscape awareness program at approximately twenty schools in Vermont and Colorado. Educators who incorporate ArcView GIS (or ArcExplorer in some cases) into their classrooms and acquire their own school site license for ArcView GIS are provided with training, support, and data. ArcView GIS is part of some of the presentations we give to schools and conferences on opportunities to integrate GIS into school curricula. ArcView GIS is also used for a Wildlife Corridor Mapping Project, at this stage primarily to georeference wildlife crossing sites.

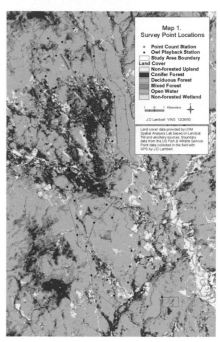

Owl Point Count Survey, Northeast Vermont. A Conservation Biology Department survey, using ArcView GIS and ArcView Spatial Analyst, data points created via GPS.

Three-dimensional overview of the Hartford, Vermont, Middle School Community Mapping Project Area, using ArcView, ArcView Spatial Analyst, and ArcView 3D Analyst.

ArcView Spatial Analyst

This extension is used routinely. Much of the work in the Conservation Biology Department is elevation and habitat related. This makes the use of grid files essential. Map queries are used to define new subsets, various analyses are performed, and maps are created.

For the Education Department, ArcView Spatial Analyst is used to create elevation shapefiles and other data sets as required for Community Mapping schools. The Wildlife Corridor Program also uses ArcView Spatial Analyst to test relationships between wildlife point data and habitat features.

ArcView Image Analysis

ArcView Image Analysis is used extensively by the Conservation Department to read and reregister image data from the Dominican Republic (winter songbird nesting habitat). It is also being used to create topographic index files for a grid of butterfly sampling sites covering the state of Vermont.

In the Education Department, ArcView Image Analysis is used extensively to register historic maps from scans. It is also used to subset images.

ArcView 3D Analyst

This extension has not been used as much as the others. It is currently used by the Education Department to create terrain models for school project areas and Wildlife Corridor Mapping Project interest areas. The use is primarily for presentations.

The Community Mapping Program will likely use this extension more as we explore ways to utilize another extension, CommunityViz™, in high school programs.

In summary, ArcView GIS and the extensions are the primary software in use by several of our conservation biologists. It has been a tremendous help in most of the conservation biology programs underway at VINS. The Education Department, primarily through the Community Mapping Program, is also using ArcView GIS extensively. The Community Mapping Program would have to be totally restructured without an available GIS such as ArcView GIS. We are pleased with the grant arrangement through the ECP program and look forward to continued participation.

VINS is also the recipient of a CTSP grant for 2000, through which we received an HP plotter, ArcPress for ArcView GIS (very useful), and ArcInfo 8 plus extensions. For the time being all our needs have been met, although we may need a second user license for ArcInfo 8 sometime in the next year. Also, the upcoming changes in ArcView 8.1 might prompt us to request a grant for upgrades.

Summary of Bicknell's Thrush Study (Kent McFarland)

Bicknell's thrush (*Catharus bicknelli*), recognized as a subspecies of the gray-cheeked thrush (*Catharus minimus*) since its discovery in 1881 on Slide Mountain in the Catskills of New York, has recently been given full species status by the American Ornithologists Union. Significant differences between the two taxa in morphology, vocalizations, genetics, and breeding and wintering distributions contributed to this designation. This classification has led to the recognition of Bicknell's thrush as one of the most at-risk passerine species in eastern North America. Partners in Flight ranked Bicknell's thrush as number one on a conservation priority list of neotropical migrant birds in the northeast. The species has been accorded "vulnerable" status in Canada and is also considered "vulnerable" by BirdLife International for the IUCN 2000 Red List.

Vermont Bicknell's thrush.

The breeding range of Bicknell's thrush in the United States is limited to subalpine spruce–fir forests of New England and New York. In Canada it is found in highland spruce–fir forests in Quebec, Nova Scotia, and New Brunswick. It has also been found in mixed second-growth forest following clear-cutting or burning in Quebec and New Brunswick. As the only breeding songbird endemic to high-elevation and maritime spruce–fir forests of the northeastern United States and adjacent Canada, Bicknell's thrush qualifies as a potentially valuable indicator of the health of montane avian populations and their associated forest habitat. Research aimed at clarifying the distribution, ecology, and population status of Bicknell's thrush in the northeast have been underway since 1992; similar studies are in progress in New Brunswick, Nova Scotia, and Quebec.

Decline of high-elevation forests in the northeastern United States during the 1960s and 1970s is a well documented phenomenon. Red spruce (*Picea rubra*) dieback has been especially pronounced, but mortality of balsam fir (*Abies balsamea*) has also been extensive and widespread, although most of this has resulted from naturally occurring fir waves.

Atmospheric deposition of acidic ions from industrial sulfur and nitrogen oxides has been strongly, although not conclusively, implicated as a causal factor in red spruce decline. Increased winter freezing injury of spruce, possibly mediated through reductions in calcium reserves, may be directly linked to high levels of acidic deposition. Despite declining trends in atmospheric sulfate concentrations resulting from mandates of the 1990 Clean Air Act amendments, acidity of precipitation in northeastern North America does not appear to be decreasing. Heavy metal toxicity from airborne pollutants has also been implicated as a contributing cause of high-elevation forest decline in the northeastern United States, particularly in the Adirondack and Green Mountains, although several recent studies indicate that lead concentrations in the forest floor are rapidly decreasing. These documented problems, combined with potential loss of habitat to global climate change, other atmospheric pollutants (e.g., mercury), ski area development, telecommunication tower construction, and proposed wind power facilities make this restricted habitat one of the most vulnerable in eastern North America.

The wintering area of Bicknell's thrush was completely unknown until 1921, when specimens of gray-cheeked thrush from the island of Hispaniola were examined and determined to be bicknelli. In addition to these and other more recent specimens from Haiti and the Dominican Republic, documented winter records of gray-cheeked thrush, probably representing bicknelli, have been obtained in Jamaica, Mona Island, and Puerto Rico. However, the precise distribution and wintering ecology of Bicknell's thrush remains poorly documented. Preliminary data suggests that the species may be limited to primary tropical forest at both high and low elevations. These forests have been heavily clear-cut, burned, and converted to other uses throughout the Caribbean as a result of burgeoning human population pressures. Forest inventories in the 1980s indicated that only 14 percent of the Dominican Republic remained covered with moist, broad-leafed forests. The consequences of such deforestation on wintering Bicknell's thrush populations are unknown but could be significant (see map below).

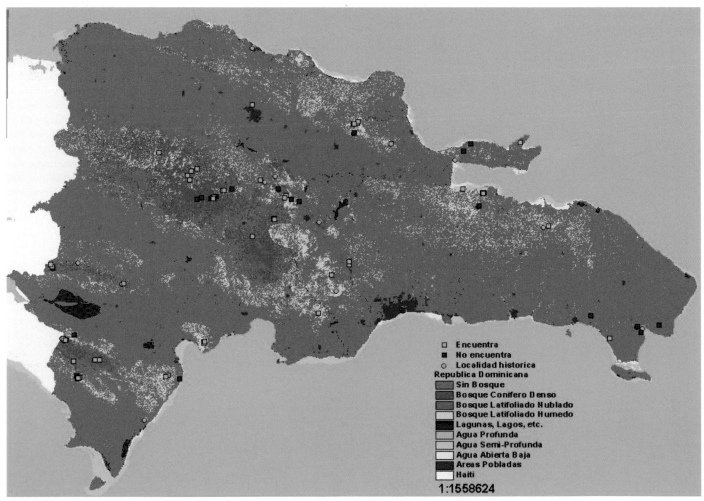

Dominican Republic habitat types and Bicknell's thrush survey sites.

Following a preliminary field investigation in 1994, we initiated long-term studies of Bicknell's thrush in the Dominican Republic during November and December of 1995. These studies were continued during the winter of 1996–97. The primary purposes of this research were to (1) visit areas of known historic occurrence of Bicknell's thrush to determine current presence/absence and habitat condition, (2) establish study sites in Parque Nacional Sierra de Baoruco to initiate a long-term monitoring program and to collect demographic and population density data, (3) survey forested areas throughout the country to document wintering distribution and habitat use, and (4) enlist the cooperation of Dominican wildlife agencies and other conservation organizations in determining the winter range of Bicknell's thrush and collaborating in future studies.

We were accompanied in the field by Dominican wildlife biologists, national park guards, and local guides. We trained them in the various field methods (e.g., taking census, mist netting, banding, and field record-keeping) used during this study. Informally, but often in considerable detail and at considerable length, we discussed conservation issues and land-use practices with Dominican natural resource officials, park guards, and guides who assisted us as well as many local citizens we met. Although the effects of such discussions are difficult to evaluate, we believe this is an important and highly effective method to raise consciousness about bird conservation at the grassroots level.

Many important questions about the ecology and stability of Bicknell's thrush populations require intensive monitoring of discrete habitat units and studies of known-identity individuals. Baseline data on population densities, territory size, movements, productivity, site fidelity, survivorship, habitat use, and the effects of human activities are needed to evaluate the conservation status of the species across its naturally fragmented, high-elevation breeding range and its winter range in the Greater Antilles. Below, as examples, we present some of the work that has been done for this project using ESRI software.

Radio Telemetry
We placed miniature radio transmitters on Bicknell's thrush adults on Stratton (1997–1999) and Mansfield (1998–1999). We used radio transmitters from two different manufacturers. All transmitters in 1997 and most (84 percent) in 1998 were from Wildlife Materials (model SOPB-2012) and weighed 0.9 gram with a battery life of 23 days. Some transmitters in 1998 (16 percent) and all in 1999 were made by Holohil Systems (model BD-2) and weighed 1.0 gram with a battery life of 60 days. In 1997 and 1998 we attached transmitters to the

base of the two central rectrices using dental floss and Super Glue. Because of an unacceptable rate of transmitter loss in 1998 and the increased battery life available to us in 1999, we attached transmitters using a harness design. We detected no obvious effects on behavior of radio-tagged birds, but we were unable to test this directly via time budgets or movements due to the species' secretive nature and the dense vegetation and rugged terrain on our study plots.

We relocated radio-tagged birds with Wildlife Materials TXR-1000 receivers and three-element, handheld Yagi antennas using homing or triangulation techniques. Homing locations were determined by identifying vocalizing birds, quietly approaching birds to pinpoint locations visually, or by circling the signal area in small habitat islands while walking the surrounding grassy ski trails. Locations were mapped in the field on detailed maps of the study areas, and observers ranked point accuracy on a scale of 0–3 (0 = exact, 1 = ~10 m, 2 = ~25 m, 3 = ~50 m radius).

Triangulation data was analyzed using LOAS™ 1.0 software (Ecological Software Solutions 2000). Field tests of bearings to transmitters resulted in a standard deviation of 200, which we used to estimate point locations. Transmitter locations with three or more bearings were estimated with the Maximum Likelihood Estimator (MLE) with a 95 percent confidence ellipse estimated using the chi squared method. We used best biangulation when MLE failed or there were only two bearings. Best biangulation calculates all intrabearing angles and selects the bearings whose angles are closest to 900 and calculates an error polygon using a 150 bearing standard deviation.

Home range location, size, and movements of Bicknell's thrush breeding season home

ranges were defined as the area used by an individual from 1 June to 31 July each season. We determined home range size and location using the nonparametric kernel method calculated with ArcView GIS and an extension called Animal Movement Analyst 2.04 (Hooge and Eichenlaub 1997). We used a fixed kernel with the smoothing factor determined by least-squares cross-validation. We calculated both the 95 percent (area the bird actually used) and 50 percent contours (core area of activity) for birds with a minimum of 30 locations. We used only those locations that were more than five minutes apart based on the general rule that locations t1 and t2 can be considered independent if the period between them is sufficient to allow the bird to move from one end of its home range to the other. Field experience suggested that thrushes could fly from one end to the other in much less time. Locations of birds known to be on the nest (e.g., brooding females) were excluded.

Home Range Overlap
We calculated a static home range interaction of neighboring thrushes from the kernel home range (KHR) using the following equations: $S_{1,2} = A_{1,2}/A_1$ and $S_{2,1} = A_{1,2}/A_2$ where A_1 and A_2 are the total KHR areas of thrushes 1 and 2, $A_{1,2}$ is the area of overlap, yielding the proportion of the home range of animal 1 overlapped by animal 2 ($S_{1,2}$) and the proportion of the home range of animal 2 overlapped by animal 1 ($S_{2,1}$). This statistic is limited in that it does not imply any mutual awareness among the tracked thrushes; however, a more rigorous dynamic interaction statistic in which birds are tracked simultaneously was not possible due to logistic and environmental constraints.

Bicknell's Thrush Home Range
We have digitized and begun to analyze radio telemetry data for Stratton Mountain. Radio telemetry data on Mt. Mansfield is entered and triangulation calculations are underway. In 1997, we employed radio telemetry to investigate how thrushes move through the ski trail-forest island complex, and assess their reactions to recreational activities. We quickly discovered that male thrushes were not holding small, discrete territories, as is generally assumed for most Nearctic–Neotropical migrants, but instead broadly overlapped .

We frequently detected several males singing and calling from the same area within a single hour. The areas of high overlap generally coincided with nest site locations. However, unlike the dunnock (*Prunella modularis*), males do not defend exclusive areas that encompass more than one female, but appear to behave more like male Smith's longspurs (*Calcarius pictus*), which defend small areas around the female. Bicknell's thrush females tend to occupy home ranges with little or no overlap, and these are much smaller than male home ranges. Our field observations suggested that females aggressively protect territories during the brief period of mating and egg laying. Further analyses of our radio telemetry data should better elucidate the dispersion patterns and movements of Bicknell's thrush, particularly in relation to its complex mating system.

Winter Range Distributional Surveys

During the winters of 1995–96 through 1999–2000 we conducted broad-scale distributional surveys in forested habitats of the Dominican Republic on the Samaná Peninsula, Los Haitises, southeastern region, eastern and northern sections of the Cordillera Central, Sierra de Baorucos, and Sierra de Neiba (see map on next page). We confirmed the presence of Bicknell's thrush at 15 of the 25 (60 percent) sites surveyed. All sites with confirmed presence consisted of primary or mature secondary broad-leaved forest or pine forest with a predominantly dense, broad-leaved understory. Most surveyed sites where Bicknell's thrush was not documented had a recent history of impact by agriculture, grazing, or fire. We found thrushes at seven of the 13 (54 percent) historic sites we surveyed. Many of these areas have been and continue to be heavily impacted through cutting, fire, grazing, and agriculture. We were unable to pinpoint the exact location of several historic sites due to their inadequate documentation or because radical changes in the surrounding landscape had made them unrecognizable.

The Loma Atravesada site on the Samaná Peninsula typifies the large-scale landscape changes that have greatly impacted some sites of historical Bicknell's thrush occurrences. Virtually all of the Samaná Peninsula is dedicated to subsistence agriculture and plantations of cacao and palm oil. In 1997, we located one of the last remaining tracts of forest larger than a few hectares on the entire peninsula. This remnant patch (25 ha.) of wet broad-leaved forest is currently in private ownership but is wholly unprotected from selective cutting of large trees for charcoal and lumber, and from agricultural clearing on its periphery. Despite these serious incursions, we found a relatively high density of Bicknell's thrush in this forest fragment. The apparently high density of thrushes may have been inflated by immigration from recently impacted surrounding areas. We failed to detect any thrushes in a much smaller (4 ha.), isolated parcel of forest on the peninsula between the towns of Sanchez and Naranjitos especially.

Field studies revealed that Sierra de Neiba was an area that was losing forest cover from harvesting and fire at an exceedingly rapid rate despite being declared a national park in 1995. Of the two Neiba localities we visited in 1997, the western section above "Vuelta de Quince" on the road to Hondo Valle showed fewer signs of conversion, with only selective removal of very large pine and hardwoods for lumber and some clearing for agriculture. In contrast, the eastern section above Apolinario centered on Monte Bonito had been extensively cleared for agriculture. In 1995, cleared areas and agricultural plots within this section of the park were estimated to occupy 30 to 40 percent of the land. In early 1997, we estimated this figure to be 70 to 80 percent, with very little forest remaining within park boundaries and none outside. We estimate complete loss of forest in this area within two to three years, although increasing fragmentation may render many forest patches too small to support forest-dwelling birds even earlier.

Boca de Yuma in southeastern Dominican Republic illustrates the difficulties we have encountered with historic records. Bicknell's thrush was found by two independent observers in the 1970s through banding captures and tape recorded playbacks. Six thrushes were captured during three days of banding (April 8–11, 1974) near Boca de Yuma, although no thrushes were captured when netting was again conducted at that site on January 7–9, 1975.

The site was said to be located "3.5 miles west of the Club Nautico entrance." A single recording of a thrush was also obtained on March 9, 1979, at an unspecified location in the Boca de Yuma area. We visited the area on December 11–12, 1995. After questioning local citizens we were able to locate the former entrance for Club Nautico. The original directions to the banding site, written as "3.5 miles," left us with uncertainty as to whether we should travel 3.5 miles or kilometers because vehicle odometers are metric in the Dominican Republic. We traveled 3.5 miles and found large agricultural fields and no forest. At 3.5 kilometers we found forest on one side of the road and recently cut forest (stumps still visible) converted to pasture on the other side of the road. A dusk census at this latter site failed to detect any thrushes. We also censused areas within Parque Del Este 1.5 miles west of Boca de Yuma and failed to detect any thrushes. The forest immediately outside of the park was heavily impacted or removed. The historic presence of Bicknell's thrush in this area remains uncertain. We plan to revisit this area in 1997–98 and survey the gallery forest along the Yuma River where humid forest currently appears to be intact.

Management and Conservation Implications

Pending full analysis of our existing data and compilation of more robust data for many aspects of this research, it is premature to provide definitive management recommendations. However, we have worked closely with several ski areas and with the Vermont Fish and Wildlife Department to provide preliminary guidelines for ski area land managers and to address several site-specific management issues. We developed a mitigation plan with Stratton Mountain for an area of new lift construction in 1999, setting aside for reforestation an area of developed trails equal in size to an area that was removed by construction activities. This exchange also served to connect several small islands of habitat, enhancing their collective value to Bicknell's thrush and other species.

We obtained ERDAS IMAGINE®-formatted land-use coverage of the Dominican Republic from a government agency in country. We imported from GPS our locations that we surveyed for Bicknell's thrush and all known historic locations. We are currently obtaining a protected area coverage to determine how well this rare bird and the remaining habitat it occupies is protected and suggest areas in need of further protection and management.

We will continue to develop management guidelines as our data analyses and experience warrant and to provide advice when requested. A detailed, rangewide conservation assessment and management placement is in the early stages of preparation but will require additional field data collection and analysis.

Comanche Pool Prairie Resource Foundation

Loren Graff   lgraff@rh.ne

The ECP grant of ArcView GIS 3.2 has been useful to the Comanche Pool Prairie Resource Foundation in our mission to improve the native range ecosystem. We can now produce high-quality maps that ranch managers are able to use to manage their rangeland. The Comanche Pool Prairie Resource Foundation acquired orthophotography to use as the base layer. Other layers that were acquired include soils, USGS topography maps, and the public land survey. Fence lines, water lines, and so forth, are added using the ArcView GIS software. Using the ArcView GIS software and the various layers, the Comanche Pool Prairie Resource Foundation assists range managers with planning and the development of intensively managed grazing systems. These grazing systems improve the range ecosystem. The software allows the managers to quickly measure lengths and acres. It also enables the managers to calculate acres of particular range sites to determine proper stocking rates. The Comanche Pool Prairie Resource Foundation is beginning to utilize ArcView GIS to assist with the management of at-risk wildlife species. Of particular interest is the lessor prairie chicken. This wildlife species is under consideration to be listed as a threatened species. The Foundation is using the ArcView GIS software to document and track areas that the lessor prairie chicken uses as "booming grounds" (mating areas). Once located, these sites can be managed to maintain the conditions that the lessor prairie chicken requires. The Comanche Pool Prairie Resource Foundation intends to continue improving GIS knowledge and capabilities. The Foundation hopes to use satellite imagery to determine biomass cover across the rangeland. This technology can be used for areawide management, individual ranch management, and wildlife management. A ranch manager could utilize the information to easily visualize what parts of the ranch are being overgrazed or undergrazed. The manager could also use the information to monitor residual cover of key wildlife habitat areas of the ranch. The Comanche Pool Prairie Resource Foundation continues to be a voluntary organization with little funding. Activities of the Foundation are conducted through the voluntary efforts of the board of trustees, as well as a few additional volunteers.

3 - Bar Ranch

At this time, we locate the booming grounds by actually driving around to look for the birds and by visiting with landowners. The lessor prairie chickens utilize the booming grounds for only a couple of months in the spring. Therefore the time frame is short to locate the areas. As mentioned earlier, these areas are typically on higher elevations, as well as on areas with shorter native grass. After the mating season, the chickens then require taller native grass in close proximity for nesting and raising the chicks. Therefore it is important to manage several different habitats near each other for the lessor prairie chickens.

Institute for Wildlife Studies GIS Strategy

Gregory Schmidt, Wildlife Biologist (GIS Coordinator)

About the Institute

The Institute for Wildlife Studies (IWS) is a nonprofit conservation organization that conducts long-term research on bird and mammal species, many of which are threatened or endangered. Our goal is to aid in strategic planning for the survival and recovery of these species. The IWS collects baseline and habitat requirement data on these species; monitors changes in population size, distribution, and density; reintroduces species to formerly occupied habitat; and investigates factors that may be contributing to population decline. We are presently involved in recovery efforts for federally listed species such as the endangered San Clemente Island loggerhead shrike (*Lanius ludovicianus mearnsi*), the threatened bald eagle (*Haliaeetus leucocephalus*), the threatened San Clemente Island sage sparrow (*Amphispiza belli clementeae*), and the island fox (*Urocyon littoralis*), which will soon be listed as endangered on four of the six California Channel Islands it inhabits. The Institute's main office is located in Arcata, California, with field offices on Santa Catalina Island and San Clemente Island in California.

The majority of our work is conducted on the California Channel Islands, but we also have ongoing projects in the South Pacific, Russia, Japan, and New York State.

Use of ESRI Software

In the year since receiving the donated software, we have established a GIS as a tool at our three offices. At least one individual at each IWS office has spent a considerable amount of time learning the basic, and in some cases the more complex, functions of each program and has trained other employees in basic operation. Our main office coordinates base coverage acquisition, creation, conversion, and maintenance. The GIS software grant came at a time when our organization began to grow at an exponential rate. The frequent addition of new projects and personnel required that our GIS contain uniform base coverages and consist of software that was easy to learn and familiar to most new employees (e.g., mainly ArcView GIS). We plan to streamline the learning process with on-site training and the development of IWS-specific applications and scripts.

Although we are still in the process of creating or acquiring base coverages for all of our project sites and modifying or validating existing base and project coverages, the GIS software has allowed us to make the most out of what is currently available. Several IWS employees regularly use ArcView GIS and the ArcPress extension. The Data Automation Kit has proved valuable for those employees who do not have the time to learn PC ARC/INFO commands but who require some of its functions for creating, editing, and converting coverages. The menu-driven Data Automation Kit has allowed the GIS coordinator in our main office to walk employees in our field offices through a PC ARC/INFO operation without having to delve into details on how to use the commands and the proper syntax. This allows an employee with little experience, or experience only with ArcView, to learn at least how to build topology, convert shapefiles to coverages, and to import and export coverages.

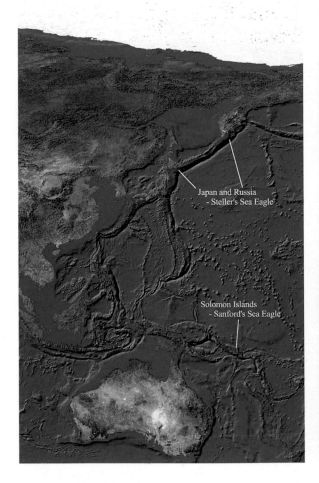

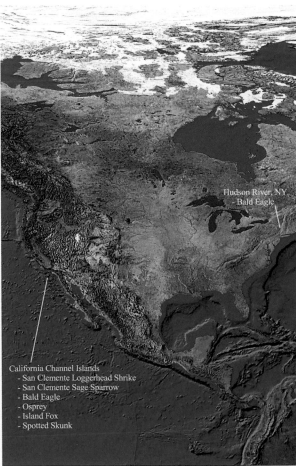

Institute for Wildlife Studies project locations and species studied. This graphic file was created using ESRI software (ArcView GIS and ArcPress) and a WorldSat Color Shaded Relief Image, both donated through the ESRI Conservation Program.

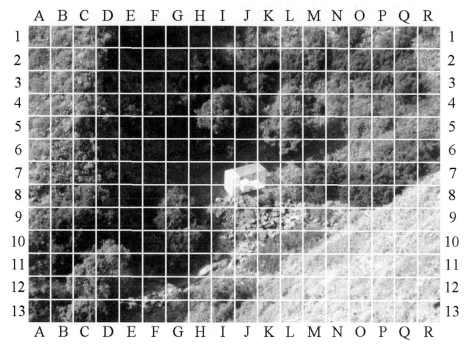

Photograph of a shrike release site with a grid overlay used to track movements of released loggerhead shrikes. Personnel placed along the canyon perimeter were able to report shrike positions to other crew members quickly and accurately using the grid cell code rather than attempting to describe the location based on vegetation.

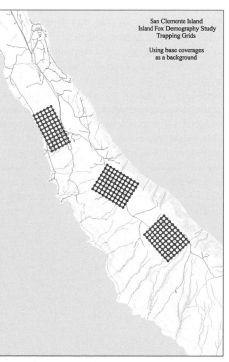

Maps of island fox trapping grids used for field navigation and logistical planning.

We are just beginning to use the ArcView Spatial Analyst and ArcView 3D Analyst extensions to create three-dimensional digital terrain models of our project sites and for three-dimensional draping of animal home range polygons over our terrain models. Once we obtain or create the necessary base coverages, we will use these programs to develop habitat suitability models for several threatened and endangered species we study. The habitat models can then be used to assist in modeling changes in animal population size and distribution with predicted changes in vegetation structure, composition, and distribution over time. Used in conjunction with other available information, we will be better able to predict the effectiveness of recovery efforts.

We have used our GIS software for project planning and implementation, for spatial analysis, and in presentations and reports. Below are examples of how we have used our GIS software within the past year in each of these categories.

Project Planning and Implementation

We used PC ARC/INFO to generate trapping grids for island fox projects on San Nicolas and San Clemente islands, California. Field personnel used the maps and the coordinates obtained from the coverages to navigate to trap sites using a global positioning system (GPS) receiver. Furthermore, the maps were used for logistical planning by providing other important features such as roads and stream courses. Several other field projects rely on GIS to work out access routes to remote sites and to determine trail construction routes.

As part of the conservation program for the island fox, the IWS has proposed the establishment of a captive breeding center on Santa Catalina Island, California. We are presently using our GIS software to aid in the selection of a suitable location for the center based on topography, vegetation, and road locations.

As part of the San Clemente Loggerhead Shrike Recovery Program, the IWS releases captive-bred shrikes from cages placed in suitable habitat at remote sites. However, many released shrikes quickly evade field personnel making it difficult to determine the outcome of the release and the ultimate fate of the birds. To better track released shrikes we created a grid to overlay on a photograph of the release site (see map left). The grid enabled field personnel to track a bird and report its position to other personnel (via radio) rapidly by using the grid cell code rather than attempting to describe the location based on vegetation characteristics.

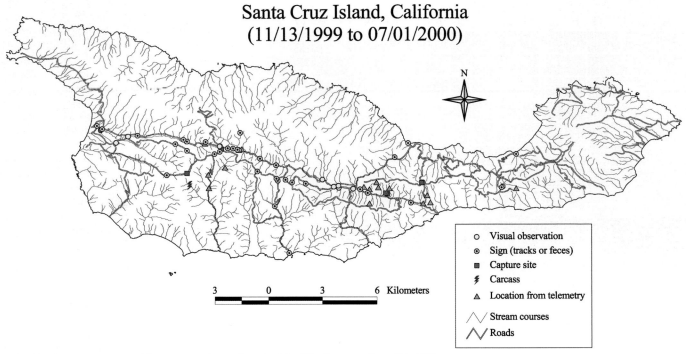

Santa Cruz Island, California
(11/13/1999 to 07/01/2000)

N

○	Visual observation
◉	Sign (tracks or feces)
■	Capture site
⚡	Carcass
△	Location from telemetry
/\/\	Stream courses
/\/	Roads

3   0   3   6   Kilometers

Island fox location map created for a presentation made to the Island Fox Recovery Team, July 2000.

Our GIS software is used regularly for quick linear measurements (e.g., road segment length, stream course lengths, and straight-line distance between two points), area calculations (e.g., vegetation polygons), and field map generation (using digital orthophoto quadrangles and/or digital raster graphic topographic maps images).

Reports and Presentations

A very important use of our GIS is in the creation of map images for technical reports and for creating slides for presentations. Prior to receiving our software grant, we relied on several programs (all nongeospatial) to generate map images for incorporation into reports and for slide production, generally resulting in low-resolution, poor-quality images. With our GIS we have standardized methods used for image generation (largely through the use of the ArcPress extension) and have greatly increased the quality of image output. By maintaining identical GIS programs in all IWS offices, we were able to standardize the image and feature coverages used as backgrounds for "heads-up digitizing" (i.e., used a mouse to input geo-spatial data rather than a digitizing tablet) and graphic file generation. The high-quality images and slides we now produce have greatly improved our reports and presentations. Moreover, all displayed features are now registered to the earth's surface rather than existing as graphic objects (representing spatial data themes) that were overlaid using nongeospatial graphics programs.

Spatial Analysis

In the past year, the majority of data collected by the IWS was at the plot level, as we were measuring habitat structural components and prey abundance and composition at sites occupied by study animals. Therefore, spatial overlay analyses at a larger scale were not necessary. However, we plan to broaden our analyses to include stand and landscape scales, especially for measuring the amount of suitable habitat present at occupied versus unoccupied sites. We will begin extensive island fox surveys on San Clemente and Santa Catalina islands this fall, which will require regular spatial analyses to map fox density, distribution, and abundance and to calculate road density and vegetation composition at occupied and unoccupied sites. We will also develop spatial habitat models to aid in estimating the number and distribution of foxes that can be maintained on each island. This information will be especially important on Santa Catalina Island where we plan to breed foxes in captivity and release them into a suitable habitat. Our goal is to release the captive-bred foxes in such a way as to maximize the probability that wild foxes will interact, form pairs, and breed naturally to augment the population that was decimated by disease in 1999.

Marine Geography

Introduction

This section of *Conservation Geography* began as an idea spawned by the growing need for a gathering place to share the work of marine researchers and professionals using GIS. As we become increasingly aware of the world's oceans and seas and the diversity they contain, we have an inherent responsibility to preserve them. This compilation of stories provides a glimpse into the many uses of GIS for marine applications being developed and used today. The people and organizations that contributed to this section are a sampling of the researchers, organizations, and professionals who are dedicated to understanding and analyzing this dynamic and changing environment.

Marine GIS has been used in many ways to gather data from ocean and seas, and integrate the information of this environment. Near shore and deepwater phenomenon such as current, salinity, temperature, biological and ecological mass, and density all play integrated roles we are only beginning to understand. Some of the areas of marine GIS development include oceanography, coastal zone management, navigation and charts, ocean industries, and conservation. In the following pages you will read of the brave marine researchers and professionals who take an active role in marine conservation, to discover and present their marine studies and findings using GIS as the primary tool.

The studies featured here include local, regional, and international marine conservation efforts. The National Oceanographic and Atmospheric Association and the Federal Geographic Data Committee Bathymetric Subcommittee have created a Shoreline Metadata standard complete with an informative marine and shoreline specific glossary and bibliography. The monitoring of dolphins in Florida Bay by the Dolphin Ecology Project uses GIS to describe dolphin density and distribution as influenced by environmental degradation. The International Marine Life Alliance is using GIS coral reef invertebrates collection for reef rehabilitation and the aquarium trade. The Bay of Fundy Marine Resource Centre as developed the national Coastal Resources Mapping Project and is mapping marine biological and ecological resources of the region. The USGS Glacier Bay Field Station has developed an extension made available over the Internet, that allows three-and four-dimensional analysis and display of volumetric, and time series, oceanographic data. People for Puget Sound has continued to develop its community driven coastal and marine survey initiatives, to become a regional leader in data analysis for marine and near shore ecosystems.

The Marine Conservation Biology Institute continues to promote the integration of conservation science into marine policy, with the Baja-to-Bering Mapping Initiative. The Heal The Bay community action group uses GIS to change policy in their region. The Center for Marine Conservation has developed a program to evaluate existing and proposed Marine Protected Areas. The Surfrider Foundation and the Beachscape program has expanded to a national network of chapters mapping coastal areas and increasing the information we have to preserve beach boundaries from coast to coast. The Gulf of Maine monitored by the New England Aquarium reveals interesting patterns between sea surface temperature and Bluefin tuna distribution. The Oceanic Research Foundation using the satellite telemetry tracking of turtles from space promotes international conservation concerns from the Gulf of Mexico to the United States. And The Woods Hole Oceanographic Institute is utilizing GIS for its moored buoy site evaluations.

The collection of authors in this section represents some of the people and organizations who have devoted their time and energy using GIS to better understand and represent what happens from the highest of high waters to the bathymetric bottom. It is our goal to assist the development of these and other studies by providing this medium to share some of the marine GIS techniques, ideas, and projects underway today. Thanks to your continued interest and support we can continue to develop this area of study and the many ways that GIS can be applied to it.

Barbara Shields, ESRI author, enjoys the coast of Southern California.

Joe Breman examines the nearshore coastal plain offshore in northern Israel.

ESRI Marine Conservation Program

The concept of the ESRI Marine Conservation Program has developed from twelve years of successful networking, education, and outreach of the ESRI Conservation Program (ECP). The ESRI Marine Conservation Program intends to evoke the interest of the marine GIS user community, and promote its development through this collaborative initiative. Thanks to the generous support of the National Oceanic and Atmospheric Administration, this collection of stories is a first effort of what we hope will become an annual forum for this expanding niche, to help reach people and influence change in the marine GIS community. The mission of the ESRI Marine Conservation Program is to support the conservation of the seas and oceans worldwide by facilitating communication and encouraging cooperation throughout the marine GIS user community. The goals and objectives of the ESRI Marine Conservation Program are to:

- Provide training and professional development opportunities to increase the capacity of the marine community in the use of GIS.

- Promote low-cost or free access to marine-related data, and encourage the collaborative sharing of marine related information.

- Publish articles annually to provoke the exchange of new ideas, methodologies, and tools relevant to the marine user community.

- Assist the marine GIS community in obtaining and utilizing GIS equipment, software, and services.

- Extend outreach to make marine-specific GIS techniques available to a wide network of marine users.

- Enhance the development and growth of the marine GIS user group by providing a forum for meeting at the annual ESRI User Conference, and hosting a website: www.esri.com/oceans

Joe Breman
Editor, Marine Geography

Monitoring Dolphin Behavior and the Effects of Restoration

Laura K. Engleby
Dolphin Ecology Project Florida

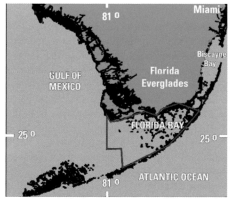

During the habitat use study, researchers observed "mud-ring feeding," a previously undescribed feeding strategy by bottlenose dolphins in Florida Bay. This feeding strategy is closely associated with the mud banks in the bay. Feeding either takes place along the edge of the bank or on the bank itself.

The ecological degradation of South Florida is as notorious as ecological degradations in Lake Eric and Chesapeake Bay. Humans have altered the South Florida landscape in ways that affect the temporal and spatial variability in water flow, nutrient loading, and productivity. Florida Bay in particular has experienced increased salinity due to the diversion of freshwater input. This increased salinity, together with elevated nutrient levels from land development sources, has stimulated algae blooms, resulting in large-scale die-offs of sea grasses, sponges, and mangroves. Consequently, declines in these habitats have caused reductions in fish populations.

Over the past fifty years, this severe degradation has been well documented by scientists. During this time, top predators such as herons, brown pelicans, alligators, and storks have declined by 80–95 percent. Sixty-eight species of South Florida's mammals, birds, reptiles, amphibians, and plants are threatened or endangered. Federal and state management agencies have responded to these declines with an intensive, $7.8 billion restoration project designed to restore the natural quantity, quality, timing, and distribution of freshwater into Florida Bay. No one knows what impacts these changes will have on a highly visible, upper trophic-level species—the bottlenose dolphin.

Regional managers of federal, state, and county agencies need information. Currently, agency managers overseeing the restoration efforts have no baseline information about bottlenose dolphin numbers, preferred habitats, seasonal movement, food requirements, or reproduction. Restoration projects require essential information on life history, density, and distribution patterns of bottlenose dolphins as well as how they relate to, and are affected by, habitat and water quality in the Florida Keys. Using researchers, volunteers, recording equipment, and a GIS, the Dolphin Ecology Project hopes to collect and distribute this information to restoration agencies for a decision making tool.

The project's GIS is useful in determining:

- The distribution and density of dolphins that regularly live around the Florida Keys and the number of dolphins that might be transiting through this area.

- The types of prey that dolphins depend on and the types of habitats that dolphins utilize to find their prey.

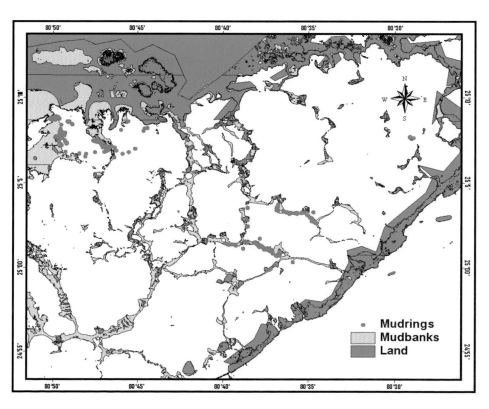

Using researchers, volunteers, recording equipment, and a GIS, the Dolphin Ecology Project hopes to collect and distribute this information to restoration agencies for a decision making tool.

Central and eastern Florida Bay mud-ring feeding sites.

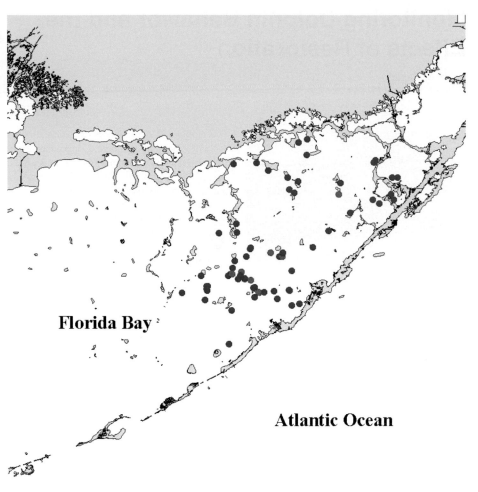

Florida Bay

Atlantic Ocean

•	Trawl locations-feeding
•	Trawl locations-nonfeeding

- The impact on dolphins produced by changes in their environment such as variations in salinity, nutrient levels, and pesticide runoff.

- By conducting surveys and focal-animal-follows from small, outboard-powered vessels, the team assesses the distribution of bottlenose dolphins and their prey in relation to water quality. Surveys conducted throughout the year reveal seasonal changes in both dolphin distribution in the area as well as dolphin habitat use. To do this, the team divides the study area into strata representing habitats of similar composition and quality. Moreover, they assess prey distribution using a small otter trawl.

The team also measures other properties of habitat qualities such as salinity, temperature, and turbidity.

GIS is instrumental in recording dolphin distribution and feeding behavior. When the team locates a pod of dolphins, they record the sighting location using a differential global positioning system (GPS) unit. They also take photographs of the dorsal fin of each dolphin in the group to recognize individual dolphins. One distinctive dolphin is selected as a focal animal and is then followed from a distance of less than one hundred meters. During each follow, the focal animal's location, habitat type, behavior, group composition, and size are recorded at three-minute intervals.

Researchers record the behavioral state of the dolphin using the following criteria:

- Travel—directed movement in a specific direction.

- Socializing—tactile contact among group members.

- Resting—slow quiescent movement or individuals remaining motionless at the surface.

- Feeding—direct evidence (prey in mouth) or strong indication of feeding (chasing prey, rapid changes of direction).

Using these criteria, the observers are able to distinguish dolphins that are actively feeding from those engaged in other activities. Once a dolphin is seen feeding, researchers then sample prey (using a small otter trawl) and water quality (temperature, salinity) using a YSI multiprobe. Thus, they obtain samples of fish abundance, species composition, and water quality at sites where dolphins are feeding. During the follow, researchers sample prey and water quality at regular intervals to obtain control data. This distribution, movement, and behavioral data collected in the field is then incorporated into ArcView GIS to test the hypothesis that dolphins feed preferentially in habitats where water quality is good and prey densities are high.

In addition to these dedicated focal-animal-follows, the Dolphin Ecology Project (DEP) examines data on dolphin distribution collected by their collaborators at the National Marine Fisheries Service South East Fisheries Science Center. Eventually, the DEP would like to overlay distributional information from the U.S. Coast Guard aerial surveys to see which habitats dolphins are using on a large scale and how their use of the habitat varies seasonally. This collaboration will disclose patterns of dolphin habitat use over a variety of spatial scales.

GIS-generated maps depict the current distribution of dolphins and prey sampling sites both for control setup and in the presence of feeding/nonfeeding dolphins. Also, maps will be generated that depict the distribution of dolphins in comparison with benthic communities and water quality parameters.

Photo courtesy of Michelle Kinzel

Over the past fifty years, this severe degradation has been well documented by scientists. During this time, top predators such as herons, brown pelicans, alligators, and storks have declined by 80–95 percent.

By employing a geographic framework for evaluating linkages between habitat quality, prey densities, and dolphin distribution, the DEP can provide sound scientific advice on the effects of restoration options on the population(s) of bottlenose dolphins in Florida Bay and in the Florida Keys. The GIS generates a comprehensive picture of the relationship between ecosystem health and dynamics of habitat use by bottlenose dolphins. This information will be valuable to managers attempting to predict, monitor, and understand the effects of ecosystem restoration efforts in South Florida.

The Dolphin Ecological Project conducted its pilot field program, which was funded by the National Marine Fisheries Service, in Florida Bay last year. Although the analysis of the first year of work is far from complete, they have made some preliminary assessment of the distribution and behavior of bottlenose dolphins in Florida Bay. Much analytical work remains to be completed, and any conclusion presented here should be regarded as provisional pending further analysis.

The surveys from the project confirm that dolphins are present in Florida Bay throughout the year. The photoidentification catalog presently includes one hundred eight identifiable animals for the eastern Florida Bay study area. A qualitative examination of the distribution of dolphin sightings and focal follows reveals that most dolphins were found in the southern portion for the study area, where the

environment is influenced by exchange with the Atlantic Ocean.

The analysis of the spatial variation of potential prey is not yet complete, but other studies have found that fish densities are relatively low in the northern Florida Bay. It is believed that the distribution of bottlenose dolphins in this area is primarily driven by the distribution of prey, which, in turn, is determined by environmental factors such as temperature, salinity, and habitat type. Due to their high visibility, bottlenose dolphins are a good indicator of the distribution of prey and, in turn, the quality of habitat and environment that support these animals.

The South Florida Ecosystem Restoration Program is the largest environmental restoration project ever attempted in the United States. Along with other critical projects that are under way in South Florida, its goal is to restore the regional hydropattern and recover endangered species and habitats. Within this affiliation, the Dolphin Ecological Project has an unprecedented opportunity to establish baseline information for bottlenose dolphins in Florida Bay before a major environmental restoration project takes place. Researchers can then evaluate the effects of restoration efforts on a top predator in this ecosystem.

The Dolphin Ecological Project outreach work consists of participation in scientific meetings and direct interaction with managers and stakeholders interested in bottlenose dolphins and their habitats in South Florida. At the conclusion of the field project, the team intends to hold several public meetings in South Florida in order to highlight its findings and promote restoration programs using the bottlenose dolphin as a flagship species for these efforts.

The work described here represents some of the first dedicated research on bottlenose dolphins in Florida Bay. The project team believes that continuation of this research will provide a rich baseline of information for assessing the response of bottlenose dolphins to future changes in the south Florida ecosystem. An increase in the density and range of bottlenose dolphins would be clear evidence of the benefits of restoration programs.

In general, public support is essential to the success of any management program. The Dolphin Ecology Project study of the bottlenose dolphins will surely be an effective tool for educating the public about the complexities of ecosystem protection and restoration.

Marine Life Alliance

Peter J. Rubec and Joyce Palacol
International Marinelife Alliance

Coral reefs and other coastal habitats throughout Southeast Asia are being destroyed by destructive fishing methods. Explosives, poisons such as sodium cyanide, muro-ami and kayakas fishing (rocks and poles used to smash corals and drive fish into nets), and illegal trawling are reducing sustainable yields to the fisheries and aquarium trades.

The Philippines has strong laws against destructive fishing. Unfortunately, the laws are difficult to enforce because there are more than 770,000 small-scale municipal and commercial fishermen and more than 4,000 aquarium-fish collectors spread across 7,000 islands. Various levels of government need the ability to collect and record accurate environmental and fisheries data, analyze the information geographically, and produce management recommendations in a timely manner. Therefore, it is necessary to balance the protection of

marine resources against the enhancement of fisheries and mariculture production. There is an urgent need to turn fishermen toward other occupations through alternative livelihood training programs.

Until 1998, the community-based programs conducted by the International Marinelife Alliance (IMA) consisted of training cyanide fishermen to use less destructive, fine-mesh barrier nets to capture marine-aquarium fish and to use hook-and-line methods to capture live-food fish such as groupers and snappers for export to Hong Kong, Taiwan, and mainland China. However, with the publication of recent scientific studies demonstrating that the use of cyanide destroys coral reefs, the IMA has increased emphasis on mapping coastal habitats by using GIS to support habitat conservation.

The training programs require coordination with municipal governments and other non-government organizations. When introducing the Destructive Fishing Reform Program (DFRP) to an area, the IMA training team discusses the initiative first with local village officials and then with the fishermen, who generally give a high priority to their economic needs. They also discuss ways to increase income either from fishing or other alternative livelihoods. The sustainable use of natural resources is addressed through the community education component of the DFRP. Local management options are considered such as limiting access, policy enforcement, or sanctuary development.

IMA staff provided assistance and GIS training to personnel in the Philippine Bureau of Fisheries and Aquatic Resources (BFAR) in the central office; field offices; and selected municipalities situated in the provinces of Cebu, Bohol, and Palawan. The training provides students with a working knowledge of GIS concepts and applications so that they may assist these government agencies with the management of the country's coastal resources. Part of the GIS training has focused on the determination of boundaries using ArcView GIS to define the spatial extent of municipal waters.

After GIS training, the next step is the development of the database. Five sites in three provinces were chosen for the GIS mapping: Olango Island and Lapu Lapu City on the Island of Cebu, Guindacpan and Talibon on the Island of Bohol, and Santa Cruz in Davao Del Sur–southern Mindanao. The IMA's GIS personnel have created maps for eleven villages (barangays) on Olango Island situated in the municipality of Lapu Lapu near the Island of Cebu, two villages in northeastern Bohol, and three villages situated in the Davao Gulf of Davao Del Sur.

The GIS work was geared to the creation of coverages to support community decision making such as deciding where to establish a marine protected area or deciding the areas most suitable for various types of mariculture. The maps for each village consist of the coastline, a composite habitat map depicting terrestrial and marine zonation, bathymetry, ocean currents, salinity, and temperature. Symbols used with the composite habitat maps depict areas where different fish and invertebrate species are caught; the location of stationary

fishing gears; areas with destructive fishing; the boundaries for marine-protected areas or wildlife sanctuaries; human use areas such as public parks, resorts, and marinas; and locations with coastal pollution.

The composite habitat maps depicting terrestrial and marine zones such as the types of shoreline, mud flats, sea grass beds, passes/channels, reef flats, and barrier reefs were created from multispectral SPOT (Système Probatoire d'Observation de la Terre) satellite imagery obtained from the Philippine National Resource Mapping Authority. The scale of the rectified maps created using ARC/INFO and ArcView GIS ranged from 1:8,500 to 1:11,500. Bathymetry contours were digitized from nautical charts. Seasonal salinity and temperature data will be interpolated using ArcView Spatial Analyst from point measurements presently being gathered using data loggers. The fisheries species distribution data was obtained from interviews with local small-scale fishermen.

A promising means by which community-based coastal resource management programs can be implemented in the Philippines is through territorial use rights in fisheries (TURFs). Community control of the means of production through TURF management has the potential of resolving user conflicts and reducing fishing effort in specific areas. Changes in the Local Government Code in 1991 and the new Fisheries Act in 1998 granted municipalities the right to license fisheries and lease mariculture sites within municipal waters. Currently, municipal governments through Fisheries and Aquatic Resource Management Councils (FARMCs) have the authority to regulate the implementation of TURFs. The potential of using GIS associated with environmental monitoring and spatial management strategies, such as zoning areas for sanctuaries, for certain fisheries, or as sites for mariculture, is now being evaluated.

Corals and other reef-invertebrate species reared on TURFs for export could become an important source of revenue for local communities. It may be possible to halt the destruction of coral reefs by demonstrating that reef fish and invertebrates can be reared profitably for the aquarium trade. This should be tied to programs to convert small-scale fishermen to TURF farmers.

The IMA has been using GIS to delineate the boundaries of marine-protected areas and TURFs for mariculture through consultation with the municipalities, FARMCs, BFAR, and the fishermen. The GIS database is being used to assist with this planning process.

The first step of the planning process is to create habitat maps that show species

Coral farm in the village of Caw-oy on Olango Island depicting the rearing of coral fragments in concrete frames deployed in suitable reef areas.

distributions and then to analyze the maps to determine which areas are most suitable for mariculture. Areas already being used for fishing are excluded from consideration; likewise, areas with siltation or unfavorable water quality from pollution are also excluded. Next, underwater surveys are conducted to determine which of the remaining locations would best support the farming of giant clams, coral fragments, or live rock. Finally, the potential of the sites are rated. Sites close to existing coral reefs are rated more highly since they already support populations of the organisms desirable for mariculture. Figure 1 depicts the location of hypothetical TURFs overlaid onto the habitat maps in the village of Sabang on Olango Island.

The IMA is working in cooperation with government organizations and the FARMCs to plan the placement of TURFs. Key habitats, such as mangroves, sea grass, and coral reefs, will be protected, while other areas are leased to members of the community. The goal is to shift fishermen away from destructive fishing by demonstrating the economic benefits of farming reef organisms while raising

community awareness of the need for conservation of coastal habitats.

GIS-produced maps support decision making concerning the allocation of lease sites used as TURFs. The FARMCs can limit access by mobile fishermen using destructive fishing methods. Members of the community are able to police the TURFs because they have control over the resources being farmed. Hence, GIS becomes a key tool to support spatial management of these vital marine resources.

Giant Clams
Generally, it takes three to five years to grow giant clams large enough to harvest for human consumption. In contrast, it takes just less than one year to rear clams to about 30 centimeters for the aquarium trade. Hence, the export of giant clams to the aquarium trade can provide more immediate economic returns than harvesting them for food.

The IMA has established a sanctuary for holding giant clam brood stock. A hatchery for giant clams will produce clam spat for distribution to coastal communities. Fishermen will be trained to rear the clams in cages situated on TURFs allocated by the FARMCs. The giant clams will be used for restoration of wild populations and for export to the aquarium trade.

Live Rock
The IMA plans to train Filipino fishermen to create artificial rock from coral sand mixed with concrete molded into various forms. The artificial rock will be deployed on TURFs in coral reef areas, where the rocks can become coated with coralline algae and then be colonized by planktonic larvae produced by marine invertebrates (sea anemones, zoanthids, crinoids) situated on nearby reefs. After about a year, farmers will be able to harvest the rock for export to the marine-ornamental trade.

Hard and Soft Coral Fragments
In an effort to rehabilitate damaged coral reefs, coral reef scientists have developed techniques for transplanting corals and culturing fragments attached to artificial substrates. The coral fragments can be propagated in TURFs situated on reef flats.

A coral farm is already well established on Olango Island (photos, previous page). Dr. James Heeger of the University of San Carlos created a coral farm in the village of Caw-oy. An environmental training center was established to train local fishermen in the basic skills of coral farming. This includes methods for selecting donor corals, applying fragmentation techniques, monitoring and maintenance during the grow-out period, and setting up cooperatives for marketing the corals.

Reef rehabilitation and ecotourism come together at a coral farm on Camotes Island in the Philippines.

All photos in this chapter courtesy of Peter J. Rubec.

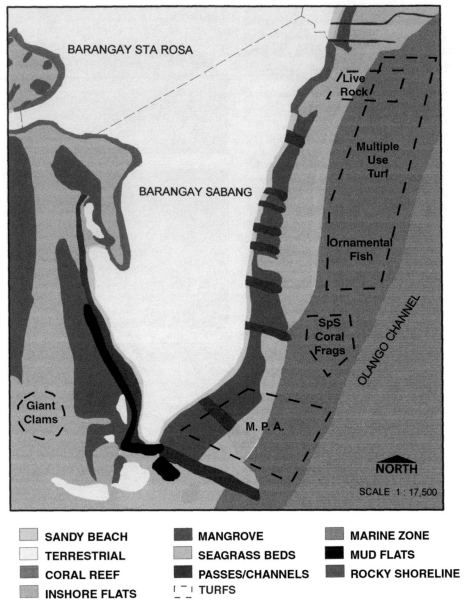

SANDY BEACH	MANGROVE	MARINE ZONE
TERRESTRIAL	SEAGRASS BEDS	MUD FLATS
CORAL REEF	PASSES/CHANNELS	ROCKY SHORELINE
INSHORE FLATS	TURFS	

Figure 1. Marine resource map of Barangay Sabang on Olango Island depicting hypothetical locations of a marine-protected area; a TURF for collecting ornamental fish; and TURFs for the culture of giant clams, coral fragments, and live rock.

The IMA recently took over the management of the Caw-oy coral farm, which will become a regional training center. The IMA has initiated projects to rehabilitate coral reef areas and expand training in coral farming to other villages on Olango Island, on the Island of Bohol, and in the Davao Gulf situated in southern Mindanao. Using corals from Caw-oy, the IMA has created a second coral farm to support reef rehabilitation situated at the village of Consuelo on Camotes Island off Cebu. Moreover, the IMA is also promoting ecotourism at the coral farm by creating underwater trails for divers.

The IMA is assisting the communities to obtain United Nations Convention on International Trade in Endangered Species of Flora and Fauna (CITES) export permits from the Philippine government. These permits will allow coral fragments, giant clams, and live rock reared on TURFs to be exported to the aquarium trade to enhance the income of local people. The IMA is a nonprofit, nongovernment marine conservation organization with offices in the U.S.A., Philippines, Vietnam, Hong Kong, Indonesia, Vanuatu, Guam, and the Marshall Islands. The goals of IMA are to conserve marine environments and protect

both biological diversity and sustainable use of marine resources by local people.

For TURF farming to become a reality, various integrated strategies need to be developed including village-level training and loan programs that assist local communities to create hatcheries and other infrastructure. Scientists need to become more involved in training programs to transfer their knowledge to fisherfolk. The use of GIS can assist with zoning coastal nearshore areas in a manner similar to terrestrial coastal planning.

Multiple Marine Uses of GIS

Martin Kay
Bay of Fundy Resource Centre

The use of GIS has long been the bastion of government as well as business. Today, marine geographical depiction through GIS is becoming a key factor in marine planning, policy making, conservation, and research. To begin the development process of integrated coastal resource management, the Coastal Resources Mapping Project for Digby and Annapolis counties was undertaken as a joint project between the Western Valley Development Authority and the Department of Fisheries and Oceans (DFO)–Canada.

The affordability of today's GIS programs has allowed the Marine Resource Center (MRC) to prioritize the development of a fully functional GIS lab. The MRC's Coastal Resource Mapping Project locally offers GIS information and technology that enable a wider use by community groups and organizations. Some examples of how the community uses MRC's data and maps include a Storm Surge Water Project in partnership with the Clean Annapolis River Project, Clam Flat Restoration/ Harvesting Plan in partnership with the Annapolis and Digby Counties Clam Management Board, Coast Guard Risk Analysis Project in partnership with Canadian Coast Guard (CCG) and, finally, the combined DFO and CCG Emergency Response plan. The MRC's GIS project supports these causes in a visually meaningful way. The MRC continues to expand its in-house GIS information and

to enlarge its accessibility to other groups involved in the region's integrated coastal resource management initiatives.

Because of the MRC's broad interest in integrated management for the Bay of Fundy and the Gulf of Maine, there was a need to expand some of the data sets for the entire Bay of Fundy regions and Nova Scotia coastal areas. Fortunately, Canada's Department of Fisheries and Oceans had already done the work. They generously contributed their digital data and basemaps to the MRC, who gladly transferred the information into their GIS. In addition, the Department of Fisheries and Oceans contributed eight extensive Fish Species Profiles as well as three Ecosystem Community Profiles. These, too, will be linked to the GIS at the MRC office and will be an invaluable source of information for mapping applications.

Another contributor is the Nova Scotia Geomatics Centre, a provincial government agency who not only provides technical assistance to the MRC but who also has donated twenty-three digital files of the new Coastal Series of maps for the Bay of Fundy. These maps, combining the data from both topographical maps and navigational charts, are the first of their kind in Canada and will be of considerable importance to both marine-based and coastal activities.

Because of its location on the Gulf of Maine, the MRC has also been very active in working with other community organizations in neighboring New England. In particular, the Cobscook Resource Centre in Eastport, Maine, has been a key partner in this cross-border networking effort. Over the last few years, its development has paralleled that of the MRC, and therefore, the two resource centers have come to rely on each other. Among their similarities is a major commitment to bring GIS capacity to marine issues at a grassroots community level. A key part of their cooperation has been to develop innovative ways for facilitating cross-border data sharing. The need for data sharing became apparent as soon as the two centers saw each other's maps, both of which showed blank areas on their opposite borders. A collaborative project is expected in the near future.

The MRC embarked on another project in early 2000 that was funded through the Canadian Rural Partnership Program, a federal government program. The goal of the project was to educate and promote GIS at a community level that included both local groups and fishermen organizations. At first, some of these groups were not convinced that having a map of their information would be of much immediate benefit to them. However, when agencies were shown the data in a digital format,

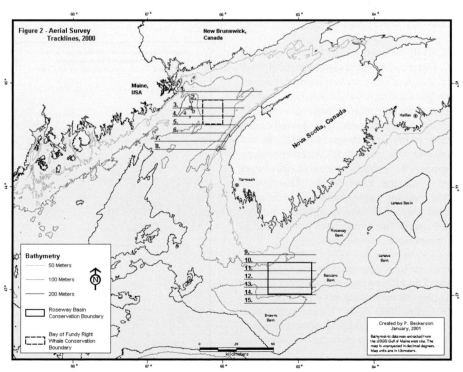

Figure 2 - Aerial Survey Tracklines, 2000

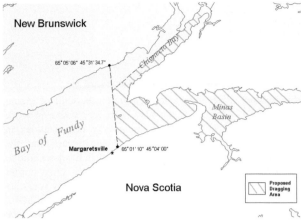

New Brunswick

65°05'06" 45°31'34.7"

Bay of Fundy

Margaretsville 65°01'10" 45°04'00"

Chignecto Bay

Minas Basin

Nova Scotia

Proposed Dragging Area

Designated proposal area for sea urchin dragging permit.

they began to realize the possibilities available to them with GIS. For example, they saw that GIS has the ability to add layers of data together, which is useful for comparative analysis. This function offered the capability to monitor fish catches and their distribution over large areas during a given period of time. Fishermen and regulatory agencies alike found this spatial representation to be of great worth.

When monitoring fish habitats, the GIS Tag Recovery Map serves as a strong analytical tool. The MRC will assist an organization in New Brunswick with the development of a map of the St. John River. The map would show color-coded areas where fish were tagged and then released and also where the fish were then recaptured and the tags recovered. The MRC also provides support to a fishermen's group known as the Fund Fixed Gear Council Ground Fish Project. Together, they are designing a framework for GIS mapping as a component of their ongoing research. Factors considered in the design are issues of data sharing and digital file licenses.

MRC's GIS supports report writing on marine life. The East Coast Ecosystem completes aerial and boat surveys every summer to determine the location of right whales, dolphins, sharks, and other marine mammals found in the Bay of Fundy, particularly around Grand Manan Island and Roseway Bank. In collaboration with worldwide conservation organizations, the primary focus of this work is to track the right whale. The marine animal data for August 2000 has been incorporated into a GIS, which, in turn, produced fifteen maps. These maps will be included in the organization's final report.

Clams are also important to the Annapolis Basin. The GIS-produced clam map is used to display the current status of particular clam beds during 2000–2001. This map has been used to assist the management board in their decision regarding harvesting practices and conservation issues.

Currently, scientist Sabrina Sturman has begun a suitability study of potential lobster locations in St. Mary's Bay. Her Lobster Larvae Settling Study will not only provide an update of previous data on the crustacean but will also serve to increase the detail of coastal zone maps of that region.

For a presentation at the annual sea urchin harvesters meeting, the MRC prepared five maps of the Digby Gut, Grand Passage, Petite Passage, and a section of coast around Delap's Cove. Based on data provided by Nova Scotia Department of Fisheries, these maps depicted bathymetry and license condition prohibitions. Smaller scale maps displaying regional overviews were also generated for the conference. Urchin harvesters will use these maps as working tools to sketch topics of discussion at their meetings, and analysts will use the maps as tools to develop a management plan.

Digby will also use GIS applications for tourism. A 500-acre site, located on the bay's neck, is being proposed for developing an ecoresort. An important component of the project's proposal includes maps of trails, infrastructure, and resources. Other controversial sites are the areas' nostalgic lighthouses. The Lighthouse Preservation Society looks to MRC for a GIS Web-based map for Annapolis–Digby counties. This is an essential step in promoting community involvement in stewardship of lighthouses before these marvelous edifices are all sold into private hands.

One more application for GIS is charting abandoned fish weirs. The Western Nova Scotia Yacht Club contributed a chart to MRC of areas in the Annapolis Basin that are affected by abandoned fish weir poles. At low tide, these poles are conspicuous at approximately three feet above the water. However, at high tide or during times of low visibility, the posts prove a serious unseen hazard to boaters. The Yacht Club wants to maintain the safety reputation of Annapolis Basin and Nova Scotia for the visiting yacht trade. Therefore,/ to prevent collision damage, the club seeks to have the fish weirs removed. Based on the information provided by the club's chart, the MRC created a GIS map to show these dangerous sites, a crucial tool for the club when it lobbies to have the hazards removed.

Finally, the MRC's GIS is important in supporting the region's ecological disaster plan. The Canadian Coast Guard is designing an oil spill response plan for the Bay of Fundy. The MRC will head the effort to complete a plan for its side of the bay. This would involve meeting with community leaders, community organizations, and businesses to map out the resources and staging areas for emergency response teams.

The Bay of Fundy Resource Centre is diligently moving forward in creating a variety of opportunities offered by its GIS. With a strong GIS support team, the center's efforts and foresight have been well placed as it provides the facilities for meeting the challenges of the Bay of Fundy area by designing viable GIS solutions for a multitude of users.

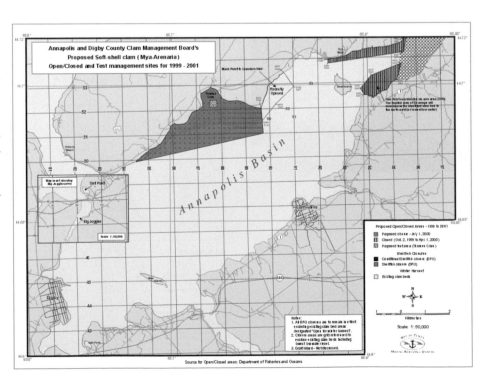

USGS Glacier Bay Field Station Develops GIS Tools

Philip Hooge, Ph. D.
USGS Glacier Bay Field Station

The USGS Glacier Bay Field Station takes GIS seriously when developing tools for its own marine geographic research. The project's scientists have come up with some worthwhile GIS marine geography solutions. Although these were originally designed for Glacier Bay research projects, the designers had the foresight to make them adaptable by others in the marine geography GIS community. These tested GIS extensions can be downloaded from their USGS Web site.

One popular extension arose in an effort to perform spatial studies of movement behavior that can be integrated into the GIS. The GIS team developed software that integrates ArcView GIS with a large collection of animal movement analysis tools. This application, Animal Movement, can be loaded as an extension under multiple operating system platforms. The extension contains more than forty functions specifically designed to aid in the analysis of animal movement, these include parametric and nonparametric home range analyses, random walk models, habitat analyses, point and circular statistics, tests of complete spatial randomness, tests for autocorrelation and sample size, point and line manipulation tools, and animation tools. This data could be collected from radio tags, sonic tags, Argos satellite tags, or observational data. The program is designed to implement a wide variety of animal movement functions in an integrated GIS environment. The program also has significant utility for analyzing other point phenomena.

Above: Split shot at Marble Island.
Below: Underwater rebreather setup for benthic mapping.

Photo by Jeff Mondragon

The bay from Marble Mountain.

Mountain and driftwood.

Landsat false color composite of Glacier Bay.

Oceanographic Analyst is an extension that allows three- and four-dimensional analysis and display of volumetric and time series oceanographic data. It is the only tool available at the current time that permits integration of full-cast oceanographic data into a GIS system. Its functions can also be used for calculating photic depth, integrating chlorophyll-a, processing weather data, exporting and importing, aggregation, and summarizing.

Two other extensions on this site worth noting are Spatial Tools and the Population Viability Analyst (PVA). Spatial Tools extends the capability of ESRI's ArcView Spatial Analyst extension by allowing additional menu-based tools for raster data manipulation including creating mosaics, merging, clipping, contour profiling, cleaning, and warping, among others. PVA is a powerful method for evaluating the probability of a population's extinction in the face of variation. PVA offers an alternative to traditional population models that are strictly deterministic. The PVA extension is designed to examine the influences of multiple types of variation including differences in the values of population parameters such as initial population sizes and birth or death rates and differences in the way those parameters vary. This extension allows the user to create life tables based on overall population parameters or to utilize a fully populated life table to conduct multiple population simulations. The program is in its development stage, and some caution is advised in interpreting results.

Tidewater Glacier.

Conservation Geography *thanks John Brooks for photographs of Glacier Bay.*

People for Puget Sound's Estuary Habitat Program

Tom Dean, Restoration Coordinator

People for Puget Sound's Estuary Habitat Program focuses on three goeographical areas: shorelines, deepwater habitat, and estuarine marshes and mudflats.

Citizen Shoreline Inventory Atlas

In 2000 we conducted two major pilots of a new system for gathering data in the Puget Sound's nearshore environment. With the Rapid Shoreline Inventory (RSI), we recruit, train, and deploy a team of volunteers to gather a set of physical and biological data on contiguous 150-foot segments of shoreline during an extreme low tide.

We use the resulting data to identify important shoreline areas for conservation or restoration action. Puget Sound's shorelines support the spawning of three species of forage fish (herring, surf smelt, and sand lance) that comprise the keystone of the Puget Sound food web, so the protection and restoration of these habitats is critical to the health of the Puget Sound ecosystem.

Marine Protected Areas Initiative

People for Puget Sound's work to protect the marine ecosystem in the Northwest Straits has centered around the Northwest Straits Initiative—formerly known as the "Murray–Metcalf Process" or the negotiations over the politically infeasible National Marine Sanctuary proposal. The most important of the initiative's five-year benchmarks is to establish a scientifically based regional system of marine protected areas (MPA). Kathy Fletcher, executive director, and Mike Sato, director of our North Sound office, both serve on the Northwest Straits Commission that was formed to implement the initiative. People for Puget Sound staff and members also participate in the seven county-level Marine Resource Committees (MRC) that feed into the commission. Our scientific and technical team has provided expertise to the commission and MRCs including conducting shoreline resource inventories in the field in two of the seven counties so far.

Our GIS work in defining groundfish habitat and populations directly assisted Skagit and Jefferson County Marine Resource Committees in their work to identify protected marine areas, and has led San Juan County's Marine Resource Committee to consider expanding its voluntary "no-take" areas to include all the waters of San Juan County.

In conjunction with establishing local presence among all seven MRCs and participating in the North West Straits Commission, we have formed a coalition of Washington and British Columbia NGOs, tribes, and user groups to focus attention on establishing marine protection zones between the U.S. San Juan Islands and the British Columbia Gulf Islands. We have used our GIS methodology and coalition to designate the Orca Pass International Stewardship Area. We have formally signed a "working agreement" with the local governments of San Juan County and the Islands Trust to gather and analyze data and to educate and involve local residents. This effort has gained the attention of Canadian and Washington agencies and the last quarterly meeting was attended by more than forty NGO, government agency, and tribal representatives.

Duwamish Estuary Stewardship Project

Our work in the Duwamish River Estuary during 2000 has been a great success for the organization. We were instrumental in completing a major new estuarine restoration project, we organized major stewardship projects at four other sites, and we created a citizen

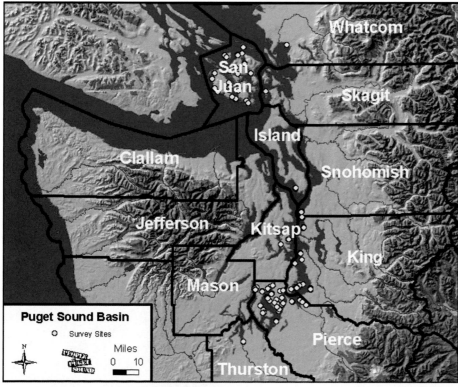

Citizens Shoreline Atlas. This atlas has become the most comprehensive citizen-based, nearshore habitat assessment of its kind in Washington state. A GIS stores the information collected by the citizen stewards. The GIS makes it possible to display the data in relationship to a series of interactive maps that are accessible at the organization's Web site, http://www.pugetsound.org/csi.

Siri Qale takes GPS readings on Maury Island, King County.

stewardship team that is gathering important monitoring data on these sites. These project sites provide critical rearing habitat for Puget Sound chinook salmon, which are listed as threatened on the Endangered Species List.

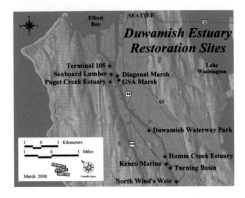

Using a submeter GPS, we created stratified basemaps for five restoration projects that were ready for monitoring. With expert help from a team of biologists and statisticians, we developed a GIS to grid the basemaps and place our sample points. We then returned to the field and navigated to those points in order to place stakes to mark the sample locations for our volunteers. The volunteers visit the sites in the months of May, July, and September to gather data on plant vigor at the sample locations. Over time, we will be able to create a GIS that maps the development, diversity, and movement of vegetation at these restoration sites.

In 2001, we will be expanding the program to cover ten project sites in the Duwamish and Puyallup river deltas. Our protocol for this volunteer monitoring program is available on request. We are currently working on a Web display for this project.

New Initiatives
During the summer, we were invited by the State Department of Natural Resources to nominate areas for their Aquatic Lands Reserve Program. We assembled ten maps that focused on concentrations of resources based on data from various state agencies. As a result of our nominations, the DNR designated the Maury Island Aquatic Reserve, which has been met with considerable public support.

As a result of this and our other efforts outlined above, People for Puget Sound has become a regional leader in data analysis for marine and nearshore ecosystems. We have been working with a wide array of local, state, and federal jurisdictions to apply the concepts we developed with the Skagit Estuary Restoration Assessment to all available data sets for the marine, nearshore, and estuarine environments in Puget Sound.

The resulting first-generation Puget Sound Assessment will be a multivariate statistical model that can be used to target restoration and conservation actions in all of Puget Sound's environments. In developing the model, we will be able to identify key data gaps and, over time, use new data to improve the model. This concept represents a large leap in our GIS goals and capacity. We very much appreciate the support of CTSP and ECP, and hope that we will be able to continue this very fruitful partnership.

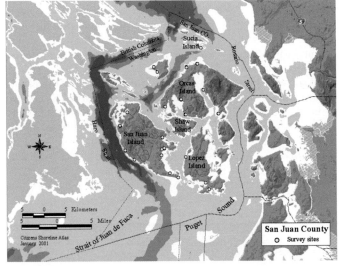

CSI Sites: Pierce County
○ Survey Sites

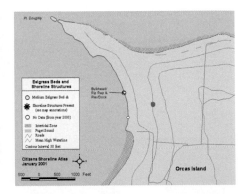

The Baja California to Bering Sea Mapping Initiative

Lance Morgan , Marine Conservation Scientist
Noreen Parks, Communications Coordinator
Marine Conservation Biology Institute

In some ways, conservation in the sea is no different from conservation on land; protecting places (and thereby the genetic, species, and ecosystem diversity associated with them) is a more comprehensive, cost-effective, and politically viable strategy than imposing separate regulatory regimes on individual species. Growing recognition of this reality has spawned a fresh paradigm for conserving marine biodiversity and strengthening fisheries management: marine protected areas (MPAs). Interest in this new approach has increased dramatically in recent years as the traditional approach of "command-and-control" regulation has failed to stem the tide of biodiversity loss and fisheries collapse.

Marine Conservation Biology Institute (MCBI), a nonprofit organization based in Redmond, Washington, with a public policy office in Washington, D.C., is dedicated to advancing the science of marine conservation biology and promoting cooperation to protect and restore earth's biological integrity. The organization has played a leading role in fostering the MPA paradigm. In January 2000 MCBI held a scientific workshop that culminated in an Executive Order to set up the government framework to establish a national system of MPAs. In accord with its mission to promote the integration of sound conservation science into marine policy, MCBI is currently focusing on developing the science-based blueprints for enacting the broad policy goals set forth in the Executive Order.

Without a Map,
Do We Know Where We're Going?

Successful place-based strategies require identifying conservation targets, so the first logical step is to produce maps of the most important places to protect. Along the Pacific coast of North America, where a movement has emerged to establish a network of MPAs extending from the waters of Baja California to the Bering Sea, no such comprehensive, region-wide map exists. In some areas researchers have collected many types of geological and geophysical data including seafloor bathymetry, sidescan sonar images, sediment and rock types, active fault zones, and submersible-based observations and measurements. Collectively this information represents one of the most extensive marine geologic databases in the world and it could be enormously useful in planning and designing MPAs. However, most of this data is not integrated in a GIS that

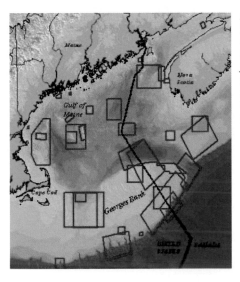

An example of previous MCBI efforts, based on the results of a scientific workshop held in March 1999: Proposed benthic priority conservation areas for the Gulf of Maine, Georges Bank, and the Bay of Fundy, as identified by marine experts from Canada and the United States, and overlain on colorized bathymetry.

could be utilized to characterize, classify, and predict the distribution of geological features and associated biological features.

MCBI has launched an initiative to integrate these data sets and supplement them with data from other less-studied areas to produce a scientifically credible map of delineated priority areas that will make the otherwise vague and abstract underwater places tangible for the stakeholders. Collaborators in this multiyear effort are the North American Commission for Environmental Cooperation, Canadian Parks and Wilderness Society, Natural Resources Defense Council, and the National Oceanic and Atmospheric Administration. The goal of the project is to generate a user-friendly GIS with interpretations that are comprehensible for lay audiences as well as the scientific community.

The challenges of generating the "B-to-B Map," as it's been dubbed, are not insignificant. It will span a vast geographic scope with a variety of jurisdictional boundaries, and at present, data is disparate and inconsistently available. The situation calls for state-of-the-art GIS capability in combination with multiscientific expertise. As all GIS enthusiasts know, layering many types of data creates a powerful tool that far surpasses the simple stockpiling of the information. The significance of one key type of marine data was recently demonstrated by the discovery that deepwater rockfish assemblages are remarkably alike in similar bottom habitats from central California to Alaska, extending from latitudes of roughly 36–55 degrees. Findings

such as this underscore the value of bathymetric and seabed classification data in assessing priority habitat identification. Because physical data can serve as a proxy to indicate the locations of certain types of bottom communities, they are highly useful complemenst to patchy biological data of varying quality. Furthermore, in areas of intense fishing activity or other disturbance to biological communities, extensive biological sampling might not lead to any useful results.

Assembling the Pieces and
Bridging the Gap

The MCBI-led Baja California to Bering Sea mapping project will strive to integrate the efforts of a number of other nongovernmental organizations and government agencies that have begun priority habitat mapping projects of smaller scope in areas along the West Coast of North America. These groups include the World Wildlife Fund, Living Oceans Society, The Nature Conservancy, Conservation International, Canadian Parks and Wilderness Society, Channel Islands National Marine Sanctuary, the U.S. Federal Fisheries Management Councils, and others. An appreciation for the value of a single robust map for the region is growing among these groups; not only will it serve as an informational resource par excellence, but it also promises to confer a sense of solidarity among the conservation allies that stand behind it.

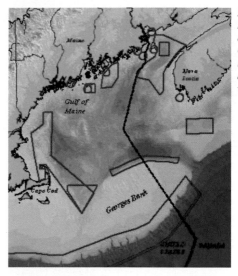

Based on the results of a scientific workshop held in March 1999: Proposed pelagic priority conservation areas for the Gulf of Maine, Georges Bank, and the Bay of Fundy, as identified by marine experts from Canada and the United States. Details available at www.mcbi.org.

Currently planned for an intermediate stage of the mapping project is a process to collect an often-ignored type of information: input from traditional ecological knowledge holders such as Native Americans, fishermen, and recreationists. These individuals will be asked to identify areas they judge to be high conservation priorities, and indicate the criteria underlying their priority assessments. Incorporating this input into the GIS will help government decision makers and NGO representatives identify areas of high socioeconomic and/or conservation value. This will enable them to set priorities and design MPAs to maximize benefits and minimize conflicts.

We believe this broad-based, inclusive process will provide a map that combines scientific credibility and political benefits to the maximum extent possible. It will generate the best possible product and create a sense of ownership or "buy-in" among the participants. If we are correct, the resulting map will serve as a foundation and stimulus for local, national, and international discussions on MPAs, fisheries conservation, coastal zone management, and ocean policy. This process is not intended to designate MPAs or to replace stakeholder processes that do so. Rather it aims to provide basic information that government agencies, conservation groups, and other interest groups need to assess the activities and degree of protection appropriate for particular marine areas.

The planned end product will be a comprehensive, easily accessible, multilayer GIS database, in ERDAS and ArcView GIS formats, that can be distributed on a platform-independent CD–ROM to a diverse scientific and nonscientific community. It will also be distributed in hard copy, and the data sets will be made available via the Web.

MCBI's efforts to promote the integration of science with marine policy began as the dream of Elliott Norse, a marine ecologist who founded the organization in 1996 and serves as president. Prior to that, Norse served in Washington, D.C., on the President's Council for Environmental Quality and edited the seminal book, *Global Marine Biological Diversity.* MCBI holds symposia and scientific workshops on emerging marine conservation issues and works to raise public and policy maker awareness on them. MCBI maintains an extensive Web site at www.mcbi.org/.

Heal the Bay: Community Action Group Uses GIS to Change Policy

Mark Abramson, Stream Team coordinator

A small group of citizens can indeed fight city hall and make a huge difference in conservation policy. Citizens of the Malibu Creek Watershed area started with a vision of restoring their beloved watershed and took action that has made statewide changes. They continue to thrive and serve as a model for volunteerism, technological application of GIS and chemical testing, and political advocacy for many conservation groups.

The Malibu Creek Watershed encompasses one hundred ten square miles of some of the most scenic natural and recreational resources in Southern California. This watershed is home to many different types of land uses ranging from the rural undeveloped Santa Monica Mountains to dense urban developments. These natural and recreational resources draw millions of visitors every year.

Three of the most recognized resources within the watershed are Malibu Surfrider Beach, Malibu Lagoon State Park, and Malibu Creek State Park. These sites serve as recreational areas for beachgoers, birdwatchers, hikers, mountain bikers, naturalists, surfers, and swimmers. This beautiful area is also home to numerous species of wildlife including the endangered California gnatcatcher, steelhead trout, red-legged frog, tidewater goby, and the California brown pelican. The problem is that as development continues to grow, so does urban runoff that negatively influences the ecology of the watershed and endangers the health of people who recreate there.

In the past, Los Angeles County treated its beaches and coastal waters as a dumpsite. Because of rising concerns about the habitat, a small group of citizens took it upon themselves to heal this ecologically troubled area that was suffering from some of the worst levels of contamination found anywhere on the nation's coastlines. This group of advocates solicited the help of the California Coastal Conservancy to initiate a conservation project that was soon dubbed Heal the Bay. A supportive conservation grant was also issued by ESRI providing GIS software, training, and technical support. Volunteer citizens were eager to help in the effort to gather and record information within the Malibu Creek Watershed.

Mapping volunteers are trained and certified. The GIS training includes GPS training, habitat assessment, and using data in ArcView GIS.

Image Hot Linked From Invasive Veg.

Image Hot Linked From Discharge Point

Stream Walk Data Collection

STREAM TEAM
MALIBU CREEK • HEAL THE BAY

Exotic Invasive Vegetation.shp
Streambank modifications.shp
Discharge & Outfalls

Artificial streambank.shp
Dump sites
Exotic invasivesw1.shp

Universal Transverse Mercator Projection
North American Datum, 1927

500   0   500 Feet

In addition to having these dedicated community workers, Heal the Bay contracts with the California State Polytechnic University, Pomona, graduate program, and the Department of Landscape Architecture to map and model the natural processes of the Malibu Creek Watershed. These students helped design the Stream Team Program including a method to enter data collected at precise locations using ArcView GIS. The graduate students also created a way that data can be exported and used by other agencies in the region for decision making purposes. ArcInfo is an integral component of Heal the Bay's GIS. Because the training program has had such great success, now other agencies, such as the National Park Service, state parks, municipalities, and other volunteer monitoring programs, have also been enrolling and applying this knowledge to their own particular conservation efforts.

Stream Team volunteers walk along the streams and creeks in the Malibu Watershed that eventually empty at Malibu Lagoon State Park and world famous Malibu Surfrider Beach. Combining GIS and chemical testing techniques, the team locates illegal/illicit connections that flow into the creeks, identifies areas that are environmentally damaged or disturbed, and tests the water quality of different locations within the watershed. By taking digital photographs, entering coordinates, and creating site reports, the team provides a full depiction of at-risk areas.

Those people who work in the GIS component use the Ashtech Reliance global positioning system (GPS). The submeter GPS unit is well suited for its task because it provides the finer level of detail that is required by the project. Furthermore, this GPS unit performs better under the tree-canopied canyons. Sometimes data gatherers must be creative in data collection because of the satellite positioning relationship to these difficult areas. The Stream Team uses 25-foot telescoping antennae poles to rise above the tree canopy and acquire satellite signals to perform their surveys.

The GPS team will look for discharge points and outfalls that are potential sources for pollutants to be dumped into the creek. These pipes are GPS recorded, then volunteers complete data sheets about the site incident. Other events that are recorded into the GIS database include unstable stream banks, exotic invasive vegetation, in-stream pool habitat, land uses that are obviously impacting the stream, dump sites, artificial stream bank modifications such

Developers' efforts to shore up the creek's sides with boulders only exacerbate the problem because this addendum serves as a funnel that intensifies the flow downstream. Because of their environmental impact, these fortifications are also mapped.

as concrete stream banks, and impairments within the stream channel. Each time one of the above-listed items is mapped, a digital photograph is taken of the exact item. The digital photographs are then hot linked in ArcView GIS to the exact location of the item mapped. Combining maps and digital imagery allows the data to be more easily understood by decision makers and the general public.

Malibu is renowned for its landslides, and data collectors make a special effort to map the creek's unstable stream banks. The sedimentation these slides generate creates a big impact on the watershed. For instance, slides affect the steelhead habitat and impede steelhead spawning areas. Frequently, unstable stream banks are correlated with impervious surfaces such as areas that have been covered with concrete. Because these surfaces no longer absorb water, the creek bears faster and larger flows downstream, which in turn scour its outside banks. Developers' efforts to shore up the creek's sides with boulders only exacerbate the problem because this addendum serves as a funnel that intensifies the flow downstream. Because of their environmental impact, these fortifications are also mapped.

A wide range of vegetation is part of the creek system that cuts completely through the Santa Monica Mountains. The team documents the watershed's riparian habitat. A research team has most recently discovered the San Fernando Valley spine flower. This indigenous flower, previously assumed extinct, exists only along a limited strand of the watershed. Intrusive exotic vegetation is also entered into the database. Identifying these areas, the GIS spatially indicates where native vegetation is being pushed out by exotic invasive vegetation. The GIS offers analytical tools for targeting exotic plants for removal and marking areas of natural vegetation.

The Stream Team also maps habitat within the stream channel. Here, the creek supports pool habitats that serve as holding places for fish and frogs. A project team discovered a rare red-legged frog in the headwaters of Las Virgenes Creek. This habitat, dutifully plotted into the database, was not only site referenced but pictorially illustrated as well. This style of presentation makes a strong impact in educating the public. Some project supporters and watershed stakeholders specifically expressed

concerns about the distribution of different species along the creek. GIS mapping is capable of establishing trends that provide a history of these movements.

In an effort to protect the receiving water area, the team maps the location of impairments created by algae and sediment to quantify the level of these impairments. Nuisance algae reflects a nutrient issue in the ecosystem. When it reaches an estimate of greater than 30 percent, the area is considered impaired. However, nuisance algae is not really quantifiable, so the team maps it as a line and uses a 10 percent range to estimate the percent of stream channel covered by algae (e.g., 60–70 percent) to indicate the severity of the impairment. This information is made available along with linked images so that decision makers can see locations and pictures of the impairment area.

Hot spots indicated by the GIS offer signposts for Heal the Bay chemistry teams to check these areas' pollutant levels. Currently, water chemistry testing is done at seven locations throughout the Malibu Watershed. The organization trains volunteers how to collect and test water. A few of the properties tested for are pH, stream flow, dissolved oxygen, nitrates, phosphorus, and ammonia. They also test for the entercoccus bacteria. By combining the GIS database with the chemistry analysis database, water quality trends can be spatially depicted, and areas with consistent problems can be readily identified.

The outcomes are published on a Web site that presents the information as a link map that acts as a report card, grading beaches A to F. For instance, by monitoring the LCU fecal pollutants, evaluators first assess the increased likelihood of people using these beaches getting sick, and then they assign the letter grade. In addition to this public information, the Web site will incorporate an FTP site where interested parties can download free data. Beachgoers and policy makers alike find the Heal the Bay Web site a vital source of information.

The success of Heal the Bay is found in its influence in public policy making. The mapped information that volunteers gather is forwarded to those government agencies responsible for protecting the watershed's resources. California Governor Gray Davis has signed a mandate for an allocation of $100 million for the improvement of Southern California coastal beaches. At the local level, the City of Malibu recently approved a resolution supporting a Malibu marine sanctuary and refuge to be created by an act of the state legislature and the governor. If passed in its current form, the refuge would protect twenty-seven miles of shoreline coast and related habitat and extend three miles offshore. It would ban such activities as commercial fishing, kelp cutting, the removal of marine flora and fauna, oil drilling, and harmful scientific experiments.

At the state level, Heal the Bay helped to draft AB411, a vital piece of legislation that was implemented as part of the California Health and Safety Code providing the public with "right-to-know" information about water quality at local beaches. This requires that those beaches serving more than fifty-thousand visitors a year must conduct weekly testing of three specific bacteria indicators: total coliform, fecal coliform, and entercoccus. Warning signs must be posted by the Health Department if State standards for bacterial indicators are exceeded. Project members have been working diligently in creating an easily understood means of beach assessment. Thus, the Heal the Bay project has grown to the point that it now grades beaches from Santa Barbara County to the Mexican border.

Not so long ago, Santa Monica Bay was a different place. In a short time, Heal the Bay has made it better. Now in its second decade, Heal the Bay continues its fight to find workable solutions to the problems threatening the future of the bay and all of Southern California's coastal waters.

The Stream Team uses 25-foot telescoping antennae poles to rise above the tree canopy and acquire satellite signals to perform their surveys. The GPS team will look for discharge points and outfalls that are potential sources for pollutants to be dumped into the creek.

Growing a National Database

Jack Sobel   Jsobel@DCCMC.ORG
Center for Marine Conservation

The Center for Marine Conservation (CMC), headquartered in Washington, D.C., has deployed GIS to compile and utilize databases on marine and coastal protected areas (MACPAs) throughout the United States. CMC's initial GIS efforts were directed toward developing the first complete database of federal MACPAs and eventually expanded to include state and regional information. CMC's MACPA database contains standard information including size, year established, and management entity for all sites and more detailed and expanded information for other sites. CMC's GIS work is intended to support public decision making for marine- and coastal-protected areas and marine conservation.

Comprehensive information about marine- and coastal-protected areas has traditionally been difficult for the public and decision makers to access because it has been scattered among a host of management agencies and institutions. These groups expressed a need for expansive geographical information. Obtaining its initial support from the U.S. Environmental

Protection Agency (EPA), the CMC initiated efforts to compile a comprehensive database of marine- and coastal-protected areas in the United States. Over the past several years, the CMC has collaborated with its partners to compile related regional and local information. CMC database was developed to allow government agencies, nongovernmental institutions, and the broader public better access to information about protected areas.

A variety of methods have been used for compiling databases. The initial federal MACPA database was created by integrating information from various sources including the EPA, the National Oceanic and Atmospheric Administration (NOAA), the National Park Service, and the U.S. Fish and Wildlife Service. Some important attribute data was also available on federal agency Web sites and publications. Good sources for data gatherers were lists of coordinates found in various federal regulations. First they entered these into MS Excel, then combined them with ArcView GIS, and finally applied a given list of attributes. These

Redtail triggerfish, an example of marine wildlife protected via creation of the Tortugas Ecological Reserve in the Florida Keys National Marine Sanctuary. Photographer, Don DeMaria and Woodfin Camp.

attributes included the protected area's name, state, designation, acreage, and the year it was established.

When available, GIS data was collected from the various state and federal agencies. ArcView GIS was used to reproject the data into a common coordinate system. When GIS data was not available, coordinates derived from federal regulations were used to create points and polygons. Coordinates were entered into MS Excel and saved as a DBF file. The DBF files were then imported into ArcView GIS, and the Add Event Theme feature was used to create the necessary polygons or points. As needed, polygons of protected areas were clipped to a common shoreline to create a clean coast. The various attributes that came with the data were replaced with a common set of attributes. Attributes were obtained from phone interviews, regulations, Web pages, brochures, and other sources. When available, official information about size was used. In some cases, attribute information was calculated using GIS. For example, ArcView GIS was used to calculate the size of some fishery closures that were created from boundary coordinates.

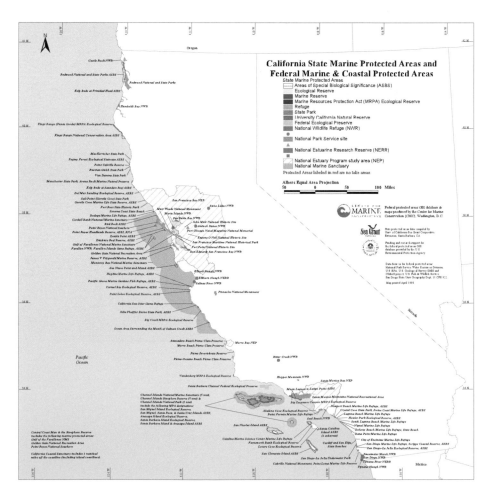

The initial federal GIS MACPA database was put on both CD–ROM and a set of large-format hard-copy maps, which were delivered to the EPA. Both CMC and EPA have used these resources for a variety of purposes and have posted MACPA maps from them onto their respective Web sites. The CMC itself has used this information effectively in meetings with government agencies, conservation groups, and the public. As part of its marine-protected area initiative, the federal government (NOAA) has developed its own MPA Web site and is developing a more elaborate MPA database. CMC has met with and shared information from its database with NOAA representatives.

The CMC had also produced GIS maps of marine-protected areas in California from a database compiled by the California Sea Grant. Conservation representatives were able to provide maps to members of the California state

legislature that showed the actual amount of marine area in which fishing is completely prohibited compared to the areas considered marine "protected" areas. The maps served as very effective visual tools in CMC's ongoing efforts to establish more no-take marine reserves along the California coast.

The CMC, in collaboration with several partners, is using the Gulf of Maine region as a prototype for developing improved regional GIS databases on MACPAs. Data in the Gulf of Maine includes state-protected areas and indicators of the extent of protection offered by each area. The CMC used phone interviews, regulations, and other information sources to obtain information on management objectives and regulatory measures that address pollution, protection of benthic habitats, protection of species, and protection of coastal habitats. The CMC also convened a panel of regional experts to develop a model

Top to bottom: Sea turtle, dolphin, Marbled Grouper and Hogfish, Riley's Hump. Photographer, Don DeMaria (copyright).

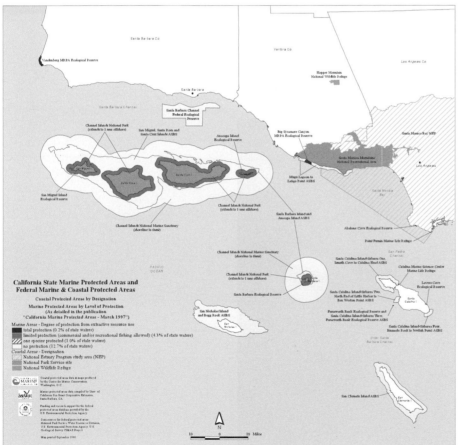

that derives a conservation value based on the size, permanence, and seasonality of each area. The results of all three analyses are contained in the expanded database, which the CMC hopes can be modeled in other priority areas. Information from the expanded database is currently being used to develop a book and poster that will be an important tool for CMC staff in New England.

CMC uses ArcView GIS and ArcView Spatial Analyst to integrate information from their initial MPA and MACPA GIS database efforts with information from other sources to help evaluate existing and proposed MPAs and develop more effective and comprehensive MPA networks for the future. These integration efforts have involved physical information (e.g., bathymetry and currents), biological information (e.g., species and habitat distribution), and human use information (e.g., fishing effort and diving locations). CMC is also increasingly integrating GIS into its marine debris, fisheries, pollution, and protected species work.

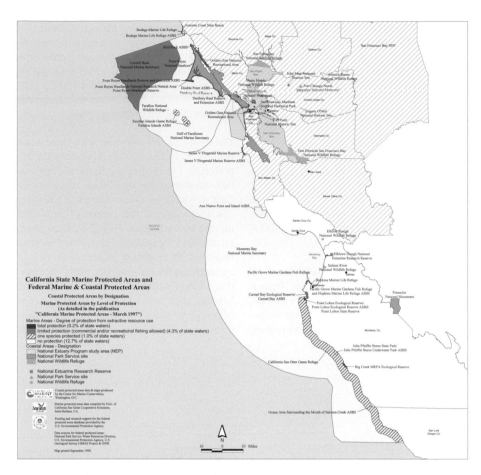

Author Jack Sobel on a research expedition to help create the Tortugas Ecological Reserve in the Florida Keys National Marine Sanctuary.

Photograph courtesy of Don Kincaid

Effective Use of Volunteer Resources: GIS in Community Activism

Chad Nelson, The Surfrider Foundation

People who live near the sea love the oceans and gladly volunteer to help preserve them. Many conservation organizations are formed as the result of a communal desire to rely on community volunteerism not only for direct cleanup projects but also to gather information and organize educational programs. Originally, the Surfrider Foundation focused on Southern California coastal areas, where they had such success in organizing the efforts of volunteers that the organization quickly expanded to 50 chapters around the nation.

The Surfrider Foundation's core competency is community-based education and activism. To strengthen and build on this grassroots educational focus, the organization must facilitate the dissemination of up-to-date, science-based information at the community level. Surfrider has been able to accomplish this by the

effective development of programs such as Beachscape, its popular community-based coastal mapping program.

Using Beachscape, volunteers document the physical characteristics, land-use patterns, pollution sources, public access, erosion, habitat, and wave characteristics of the nation's coastlines. The aim of this program is to mobilize the Surfrider Foundation's vast national network of local chapters and volunteers to characterize local coastal areas at a scale smaller than is currently available from most data sets.

The Surfrider Foundation's members represent an enormous workforce of interested citizens who have knowledge of their local community. Tapping this community resource enables the Beachscape program to develop

data sets at a scale that would be prohibitively expensive for traditional, contract-based data collection projects.

The project uses geographic information systems to store, analyze, and publish data about the influence on coastlines by both natural and human-influenced conditions such as the locations of outfall pipes, hard structures, erosion "hot spots," accumulation of marine debris, and beach accesses (see map below). The Beachscape program illustrates cumulative impacts and provides a basis for evaluation of coastal projects and management proposals. Thus, this program empowers local citizens with the information and skills they need to be effective advocates for coastal resource protection in their communities.

Beachscape is implemented through the Surfrider Foundation's national network of chapters, who collect existing GIS data, nondigital maps of beach features, and volunteer to field map the beach. Beachscape has three levels of implementation: Basic, Intermediate, and Advanced.

Basic Beachscape

Basic Beachscape is the project's simplest level of beach mapping. Basic Beachscape involves using a one-page form to map a local beach. Accompanying the form is a mapping guide that explains how to fill it out. Available to all chapters, Basic Beachscape is an "entry level" activity that is relatively simple to implement. Team members can enter the collected information into a GIS database as well as keep it in a binder at the chapter level.

The methodology for Basic Beachscape is relatively simple. Using a USGS 7.5-minute topographic quadrangle (topo quad) the chapter members delineate their coastal area into discrete beach sections. These sections are numbered and named. Volunteers are then sent to the beach to collect beach attribute information using the Basic Beachscape form. The

Beachscape
Laguna Beach Surfrider
Stormdrains

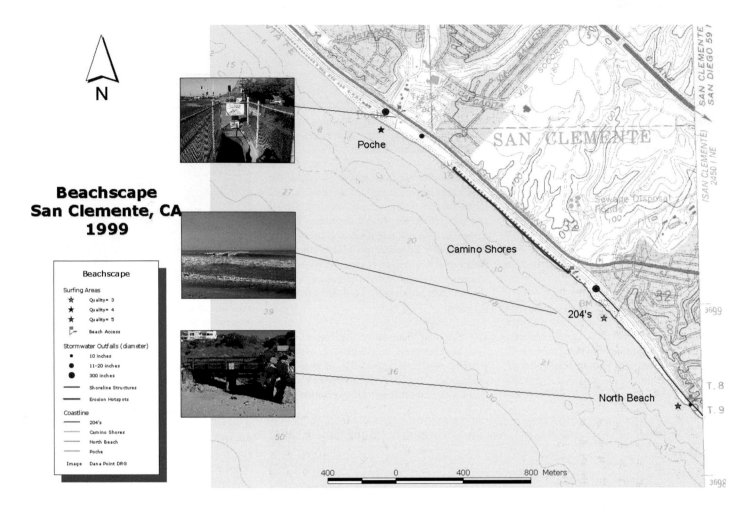

**Beachscape
San Clemente, CA
1999**

Beachscape

Surfing Areas
★ Quality= 3
★ Quality= 4
★ Quality= 5
⚑ Beach Access

Stormwater Outfalls (diameter)
• 10 inches
● 11-20 inches
● 300 inches
— Shoreline Structures
— Erosion Hotspots

Coastline
— 204's
— Camino Shores
— North Beach
— Poche
Image Dana Point DRG

information collected is then entered into a Microsoft Access database. A beach delineation theme is generated in ArcView GIS by creating line segments from a coastline derived from the 7.5-minute USGS quadrangle. The attribute information can then be linked to the beach segments in ArcView GIS.

Intermediate Beachscape

Intermediate Beachscape is a more complex mapping program. It involves mapping specific features on a USGS topo quad and filling out attribute forms specific to the feature. This project requires a Beachscape coordinator at the local chapter level and entails some training of the coordinator and users. The collected information is entered into a database and digitized in a GIS. The data is made available via compact disc, paper maps, and the World Wide Web.

The methodology for Intermediate Beachscape involves field mapping conventions, database design, heads-up digitizing, and project and layout development in ArcView GIS. Field mapping involves marking the locations of specific beach features on a USGS topo quad using a standardized mapping protocol. The team records feature attribute data by filling out a paper feature attribute form and photographing the feature.

If the chapter has a differential GPS unit, this is used to map the feature more accurately. Once the fieldwork is completed, the data is reviewed for an assessment of accuracy and quality by the chapter coordinator. The maps and data forms are then sent to the Surfrider Foundation national office to be entered into the GIS. The attribute data is entered into an Access relational database. Next, the feature data from the maps is heads-up digitized using the USGS digital raster graphic (DRG) of the topo quad. A separate theme is created for each beach feature.

Finally, the database is imported into ArcView GIS and linked to the features in the theme. The photographs are also included and can be linked using the hot link. ArcView GIS layouts are also created to illustrate a basemap for the beach. This information is then made available to the chapter on compact disc, printed maps, and over the Web. The Surfrider Foundation intends to make all data collected for Beachscape publicly available.

Advanced Beachscape

Advanced Beachscape, still in the prototype phase, is the third tier of Beachscape. It is the most complex phase and requires an increased level of commitment from associates. Advanced Beachscape involves Beach Stewards who will "adopt" a stretch of coast

and map it seasonally or even monthly. These Beach Stewards' mapping efforts are more intense than those of members who use the Basic Beachscape program. The Advanced Beachscape effort will include mapping coastal physical processes (analogous to the U.S. Army Corps of Engineers LEO study) and wildlife sightings. This information will also be included in the database and entered into the GIS. Naturally, Beach Stewards will participate in a more intensive training session.

Surfrider publishes its findings in the State of the Beach report, which summarizes the mapping information gathered by Beachscape volunteers. The information collected for this report, along with the initial data collected by Beachscape volunteers, provides a baseline snapshot of the current state of America's beaches. This information is not only useful in itself, but it will allow Surfrider to track trends in important coastal variables such as the amount of beach access and the extent of coastal armoring.

Surfrider Foundation's program has already affected conservation successes, and the implementation of Beachscape is going well. Selected as the first chapter to pilot the prototype, the Ventura County chapter has successfully mapped the entire sixteen miles of coast in their county. During the summer, they held several volunteer beach-mapping events, which served to educate the public in Ventura about the importance of monitoring beaches and also about some of the negative consequences of poor coastal management. During these sessions, participants collected a valuable baseline of beach information for the County. More than ninety storm drains, a prime source of ocean pollution, were identified.

The Boston chapter successfully mapped beach access points (or the lack thereof) in their state. They used these beach access maps in a hearing to fight for more public access to Massachusetts coastline.

In the midst of one of the longest beach closures in the city's history, the city of Huntington Beach and Orange County officials were aided by a Huntington chapter member who used a Beachscape map of storm drains to alert officials about storm drains. The Huntington/Long Beach chapter also used a map of city storm drains and the results of a Santa Monica Bay epidemiological study to create a poster that warns surfers not to surf or swim near storm drains.

Having defeated a potential damaging bike path and seawall proposal in San Clemente, California, the San Clemente chapter is now working with the city to take a proactive look at coastal and erosion management in San Clemente. Ninety percent of San Clemente has already been mapped using Beachscape.

Through the use of its effective volunteer programs, the Surfrider Foundation has improved the availability of data along the coast, thus strengthening coastal management. The foundation continues its progressive mapping program efforts. For example, Surfrider, along with its partnership organizations, is also attempting to create data standards that promote further sharing of spatial data.

⊕

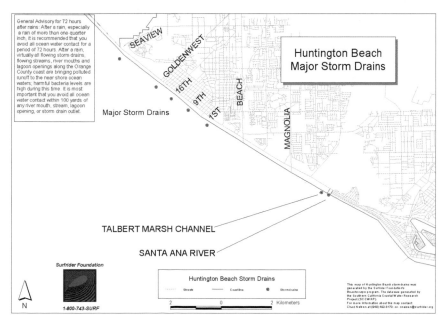

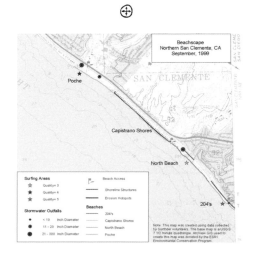

Tuna Distribution in Relation to Physical Features in the Gulf of Maine

Rob Schick, New England Aquarium

A bluefin tuna school as seen from a spotter plane.

In the New England Aquarium Edgerton Research Lab, scientists use GIS to help analyze the distribution patterns of several species in the Gulf of Maine, including the distribution of bluefin tuna (*Thunnus thynnus*) in relation to various physical features. Specifically, they are interested in exploring the spatial correlation between distributions of bluefin tuna as determined by targeted aerial surveys and sea surface temperature (SST) fronts, or temperature discontinuities, in the surface layer of the Gulf of Maine.

From 1993 through 1997, fishery-linked aerial surveys were organized and carried out by Dr. Molly Lutcavage of the New England Aquarium. Together with spotter pilots from the bluefin tuna fishery, Lutcavage located and photographed more than 2,000 schools of giant bluefin tuna. The goal of the surveys was to locate, document, and count the bluefin tuna schools present in the Gulf of Maine over the summer periods from July through October. Working with the fishing industry, the pilots logged more than 190,000 miles of survey effort, documenting some 1,800 schools in 1994, 1995, and 1996 (Lutcavage, et al. 1997). Starting in 1994, Molly outfitted the planes with GPS, cameras, and a laptop programmed with a data acquisition system that allowed pilots to use the interface between the laptop and the GPS to precisely document the location of schools. They then photographed the schools that were seen at the surface and later counted the number of fish seen in each school. This data forms the basis of the distribution data used in the related GIS work, which also attempts to determine where fish are located in relation to specific environmental variables such as SST fronts.

To build the environmental database in conjunction with the tuna data, scientists made use of the Distributed Oceanographic Data System (DODS), a software interface created by the University of Rhode Island (www.unidata.ucar.edu/packages/dods/). DODS makes available a wealth of oceanographic data from a variety of different sources via the Internet. For example, currently via DODS, frontal data is available from Florida to Nova Scotia, and the New England Aquarium has been able to acquire frontal data for each day that tuna were sighted in the Gulf of Maine between 1993 and 1996.

Using the data from DODS, the team created vector coverages of the fronts using ArcInfo GENERATE command. They further manipulated the data to create several output grids including a grid of an individual day's worth

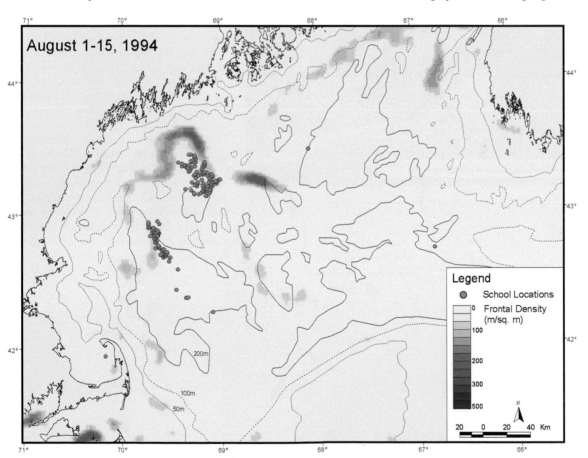

August 1-15, 1994

Legend

- School Locations
- Frontal Density (m/sq. m)
  - 0
  - 100
  - 200
  - 300
  - 500

20 0 20 40 Km

Two-week window of tuna locations and sea surface temperature frontal density in the Gulf of Maine (August 1 through 15, 1994). Locations of tuna schools were documented from aerial surveys. The frontal density layer was created by converting a vector file of all fronts seen in early August to a density grid using the line density command. (Unpublished data, M. E. Lutcavage, R. S. Schick, and J. L. Goldstein; New England Aquarium).

of fronts, a daily grid representing the distance to the nearest front, a density grid of all the fronts seen in two-week intervals, and seasonal average density grids for each year. These grids were created using a combination of GRID commands including LINEDENSITY, EUCDISTANCE, and MEAN. Finally, the data was projected to a common Lambert Conic Conformal projection. All of the above steps were automated using an ARC Macro Language™ (AML). (All the code required to perform these tasks is available from the author.)

To begin investigating the quantitative relationship between these fronts and the bluefin tuna distribution, the team built a separate AML that made use of the GRID command SAMPLE. They extracted the x,y locations of each school (in projected units) at a daily or sometimes subdaily temporal resolution. This AML used the x,y values to query four different grids: depth, slope, distance to the nearest front, and frontal density. The output from each AML was pasted into one large database for further statistical analysis. Currently, the New England Aquarium uses spatial statistics to analyze this distribution. Initial exploration into the data set indicates a great deal of variability in school locations in relation to fronts, but certain strong frontal events seem to lead to larger aggregations of tuna.

These results are interesting at several different levels. At a basic biological and ecological level, they show the environmental conditions in which tuna are seen in the Gulf of Maine. Beyond simple descriptions of tuna locations, the spatial relationship changes with time and with large changes in the physical oceanography of the Gulf of Maine. Bluefin are highly mobile predators who are in the Gulf primarily to feed, thus any interpretation of a static map of bluefin distribution must be made with appropriate caution. Because bluefin are capable of ranging across the entire Atlantic Ocean, surely they are capable of redistributing themselves within the Gulf. Using GIS, scientists can make inferences at a subdaily resolution, allowing for highly detailed investigation into bluefin biology.

New England Aquarium is excited about the application of GIS and is anticipating new discoveries and scientific findings for bluefin tuna and for other pelagic species, such as right whales (*Eubalaena glacialis*), in the Gulf of Maine and beyond.

References:

Lutcavage, M. E., J. L. Goldstein, and S. D. Kraus (1997) "Distribution, relative abundance, and behavior of giant bluefin tuna in New England Waters, 1995." International Convention for the Conservation of Atlantic Tunas Coll. Vol. Sci. SCRS/96/129.

Tracking Sea Turtles: Conservation from Space

Michelle Kinzel, Oceanic Resource Foundation

During August and September 2000, on an isolated, subtropical beach in Veracruz, Mexico, a small team of researchers, sea turtle biologists, and local volunteers put into action a satellite telemetry project, SAT TAG 2000, that had been months in the planning. In an effort to promote international conservation and support research efforts in Mexico and the United States, Oceanic Research Foundation secured funding and support for this satellite telemetry study. SAT TAG 2000 was established and conducted in response to the urgent need to develop international measures of conservation for all species of sea turtles, as well as their habitats.

The objectives of this study were to determine the migratory corridors and habitat usage patterns of green sea turtles, Chelonia mydas, which nest on the beaches of Lechuguillas, Veracruz, Mexico. The long-term goals of this study included providing the scientific findings to conservationists and policy makers in an effort to increase protection of this endangered species and its crucial habitats.

and wide reaching as they often traverse expansive ocean basins and occupy the territories of many sovereign nations. Most species will nest on the beaches of one nation and then travel across oceans or international borders to feed in another. These ancient reptiles represent keystone species and provide a focal point upon which to initiate ecosystem conservation and recovery programs.

There are six species of sea turtles that inhabit the Pacific or Atlantic oceans of the United States: the green turtle (*Chelonia mydas*), hawksbill (*Eretmochelys imbricata*), Kemp's Ridley (*Lepodochelys kempii*), leatherback (*Dermochelys coriacea*), loggerhead (*Caretta caretta*), and Olive ridley (*Lepodochelys olivacea*). At present, all six species are protected under the Endangered Species Act and listed as either threatened or endangered by the U.S. Fish and Wildlife Service. All but the Kemp's Ridley turtle migrate to Mexican waters to reproduce and deposit their eggs along sandy beaches.

The sea turtles' reliance on a terrestrial nesting site and return to their natal beaches makes them vulnerable to being caught and killed for their meat, shells, and eggs.

The females do not incubate the nests, choosing instead to roll the dice in a biological reproductive numbers game that will have the females laying up to seven hundred eggs per breeding season. Thus, the eggs are highly vulnerable and sought out by a myriad of predators including feral dogs, nocturnal mammals, beach crabs, and man.

Both male and female adult turtles are easily caught in the open ocean when they surface to breathe. Also, a significant cause of mortality is from incidental by-catch of several coastal fisheries. Add to these factors commercial development along nesting beaches, destruction of foraging habitats, ingestion of plastic debris, entanglement in floating debris, boat propeller injuries, boat collision mortality, toxic pollution mortality, and lighting disorientation of hatchlings and it is easy to see that these slow-maturing animals face a perilous and treacherous 25- to 30-year struggle to reach sexual maturity.

Nesting beach conservation efforts began in earnest in Mexico in 1964, and organized protection of sea turtle populations and nesting beaches has been occurring ever since. Several environmental and research organizations monitor and protect reproducing female turtles and their nests. The Mexican government has been instrumental in the establishment and funding of several such organizations. Despite the support and conservation efforts of the United States and Mexico, however, sea turtle populations are either declining in numbers or only marginally recovering.

Zyanaya, entering the Gulf of Mexico with her newly attached satellite transmitter.

Sea turtles are mysterious and alluring sea creatures that have inhabited the earth for millions of years and have been termed living dinosaurs by sea turtle biologists. Sea turtles have also been referred to as ambassadors of the sea because of the integral roles they play in the ocean ecosystems of multiple nations during their life cycles. Due to the migratory nature of sea turtles, their habitats are varied

Sea turtles face an onslaught of threats during every stage of their life cycle. Female sea turtles spend 99.97 percent of their lives in the ocean, returning to the terrestrial environment only briefly as they lay their eggs to propagate the next generation. This journey onto land is a brief event but one that makes the turtles susceptible to human poachers and the rare mammalian carnivorous predator such as the jaguar.

Recent increased protection has been granted to sea turtle species along their migratory routes, which had suffered severe population losses from mortality caused by incidental by-catch in fishing nets. The U.S. government has mandated the closure of shrimp fisheries during the crucial migratory season, March and April, along the coastal waters of Texas. In addition, because U.S. fisheries have been identified as impacting sea turtle species, the Commerce Department mandates commercial fisheries to use Turtle Excluder Devices when harvesting the seas. The data from projects such as SAT TAG 2000 helps conservation managers make sound recommendations for the management of fisheries in areas inhabited or utilized by endangered sea turtle populations.

To effectively protect sea turtle populations and implement conservation efforts, feeding grounds locations, nesting beach occupancy patterns, and migratory routes is needed. By employing GIS to spatially depict sea turtle occupancy and patterns of usage, biologists can better assess potential threats to these animals and propose effective conservation measures to eliminate or reduce these threats.

Historically, this information has been sparse and collected by monitoring the females and eggs during their brief residences on nesting beaches. Previous biological studies have relied on monitoring adult female turtles by attaching metal or plastic flipper tags, injecting Passive Integrated Transponder (PIT) tags into flipper muscle, or creating "living tags" by surgically moving pieces of the carapace and plastron. The flipper tags are cost effective, while the Passive Integrated Transponder tags involve the injection of a microcomputer chip and require expensive equipment both to insert and later read the bar code. The microprocessor chip has an individual bar code, and researchers must use electronic scanners to read the tags. This method, as well as the creation of living tags, is costly and relatively time intensive.

These methods essentially mark the turtle and can provide information on age classes, transoceanic or long-range movements, nesting frequency, and nesting beach site fidelity. However, these methods of identifying and studying the turtles do not provide the detailed information on migratory routes, feeding site occupancy, or habitat usage that is crucial to conservation efforts.

The modern era of microelectronics has provided unique opportunities to study sea turtle movements and behaviors at sea. Sea turtle biologists began using satellite telemetry technology in the 1980s. Early tracking studies used the Nimbus climatological system and provided data on the movements of loggerhead turtles in the Georgia Bight. The first confirmation of a trans-Pacific movement by a loggerhead turtle was made from the recovery of a flipper tag. This information, while enlightening, was limited in detail to the starting and ending points of the migratory route. By using satellite tags, one researcher was able to elucidate an important part of the migratory puzzle for the scientific community. In 1996, Dr. Wallace J. Nichols was able to confirm the migratory route from Baja California to Japan via a turtle tracked with satellite telemetry. Nichols was able to determine exact routes of migration and swim speeds with the use of this new technology.

Two sexually mature female sea turtles were chosen for this study. The female turtles were found during a night patrol of their nesting beach and were detained following the deposition of their egg clutches, which were safely transported to a protected and monitored corral. The turtles were placed in wooden boxes to minimize their movements as researchers attached the small transmitters to the carapace using a silicone polymer. The transmitting devices were secured with fiberglass strips and painted with an olive-colored marine paint to help camouflage the units with the carapace. The females, Zyanaya and Roberta, having been tagged with steel Monel flipper tags, embarked on a journey that would contribute to our sparse but quickly growing knowledge of green sea turtles and their movements between natal nesting beaches and feeding grounds.

Even before Zyanaya and Roberta were released, the tracking process began. The transmitting units measure 13 cm by 4 cm by 2 cm, approximately the size of a television remote control device, and are small relative to the size of the turtles' carapaces, which measure approximately one meter long. Each transmitter contains a microprocessor, which is preprogrammed with a duty cycle chosen by the researchers. To maximize duration of transmissions, the transmitters are programmed to send their signals using low wattage output, ranging from 0.5 to one watt, for a predetermined duty cycle, which is followed by an off-duty cycle. For this study, a duty cycle of six hours on, followed by eighteen hours off, was chosen.

The transmitters send their signal, which is detected by one of the four polar orbiting ARGOS satellites, which have been designated for tracking wildlife. Satellites receive data during their overpass in the region of the transmissions. Each satellite circles the earth every one hundred and one minutes, and the satellites are in position to receive data from any one location for approximately ten minutes.

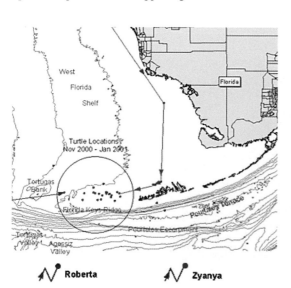

The satellite tage recorded these routes for the two green sea turtles in the Gulf of Mexico.

Because the quality of signal transmission is dependent on the angle of the satellite over the horizon, and the number of overpasses in one location, the best readings are collected near the poles. Data collected from the equator has a higher frequency of lower location class readings.

To determine the location of the transmitter, a satellite must receive multiple readings from the transmitting unit. An average reading takes between three and five minutes. This can prove to be problematic with an aquatic species such as the green sea turtles, which usually only surface to breathe and do not remain at the surface for extended periods of time. As a result, readings are received intermittently, with the results of this study showing data input approximately every three days.

The signals are categorized into location classes based on the number of readings from that particular location. The accuracy of the latitude and longitude position readings can be estimated based on the location class. The location classes, from poorest to best, include Z, B, A, 0, 1, 2, and 3. Location classes of Z indicate that the transmitter is sending a signal but the location cannot be determined. Therefore, readings from location class Z are rejected for tracking and mapping purposes. Signals from location classes B and A are somewhat reliable, but they are not as accurate as the classes zero through three. Researchers prefer to use location classes of zero or greater, which are estimated to be within 350 meters of accuracy, for plotting maps and analyzing rates of travel. For the two turtles tagged in this study, the majority of location classes were rejected because only 21.21 percent of the readings for Zyanaya, and only 18.92 percent of the readings for Roberta, were of location class zero or greater. Zyanaya covered a route of 985 kilometers, at an average rate of .631 km/hr. Roberta traveled a bit slower, at .362 km/hr for a total of 1,430 kilometers.

With the recent accessibility and implementation of satellite technology, more detailed information is now available to biologists, conservation managers, and policy makers. Satellite tracking allows for data retrieval via a transmitted signal and does not rely on the recovery of the tagging instrument. By combining the application of a transmitting device to the body of an animal with a means of signal retrieval, such as an ARGOS satellite that orbits the earth, researchers can track the specific movements of the animal over long distances.

This field project is a cooperative effort between Oceanic Resource Foundation of the United States and the Mexican environmental organization CRIP–VER. Researchers have successfully attached ST-18 Telonics transmitters to two female turtles at the end of their nesting and egg laying season. ARGOS satellite data has been transmitted since September 2000. GIS mapping has been used to plot the latitude and longitude readings into map tracks of the migratory routes.

The two female sea turtles followed different routes in their movements across the Gulf of Mexico. The first turtle remained close to the nesting beach for ten days before traveling south and crossing the Gulf of Mexico near the northern end of Cuba. The second turtle sent signals from locations on or near the nesting beach for twenty-one days before heading north and remaining in relatively shallow water near the coast before heading across the Gulf of Mexico near southern Texas. Both tagged turtles have taken up residence in feeding grounds called Tortugas Bank off the Florida Keys. The exact location of the two turtles will be monitored for the life of the batteries, predicted to be one year. Future plans include monitoring the turtles' swim speeds, dive profiles, and zones of habitat occupation.

To assess and define home ranges and zones of occupancy, the current tracking data will be analyzed using an extension of ArcView GIS called Animal Movement. This extension is designed to implement a wide variety of animal movement functions in an integrated GIS environment. The program also has significant utility for analyzing other point phenomena. The GIS maps of the sea turtles' transoceanic movements and feeding ground

locations will be compared to aerial photographs to assess habitat usage.

The Oceanic Research Foundation hopes to expand on its SAT TAG 2000 project by attaching up to ten transmitters to green sea turtles in the next five years.

SAT TAG 2000 was accomplished by the efforts of many people and organizations. Greg Carter of the Oceanic Research Foundation initiated the international project in collaboration with the Mexican organization, Centro Regional de Investigacion Pescevera–Instituto Nacional de La Pesca en Veracruz (CRIP–INP–VER), and marine biologists from Mexico and the United States. Graciela Tiburcio Pintos served as a biologist on the project and coordinated operations and field-work in Mexico. Rafael Bravo Gamboa, an engineer with CRIP, served as camp manager and coordinated all logistical operations at the tagging site. Michelle Kinzel served as a biologist during fieldwork. Greg Carter monitored the ARGOS satellite data and prepared the maps of the turtles' oceanic movements.

Please log on and discover the locations and movements of these magnificent creatures: www.orf.org/turtles.html

All photos by Michelle Kinzel

Zyanaya, making her way to the ocean after being fitted with a satellite transmitter.

Moored Buoy Site Evaluations

R. A. Goldsmith and Dr. A. J. Pluddemann
Woods Hole Oceanographic Institution

The Woods Hole Oceanographic Institution Upper Ocean Processes Group routinely deploys instrumented surface moorings in support of upper ocean and air–sea interface science programs. These moorings consist of a surface buoy equipped with meteorological sensors, a subsurface mooring line outfitted with oceanographic instruments, an acoustic release, and an anchor. A recent project calls for a one-year surface mooring deployment in the northwest tropical Atlantic, on the southwest flank of the Researcher Ridge (approximately 15° N, 51° W) in 5,000 meters of water.

Deep ocean mooring deployments provide several challenges. One of these is the trade-off between the details of the mooring design and the tolerance for error in deployment depth. Moorings are typically of compound construction, consisting of chain, wire, and synthetic elements. Principal design considerations are the compliance (degree of "stretch"), resonance (the natural frequency of vibration), and scope (ratio of mooring length to water depth). These three elements are "tuned" to achieve the desired static and dynamic response characteristics. Careful tuning can provide a more robust mooring, but in order for the tuning to be effective, the actual deployment depth must match the design depth to within a few percent (e.g., within 100 meters for a 5,000-meter design depth).

This implies that the bathymetry surrounding the chosen deployment site must be well known and that the mooring placement must be relatively precise.

The availability of accurate global bathymetry on scales of a few hundred kilometers provides a valuable starting point for determination of the mooring site. However, information on smaller scales is desired for planning the deployment, and a shipboard bathymetric survey is typically done to refine the site.

Due to difficulties in placing the mooring precisely, it is desirable to have a target area of 1 kilometer or more. The mooring is deployed buoy first and strung out behind the ship while steaming forward slowly (e.g., 1 knot). When the anchor is released, it does not drop straight down. Instead, the drag of the mooring line and instrumentation serves to "pull" the anchor toward the buoy during its descent. Anchor "fall-back" is typically about 5 percent of the water depth but varies with the mooring design and instrument load. Wind and currents influence the angle at which the

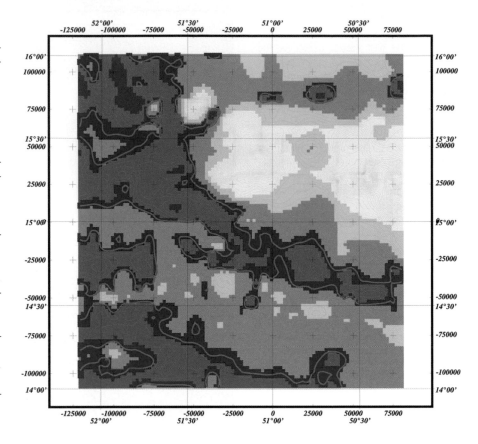

Researcher Ridge

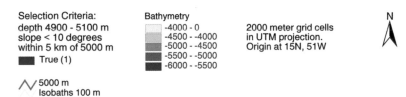

Selection Criteria:
depth 4900 - 5100 m
slope < 10 degrees
within 5 km of 5000 m
■ True (1)

Bathymetry
☐ -4000 - 0
▨ -4500 - -4000
▨ -5000 - -4500
▨ -5500 - -5000
■ -6000 - -5500

2000 meter grid cells
in UTM projection.
Origin at 15N, 51W

N

∧∨ 5000 m
Isobaths 100 m

mooring streams behind the ship, and thus the direction in which the anchor falls back. Finally, vagaries of the ship's speed and track through the water during the deployment process mean that the exact anchor release location is not predictable. Overall, these factors introduce uncertainty of at least several hundred meters in the actual anchor position relative to the target position.

To minimize the time necessary for the shipboard survey, and to maximize the likelihood of meeting the design depth tolerance, it is useful to focus on regions surrounding the target depth where the bottom slope has a local minimum. Clearly, this becomes more important when the local topography is steep (e.g., the flank of the Researcher Ridge). To facilitate the mooring deployment described here, a GIS application was used to identify portions of the

southwest flank of Researcher Ridge within +/− 100m of the 5,000-meter isobath and with local slopes less than 10 degrees.

This type of topographic site feasibility study is done all the time in civil engineering applications, but this small project provides a good example of how existing GIS software and techniques can be adapted for use with oceanographic research. As a starting point, researchers used the two-minute bathymetry derived from satellite gravity measurements (Smith and Sandwell 1997). As this was a deep, mid-ocean location, the approximately 3.7-kilometer horizontal resolution was adequate for our investigation. In some areas, higher resolution bathymetry data may be available from hydrographic or multibeam surveys and could be used if available.

Because the general area was only about two degrees (200 kilometers) square, and the bottom slope was a factor in the siting criteria, the team converted the grid to two-kilometer cells and used a Universal Transverse Mercator (UTM) projection. Although the group used ESRI's ArcView GIS software, most GIS applications will readily compute the slopes of the surface. Another common GIS procedure, the generation of buffers, was employed to define a zone within five kilometers of the 5,000-meter isobath. When these preliminary steps were completed, it was a straightforward procedure to identify all the areas meeting the defined criteria: within five kilometers of the 5,000-meter depth, between 4,900 and 5,100 meters depth, and less than 10-degree slope. These areas are shown in red on the accompanying figure, previous page.

A tentative site was identified very close to the chosen grid origin at 15° N, 51° W. In this case, the site was in the vicinity (25 kilometers south) of what had been identified by a simple visual inspection of a nautical chart. An enlarged view of the area is shown in the accompanying figure. A proposed cruise track was drawn in to help conduct the detailed survey and a chart was prepared showing both the UTM grid and the geographical coordinate used for navigation (see figure , right).

The entire exercise was done using existing GIS software, tools, and commonly available data. The procedure lends itself to being done at sea in an operation environment. Here, it not only was helpful in refining the siting of the mooring location but also in identifying several alternative sites that might not otherwise have been so readily apparent.

References:
Smith, W. H. F., and D. T. Sandwell. "Global seafloor topograph from satellite altimetry and ship depth soundings," *Science,* v. 277, pp. 1957–1962. 26 Sept. 1997.

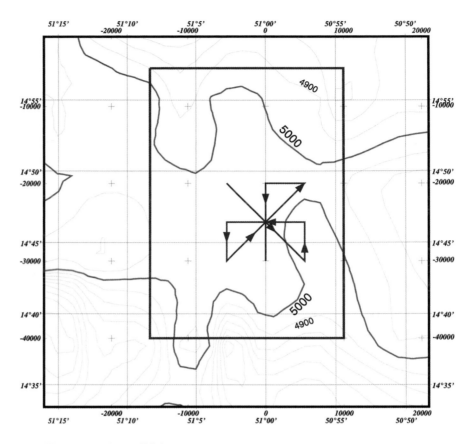

Researcher Ridge

Proposed survey track

➤

⋀ 5000 m
 Isobaths 100 m

2000 meter grid cells
in UTM projection.
Origin at 15N, 51W

N
▲

Shoreline Metadata Profile

David Stein
FGDC Marine and Coastal Spatial Data Subcommittee

The objective of the Shoreline Metadata Profile of the Content Standards for Digital Geospatial Metadata (CSDGM) is to capture the critical processes and conditions that revolve around collecting and creating shoreline data and data that intersect with our Nation's shorelines. The Metadata Profile is a Federal Geographic Data Committee (FGDC) Standard designed to standardize and clarify some of the complexities of shoreline data by providing shoreline specific metadata elements (fields), a comprehensive shoreline glossary, and a bibliography.

Because the CSDGM only allows for the documentation of generic geospatial data, the Federal Geographic Data Committee's (FGDC) Marine and Coastal Spatial Data Subcommittee (formerly the Bathymetric Subcommittee) felt that it was necessary to develop a metadata profile that addressed the shoreline. Shoreline data are important for coastal zone management, environmental monitoring, resource development, jurisdictional issues, ocean and meteorological modeling, engineering, planning, and many other uses. A published standard by a responsible agency will provide the affected community with a basis from which to assess the quality and utility of their shoreline data. The shoreline is an integral component of the geospatial data framework and the Coastal National Spatial Data Infrastructure.

The Shoreline Metadata Profile is intended to serve the community of users who are involved with geospatial data "activities" that intersect the U.S. Shoreline by providing the format and content for describing datasets related to shoreline and other coastal datasets. The Shoreline Profile is primarily oriented toward providing the elements necessary for documenting shoreline data and reaching a common understanding of the shoreline for national mapping purposes and other geospatial and geographic information systems (GIS) applications.

The Shoreline Metadata Profile is comprised of five major sections: "Mandatory Elements of the CSDGM," "Modified Metadata Elements," "User Defined Metadata Elements," "Glossary," and "Bibliography." The "Mandatory Elements of the CSDGM" section states that all of the elements (fields) of the CSDGM are available for use in the Shoreline Profile. The "Modified Metadata Element" section describes how elements from the CSDGM were modified to accommodate the complexity of shoreline data. There were two types of changes made to some existing elements. They are conditionality changes and domain changes. Conditionality refers to whether a response to an element is mandatory, mandatory if applicable, or optional. A domain is the list of possible responses to an element. In the case of a domain change, the list of possible

responses to an element were either extended or restricted. The "Extended Element" section defines the new elements that were included in the profile. Examples of new elements include tidal datums, marine weather conditions, and description of shoreline position. Finally, the "Glossary" and "Bibliography" were developed as informative annexes to the Shoreline Profile, and were designed to provide a broad-based understanding of shoreline and related issues.

The U.S. Department of Commerce, National Oceanic and Atmospheric Administration (NOAA) Coastal Services Center will maintain the Shoreline Metadata Profile for the Federal Geographic Data Committee. Questions concerning the Shoreline Profile can be addressed to the Executive Secretary of the FGDC Marine and Coastal Spatial Data Subcommittee.

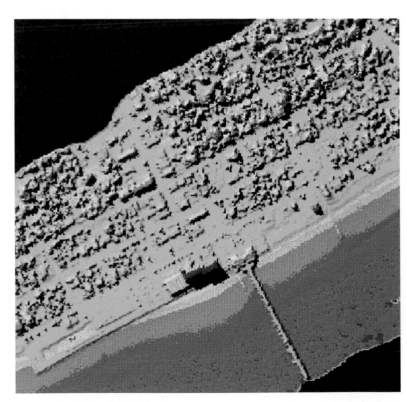

Using Technology to Resolve Coastal Resource Management Issues

New technology, data, and management techniques can help coastal communities maintain the delicate balance between the desire to grow and the need to protect natural resources. Unfortunately, a lack of expertise and high start-up costs often limit the use of helpful tools.

The National Oceanic and Atmospheric Administration's (NOAA) Coastal Services Center is a federal government agency that helps local and state coastal managers overcome these obstacles. Many of the Center's efforts use GIS extensively, as shown in the sample projects profiled below.

A GIS for the Ocean

Created for the southeastern United States, this project was the first regional GIS populated to meet the needs of ocean users. The Web-based ocean mapping and information system allows users to manipulate coastal data; simplify the spatial complexities of jurisdictional and regulatory boundaries; access background and supporting information; and download spatial data, metadata, and legal summaries. These tools and information are invaluable when making decisions regarding complex marine issues such as shipping channel, dredging, marine reserve siting, and enforcement and conservation efforts.

Ocean Planning Information System
www.csc.noaa.gov/opis/

Using Information to Diffuse Controversial Situations

Deciding where to place a proposed marine reserve can be controversial, as the implications for various user groups can vary greatly. For the Channel Islands National Marine Sanctuary in California, a GIS was used as the base of a newly created intuitive visualization tool. Various members of the community were able to visualize complex data and information and, therefore, better understand the multitude of "what if" scenarios that surrounded each point of view.

Spatial Support and Analysis Tool
www.csc.noaa.gov/pagis/html/cinms_act.htm

Bringing Nautical Chart Data into a GIS

The type of information found on a nautical chart represents basic data for most coastal resource managers. To unleash this data source, the NOAA Coastal Services Center worked with Maptech, Inc., and ESRI to create an ArcView extension that allows users to view NOAA nautical charts within ArcView. These charts, displayed as images, can provide a useful background layer for ocean and coastal GIS projects. This extension is offered to users free of charge.

Chartviewer
www.csc.noaa.gov/products/chartview/

Defining Marine Boundaries

In the past, the reach of a cannon shot determined many international maritime boundaries. With GIS and the global positioning system (GPS), precise maritime boundaries can be set, but many boundaries were designated prior to the existence of this technology. Conflicting boundary views cause a host of problems including issues regarding the enforcement of laws, questions about ownership rights, and disputes regarding resource management and mineral extraction efforts. The NOAA Coastal Services Center is working with several organizations to translate legal boundary definitions into a digitally compatible format. This work represents one of the Center's newest projects.

Marine Digital Boundaries

NOAA Coastal Services Center
LINKING PEOPLE, INFORMATION, AND TECHNOLOGY

www.csc.noaa.gov

Mapping Nature's Diversity
The Wildlife Conservation Society

Golden bamboo lemur (Hapalemur aureus), first discovered in 1986, is an example of significant new species being found in modern times even as habitats vanish, as well as an example of one of the most endangered primates in the world.

Photograph courtesy of Frans Lanting, © 2001 Frans Lanting

International GIS Support for Wildlife Conservation

Karen B. Willett and Eric W. Sanderson
Geographic Information and Analysis Program, Wildlife Conservation Society

Introduction

The Geographic Information and Analysis Program at the Wildlife Conservation Society (WCS) signed a mentorship agreement with ESRI Conservation Program (ECP) on February 18, 1999. The goal of our participation in the ECP mentorship program is to increase the ability of WCS conservation scientists to create, manipulate, and show geographic information to enhance their on-the-ground conservation efforts. Over the years since the beginning of the agreement, WCS has completed two training workshops (training a total of twenty-eight WCS field scientists), assisted another eight WCS conservationists to apply for and obtain ESRI software through the ECP, and, in general, supported the geographic information and analysis needs of WCS's conservation programs worldwide. Our mentorship agreement with the ECP has substantively contributed to the ability of our staff members to use geographic information to meet their conservation goals.

This report includes an overview of the first two GIS training workshops and examples of GIS analyses accomplished by WCS ECP recipients.

Photo courtesy Charles Convis.

Training Workshops

A key component of the ECP mentor program involves teaching grant recipients about basic GIS concepts and training them to use the software to the degree that they can work effectively on their own computers when they return to the field. Our long-term plan is to provide two training workshops per year for three years, alternating between New York and one of our three geographical regions (Asia, Latin America, and Africa). Workshops were offered in New York during September/October 1999–2001, Asia (Indonesia) in February 2000, and Latin America in February 2001.

The goals of the workshops are to:

- Build a conceptual understanding of geographic information systems and the geographic and computer science principles that underlie the successful application of GIS for conservation.

- Develop basic skills in ArcView GIS including data integration, analysis, and presentation.

- Provide a substantial start toward analyzing project data with GIS.

First Training Workshop, New York

From September 20 to October 1, 1999, the Geographic Information and Analysis Program held its first Basic GIS Training for Wildlife Conservation in New York. The course consisted of one week of structured tutorial and one week of project work based on information the participants brought from their conservation sites. Prior to the participants' arrival, we facilitated their acquisition of ArcView GIS (through ECP), ensured they obtained an ArcView GIS-capable computer for use in the field, and assisted them in developing a learning project based on conservation-related data that they had obtained from their study site. We also prepared a workshop book with exercises, lecture notes, and supplemental material that each participant took home.

The structured tutorial covered GIS basics including when and why to use GIS, GIS data sources, analytical methods, and map production using the ArcView GIS software. We used a combination of lecture and lab exercises to convey the material. Participants worked through a lab exercise that took them through all stages of a GIS analysis including data capture, data integration, data analysis, and map presentation. They went step by step through a GIS project aimed at estimating the number of people in the Bronx Zoo. We made "field observations" at locations determined with differential GPS; collected data from other sources; transformed the data into a consistent coordinate system; and conducted an analysis using buffers, intersections, and area calculations. The participants each prepared map layouts illustrating their analysis.

The second week consisted of analysis of a conservation data set that participants brought with them, using what they had learned in the first week. We facilitated their analysis of their own data. Because most of the difficulty in GIS is in application, the second week gave the participants hands-on time with their own data sets and allowed us to be sure that when they return home, they will have a start on their own projects on their own computers. In addition, we read and discussed five papers (one each morning) highlighting different applications of GIS published in the scientific literature.

Second Training Workshop, Bogor, Indonesia

From February 7 to 18, 2000, the Geographic Information and Analysis Program held its second Basic GIS Training for Wildlife Conservation in Bogor, Indonesia, hosted by the WCS Indonesia Program (WCS–IP). Fourteen members of the WCS Asia Program attended the training. The course structure was similar to the first course, though we revised the lectures with additional information and changed the lab exercise to determine the road density in protected areas in Uganda. We also taught this course using the latest version of ArcView GIS then available, version 3.2.

GIS is clearly becoming an integral part of WCS conservation activities around the world.

Laikipia Predator Project, Kenya

Laurence G. Frank

The Laikipia Predator Project, funded by WCS, works on the conservation of lions, hyenas, wild dogs, and other large African predators outside protected areas. Large carnivores worldwide are being driven to extinction because they kill livestock. We are doing intensive studies of carnivore biology and related livestock husbandry in Laikipia District, Kenya, a very pro-conservation livestock-producing area. We use radio tracking (both aerial and satellite), relating predator movements and ecology to alternative land uses that range from intensive agriculture to traditional pastoralism and commercial ranching. Lessons learned here will be applicable throughout much of Africa where large predators still exist.

Nearly all studies of large African predators have been conducted in protected areas where predators rarely come into conflict with humans. The Laikipia Predator Project is the first attempt to manage and conserve lions, hyenas, leopards, cheetahs, and wild dogs in a livestock-producing area where depredation on domestic animals causes many predators to be killed each year. In spite of these conflicts, most Laikipia landowners are strongly interested in conserving the large predators, for both aesthetic and economic reasons; ecotourism is rapidly becoming a major source of income for the region, and tourists demand predators.

The Laikipia Project commenced in 1997 with an extensive survey of both commercial landowners and traditional pastoralists. We collected data on the distribution and abundance of predators, landowner attitudes toward them, and the relative economic costs of depredation compared with other costs of producing livestock. Based on that work, several conservation foundations contributed funds to start a study of the main predators in Laikipia, with a view to producing a management and conservation plan to ensure long-term survival of predator populations. A central goal is to determine the livestock management practices that will be most cost effective in minimizing depredation losses and that can be applied widely where African predators still occur.

At the start of this study, there was no ecological data on numbers, reproduction, mortality, social group characteristics, and movement patterns of any of the predators anywhere in Africa outside of parks. In nonprotected areas, all ecological parameters are affected by human-caused mortality. The first priority, therefore, was to collect basic ecological information on the existing predator populations in Laikipia, concentrating initially on lions, spotted and striped hyenas, and wild dogs.

Lying just north of Mount Kenya National Park and south of Samburu National Reserve, Laikipia District is an extremely important area for wildlife conservation. Containing more than one million acres of semiarid ranch land, it is the only area in Kenya in which wildlife numbers are actually increasing rather than decreasing. Most landowners are committed to preserving the ecosystem while developing a sustainable rural economy. There are substantial populations of large carnivores, as well as the full range of African ungulates. It is still a remarkably intact ecosystem that presents great opportunities and challenges for conservation. It also presents unique opportunities for testing different land-use, livestock, and wildlife management systems that would be broadly applicable in much of eastern and central Africa where wildlife still coexists uneasily with livestock.

Laikipia may be unique in that commercial livestock interests are actively trying to coexist with large predators. Virtually everywhere else in the world, ranchers deal with depredation problems by eradicating large carnivores. Here, however, ranchers attempt to minimize their impact on predators, killing them only when unavoidable. Several groups of traditional pastoralists are starting to move into ecotourism, and they, too, are interested in preserving wildlife, having set aside significant areas exclusively for wildlife. This is an extraordinary area offering an extraordinary opportunity to learn how to live with large African predators.

We acquire animal movement data through aerial and satellite radio telemetry, using VHF transmitters, ARGOS PTTs, and GPS-ARGOS PTTs. Thus far, we have used ArcView GIS for mapping animal movements against property boundaries and land use classifications.

We have collars on twenty-four lions, eleven spotted hyenas, and six striped hyenas. We will continue collaring more animals and in 2000 will start a project on wild dogs in neighboring Samburu District, adding another six thousand square kilometers to the study area.

Semiannual aerial counts of all large herbivores (wildlife and livestock) are performed by the Mpala Research Centre and stored in a GIS format. Future analyses will focus on predator habitat use as a function of both livestock and wildlife distribution and the distribution of human activities (ranching, pastoralism, agriculture, population centers, tourism, and protected areas).

This is the first effort to integrate predator conservation with agriculture in Africa. Very few national parks are large enough to guarantee long-term population viability for the large predators; large areas outside parks must also be safe for these animals if they are to persist in Africa. We are supported by major conservation organizations and have excellent support and cooperation from landowners in the study area. This project is expanding rapidly. Donors, project staff, and landowners all view it as a long-term conservation effort that serves as a laboratory in which to refine predator conservation techniques that will be broadly applicable in much of eastern and central Africa.

We work in a large area and all data must be interpreted in geographic terms: predator distribution and movements, livestock and wildlife distribution, human land use practices and densities, and so forth. For instance, from our early data mapped in ArcView GIS using land use coverages in GIS, it is clear that lions avoid densely settled areas, both agricultural and pastoral, and confine themselves to commercial ranches that have a very low human density.

Use of ArcView GIS in the Africa Program

Dr. Andy Plumptre
Wildlife Conservation Society, Africa Program

ArcView GIS has been used by the Wildlife Conservation Society's (WCS) Africa Program extensively over the past year. There have been two main projects that have benefited from the support that ESRI has provided.

Mapping the Distribution of Gorillas in the Bwindi Impenetrable National Park, Uganda

WCS carried out a census of mountain gorillas in the Bwindi Impenetrable National Park in southwest Uganda. Six teams of researchers spent eight weeks in the forest surveying every 500- by 500-meter block of the forest using line transect and reconnaissance walks. Whenever a fresh gorilla trail was encountered, it was followed until three consecutive night nest counts were made to estimate the number of animals in the group. At the same time data on other species and human impacts was collected in different sectors of the park. ArcView GIS has been useful in mapping the spatial distribution of species in the park and the spatial distribution of human impacts. Mapping the human impacts has been particularly useful for showing the park managers where patrols should focus more effort.

The results were quite striking. For all primates other than gorillas, there was a definite preference for habitat at the edge of the national park. The map (bottom left) shows higher primate density with darker colors.

Mapping the gorilla groups with the relative densities of total human signs showed that gorillas tended to avoid areas of high human impact but could tolerate low levels of impact (see map right).

Identifying Areas of Conservation Importance Within Nyungwe Forest Reserve, Rwanda

In 1999 a survey was made of the Nyungwe Forest Reserve in southwest Rwanda. This forest is one of the largest remaining blocks of forest in the Albertine Rift, an area of high endemism and species richness in Africa. There has been some pressure to allow local communities access to the forest to collect honey, medicinal plants, and possibly other materials. It was decided to try to identify which areas are most important for conservation in the forest and thereby allocate access areas to local communities in sites of least conservation value. Surveys were made of mammals, trees, and birds in thirteen sites in the forest. From the data collected we could calculate species richness, diversity, and number of Albertine Rift endemic species for each

Gorilla Groups and Total Human activity

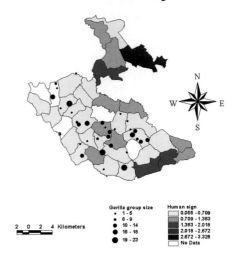

site. These could be used to prioritize sites for each taxon. Then we ranked the sites from 1 through 13 for each taxon and summed ranks across each taxon to identify the important sites for conservation of all three taxa. However, it was noticed that several high-ranking sites contained a lot of the same species and some low-ranking sites had species of conservation importance that would be omitted. Consequently we carried out a complementarity analysis weighting by endemic species as follows:

- Choose the site with most Albertine Rift endemic species and list all species that occur at this site.

- Select the next site that adds most new Albertine Rift Endemic species (and, in case of a tie, most new total number of species).

- Continue this process until all species are accounted for.

This was done for each taxon separately and then the rankings of the sites combined to provide an overall zoning plan for the forest.

Training taking place in Uganda using GPS units for the census of the gorillas.

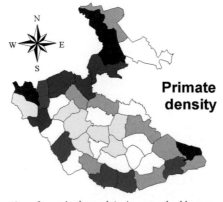

Signs of human impact in Bwindi—in this case illegal cutting of trees in the park (using a method known as pitsawing).

Indonesia Species Conservation

Margaret F. Kinnaird, Conservation Ecologist

Over the last year, we have been able to apply our ESRI grant and ArcView GIS software toward several diverse projects ranging from the effects of habitat fragmentation on the endemic Sumba Island hornbill, habitat classification for Seram cockatoos, and trends in deforestation in the Bukit Barisan Selatan National Park. Our skills have expanded significantly since the ESRI grant was received although our needs have stayed constant (and high!). We feel we have been able to expand our project funding, especially for the Bukit Barisan Selatan National Park work because we have been able to incorporate GIS analysis into grant proposals. Here we briefly outline goals, methods, and results for each project.

Effects of Habitat Fragmentation on the Endemic Sumba Island Hornbill

This study examined the interaction between fragmentation and resource availability across the entire range of a large frugivorous hornbill endemic to the Indonesian island of Sumba. Over the last several decades, Sumba's forests

have been cleared for agriculture and pastureland, and this conversion has resulted in the loss of more than 60 percent of original forest cover. What remains is a matrix of forest islands, open woodlands, and grasslands. The remaining closed canopy forests—habitat of the Sumba Island hornbill—are broken into approximately thirty-four fragments. We surveyed hornbills once in all forest patches and surveyed six of these patches intensively for one year. Because we were interested in the influence of forest patch size, shape, and distribution relative to food availability, we used ArcView GIS to digitize topographic maps and evaluate the necessary information on forest patches. We also buffered forest patches in ArcView GIS and used information on hornbill dispersal distances to ask just how fragmented the Sumba landscape is from the bird's viewpoint. We found that hornbills occur primarily in forests of more than 1,000 hectares in size and are responding to the fluctuation in the supply of ripe fig fruits in these larger forests. Forests less than 100 hectares in size are used

occasionally by hornbills, but they tend to not have resident populations. Ultimately, the value of small forests may depend on their spatial arrangement relative to large forests. We found that when we buffered the forests with a five-kilometer dispersal distance (2.5 kilometer buffer), what were once thirty-four forest fragments are now reduced to fifteen fragments. We concluded that small forests located within ranging distance of larger forests may not be perceived as isolates to hornbills. This data has profound importance for the conservation and management of the Sumba landscape, especially as Sumba's array of parks and protected areas are just being developed. Originally, small patches were thought to be relatively useless, and with our ArcView GIS analysis, we have been able to demonstrate their value when examined within the overall landscape.

Habitat Classification for Seram Cockatoos

The Seram cockatoo, found only on the island of Seram, is the largest and most striking of the white parrot species present in Indonesia. Due to habitat loss and heavy hunting for the pet trade, the Seram cockatoo has suffered a population decline and was believed by many to be on the verge of extinction. In response to international concerns for the species and the need to develop a rational plan for their long-term conservation, we began field studies to estimate the density of cockatoos in lowland forests and classify vegetation and forest structure where cockatoos are present and absent. We also developed a GIS database for assessing the status of land use relative to cockatoo distribution and abundance. We conducted systematic surveys of cockatoos and vegetation at seven sites across the western half of the island and acquired digital images of Seram from the World Bank, but original sources of data were from the World Conservation Monitoring Center (WCMC), Cambridge; RePPPRoT maps (Ministry of Transmigration); and Inventarisasi Tata Guna Lahan (INTAG), Jakarta. Information collated included elevation, vegetation types, land-use categories, and location and extent of logging concessions. We used ArcView GIS to integrate and overlay the digital data. We derived estimates of Seram cockatoo numbers throughout the island by extrapolating survey estimates to all land area identified (with ArcView Spatial Analyst from DEM) as being under 500 meters (preferred cockatoo habitat). This data suggests that the species is not on the brink of extinction. The most conservative estimate of population size is approximately 57,000 birds, a number we consider viable in

Right: Elephants photographed in BBSNP using passive infrared cameras.
Below: A tiger camera-trapped in BBSNP.

the long term if, and only if, habitat loss and hunting are brought under control. GIS analysis showed that, overall, habitat loss on Seram has not reached a level where it can be perceived as a serious threat to cockatoos. Fourteen percent of the island is set aside in protected areas, a large amount relative to other Indonesian islands. Threats remain, however. GIS analysis also showed that nearly 50 percent of Seram is held within logging concessions and 912 square kilometers of parks and protected areas are claimed by logging concessions. Unsustainable logging practices that destroy the forest canopy will dramatically reduce habitat available for cockatoos, especially if large nest trees are harvested. Management of these concessions will determine the future of a large proportion of the cockatoo population.

Trends in Deforestation in the Bukit Barisan Selatan National Park

Our largest, ongoing project concerns the evaluation of deforestation in and around the Bukit Barisan Selatan National Park (BBSNP) and the impact of deforestation on the large mammalian fauna. BBSNP is 3,500 square kilometers in size and, unlike other Sumatran parks, it still has significant areas of lowland forest and contains important populations of Sumatran tigers, elephants, and rhinoceros as well as Malay tapir. Although large, the park has an unfortunate long, narrow shape resulting in extensive boundaries subject to hunting, logging, and agricultural encroachment. To develop appropriate management strategies for the conservation of BBSNPs mammalian fauna, we initiated a study to determine the population status and distribution of mammals in BBSNP and evaluate the quality of BBSNP as habitat for various mammal species.

We sampled wildlife using automatic cameras attached to passive infrared motion sensors. These cameras were dispersed in 20 square kilometers sampling blocks at 10 kilometers intervals for the length of the park. Blocks were oriented from the edge of the park to the center and within each block, we assigned one camera per square kilometer to random UTM coordinates. Cameras were operated twenty-four hours per day for thirty days. At each camera location, we collected habitat measurements in a 10-meter radius around the camera. We counted saplings; counted and measured all trees greater than 10 centimeters DBH; took measurements of understory density and canopy closure using densitometers; and recorded presence of rattan, palms, lianas, bamboo, and wild ginger (an indication of disturbance) by quadrat. This data was then used in a hierarchical cluster analysis and Discriminate Function Analysis to identify major habitat types.

We acquired Landsat images of the park from 1985, 1989, 1992, 1994, 1997, and 1999. These images were georeferenced and used to analyze patterns of deforestation in the park and in a buffer zone of 10 kilometers around the park. We used the 1994 image extent, totaling 70 percent of the park, as the baseline for all analyses to make consistent interyear comparisons. For each image, we digitized on-screen six major habitats including forest, agriculture, burned areas, grasslands, unknown/nonforested areas, and village enclaves. We began with 1985 and classified habitats for this image. This habitat map then became the basemap for the following year (1989) and forest loss was measured between the two years. This process continued until we had a final habitat map for 1999. The 1999 habitat map was checked for accuracy by overlaying results of the on-ground vegetation classification and asking what percentage of vegetation points were properly classified according to the GIS habitat classification. Forest and nonforest habitats were classified correctly for more than 70 percent of all vegetation points. We also used ArcView Spatial Analyst to examine slope and elevation by habitat type. We overlaid wildlife data on the final habitat map and measured distances between photo sites for various mammal species and park and forest boundaries.

Briefly, our preliminary results show that relatively little land remains forested in the 10-kilometer buffer surrounding BBSNP (39.3 square kilometers or 1.67 percent of the total area) and that BBSNP itself has lost more than 28.3 percent of its original land area (661 square kilometers) at a rate of 1.81–3.1 square kilometers forest loss per year. Forest loss is primarily due to conversion to agriculture and as expected, forest loss is highest in lower elevations (less than 500 meters) of the park and on flatter slopes. From 1989 to 1999, forest loss increased by elevation and slope as the lower, flatter lands became more and more scarce. These low, flat lands are areas of highest biodiversity and therefore represent the most critical losses to the park.

We examined the distribution of tigers, elephants, rhinos, and tapirs relative to park and forest boundaries by first comparing observed and expected distributions of photos (see examples at right). Expected distributions were calculated as proportional to the camera distribution. Next, we calculated the residuals between observed and expected photo captures and looked for natural breaks in the distribution of residuals (from positive to negative) using Jenk's optimization method to calculate the Goodness of Variance Fit (GVF). The GVF identifies the optimal break in data categories when using alternative data classifications on the same set of numeric data. Once we identified the natural break for each species, we tested for avoidance of park and forest boundaries using a two-sample Chi-square (above and below the natural break). Preliminary results show that tigers, elephants, and tapirs tend to avoid park boundaries while rhinos showed no avoidance, although sample sizes for rhinos were too small to draw any conclusions. This avoidance is likely due to the fact that park boundaries seldom coincide with forest boundaries due to illegal conversion of forest to agriculture within park boundaries. Only tigers and elephants showed significant differences between photos close to the forest edge and away, suggesting that these species avoid the forest edge. Edge avoidance by tigers and elephants may be due to a behavioral aversion to humans or hunting pressure.

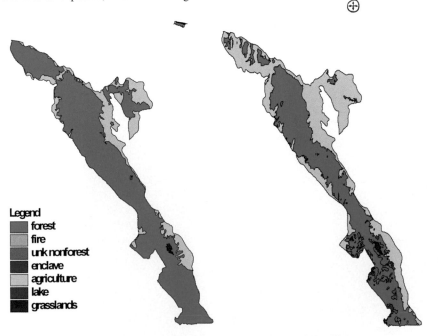

Land-use patterns and forest loss in BBSNP between 1985 (left) and 1999 (right).

Legend
- forest
- fire
- unk nonforest
- enclave
- agriculture
- lake
- grasslands

Indonesia Conservation GIS

Robert J. Lee

Background

The objectives of the Wildlife Conservation Society–Indonesia Program in Sulawesi (WCS-IP, Sulawesi) is to obtain clear and complete ecological information on protected areas in Sulawesi, and provide technical assistance in the form of training and management solutions to local stakeholders including the Indonesian government, local NGOs, universities, and communities. Our activities have focused on the following activities: 1) island-wide biodiversity

In all of our work, the use of ESRI's ArcView GIS and ArcView Spatial Analyst has been essential for understanding the ecology of wildlife as well as formulating proper management solutions to problems related to unsustainable use of natural resources and protected areas. ESRI software has facilitated the presentation of complex and sometimes difficult data to local stakeholders—scientific and nonscientific—in a clear and cogent manner. For example, the relationship between geographic features, human activities, and wildlife distribution has always been a difficult set of data to present before the use of GIS. With GIS, people who have not been formally educated can immediately see these relationships.

Additionally, one of our major foci is increasing technical capacity of national conservation professionals and students. We are training Indonesian conservation NGO members and university students in GIS with ArcView GIS.

North Sulawesi

WCS has completed ecological surveys of protected areas of North Sulawesi collecting data on wildlife distribution and composition on flora, many of whom are globally rare and threatened, and fauna and habitat types. All field data are recorded on GPS units. We have also obtained images of Landsat 7 for the province of North Sulawesi. We have digitized the following data: all major geographical features (e.g., rivers and lakes, topography, administrative boundaries such as districts and regencies, roads, villages). At present we are ground-truthing habitat type data. Based on this, we plan to produce a complete habitat map of the province.

This work will serve several functions:

- Reclassification of land use data for the government, particularly with respect to protected areas and forests from which we plan to produce an atlas.

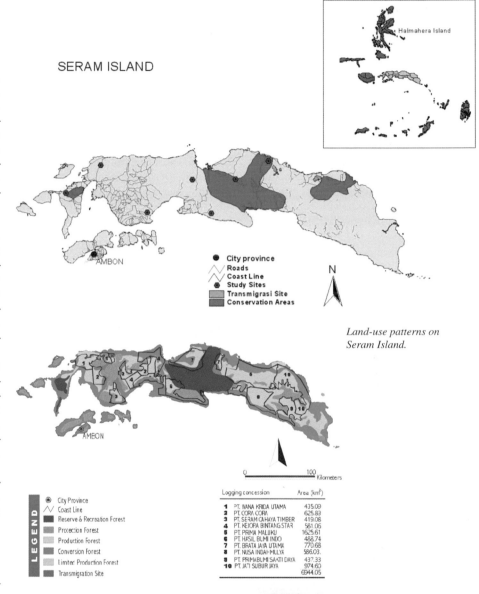

SERAM ISLAND

- City province
- Roads
- Coast Line
- Study Sites
- Transmigrasi Site
- Conservation Areas

Land-use patterns on Seram Island.

	Logging concession	Area (km²)
1	PT. WANA KRIDA UTAMA	435.09
2	PT. CORA CORA	625.83
3	PT. SERAM CAHAYA TIMBER	419.08
4	PT. KEIORA BINTANG STAR	581.06
5	PT. PRIMA MALUKU	1625.61
6	PT. HASIL BUMI INDO	488.74
7	PT. BRATA JAYA UTAMA	770.68
8	PT. NUSA INDAH MULYA	586.03.
9	PT. PRIMABUMI SAKTI DAYA	437.33
10	PT. JATI SUBUR JAYA	974.60
		6944.05

LEGEND
- City Province
- Coast Line
- Reserve & Recreation Forest
- Protection Forest
- Production Forest
- Conversion Forest
- Limtec Production Forest
- Transmigration Site

- Analysis of areas used by wildlife within forests to understand relationship between wildlife and humans (e.g., distance between settlements and forests, hunting and logging activities). We are analyzing and writing up scientific articles (e.g., bat ecology of North Sulawesi, distribution and population status of large mammals in Bogani Nani Wartabone National Park, ecology and status of Sulawesi macaques in North Sulawesi).

Southeast Sulawesi

WCS has completed ecological surveys of four of the eight protected areas in Southeast Sulawesi. At present, we are digitizing all major geographical features and plotting in the ecological data we have collected.

Vegetation mapping by WCS collaborators in Sulawesi, Indonesia.

Museo de Historia Natural Noel Kempff Mercado, Santa Cruz, Bolivia

Dr. Damien I. Rumiz   confauna@scbbs-bo.com

A grant from ESRI was awarded to Damian I. Rumiz, thanks to the support and mentorship of Eric Sanderson of the Wildlife Conservation Society. Rumiz justified this application from ConFauna (a conservation program funded by Wildlife Conservation Society at the Museo de Historia Natural Noel Kempff Mercado, Santa Cruz, Bolivia) on the need to facilitate and encourage the use of GIS tools in applied research and wildlife conservation in Bolivia. The immediate objectives of the project were to generate maps and to conduct spatial analyses of habitat and wildlife species distributions, particularly in the case of a study of the maned wolf (*Chrysocyon brachyurus*) in the Huanchaca plateau in northern Santa Cruz. Lila Sainz and Rumiz were co-principal investigators in this study, funded by the Noel Kempff Mercado National Park and TNC, who assessed the distribution and habitat use of the maned

wolf in that park. The study was expanded with extra funding from WCS and was benefited with the grant from ESRI and support from the Museo NKM.

The use of this GIS tool showed us a variety of new potential applications within our goals of research for conservation. This brief report describes the main training and research achievements accomplished with the use of ArcView GIS in ConFauna during 2000.

Methods and Results

We were able to install and start using the ESRI software at the end of 1999.

During six field trips spanning different seasons, we assessed the occurrence of maned wolf signs such as tracks, scats, food remains, beds, and scent (see photo next page) in

four sites of Noel Kempff Mercado National Park, Santa Cruz, Bolivia. We recorded sign locations with a GPS and defined the habitat type in which they occurred according to the twenty-two vegetation units recognized in a digital map of the park. We described the signs and took a series of measurements from five hundred wolf footprints to attempt a statistical separation of the individuals that produced them. With the help of Zulma Villegas and Geovana Carreno, GIS specialists of the Geography Department of the Museo NKM, Lila learned to georeference and correct the available digital maps and images for the area. She first adjusted the forest–savanna borders (according to GPS locations she recorded in the field) in the Landsat images using ERDAS and later in the vegetation map using ArcView GIS.

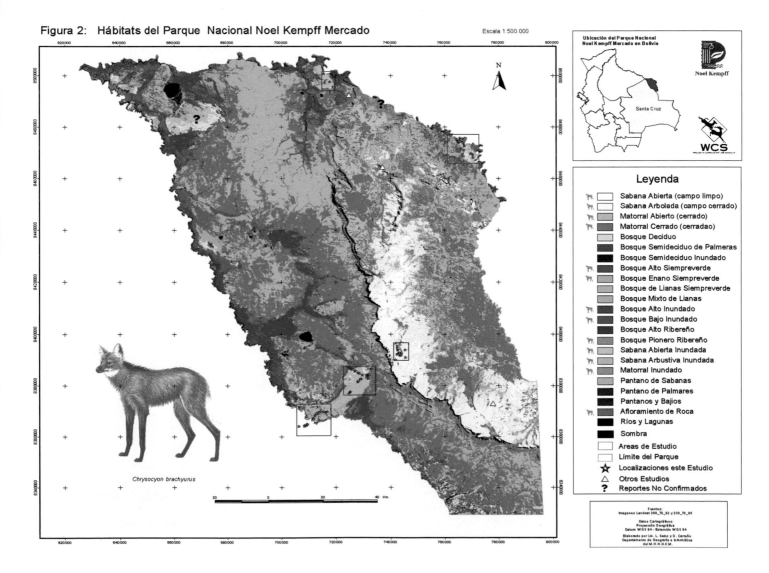

Figura 2:   Hábitats del Parque Nacional Noel Kempff Mercado   Escala 1:500.000

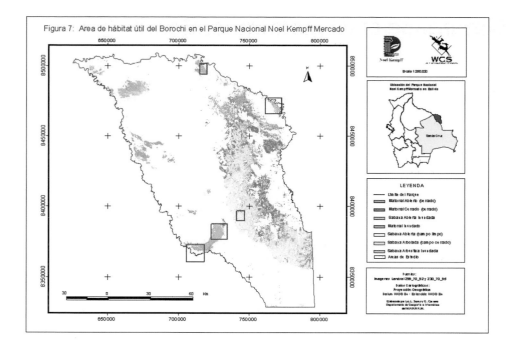

Figura 7: Area de hábitat útil del Borochi en el Parque Nacional Noel Kempff Mercado

We recorded thirteen habitats as used by the wolves in our study sites and estimated their total area. In another map, we showed only the potential habitats for the wolves (see map above). Based on these values and other parameters, we identified nonvisited sites that are potentially inhabited by the wolves. We also estimated the potential population size for this threatened species in the whole park.

Wolf signs occurred in thirteen habitat types, eight of savannas and shrublands and five of forests adjacent to open habitats. Open habitats, useful for this species, cover 360,000 hectares or one quarter of the park. Wolf signs

were easy to identify in the field but were of limited use to estimate population abundance or frequency of habitat use. Clustering of tracks showed some potential to identify individuals but needs to be improved. We recommend recording animal signs and assessing their distribution with the help of digital maps to monitor populations of this and other threatened species of the park.

Future Plans

With the use of ArcView GIS we improved the quality of our research and expanded the possibilities for data analysis. One of us became moderately proficient in the use of the software

and another started to use it. Other students are planning to include similar analyses in their theses. We will apply a similar approach to study the distribution of another threatened species, the pampas deer (*Ozotoceros bezoarticus*) in the park. We are learning about other applications and tools offered by ESRI and plan to apply for new support in the near future.

Mamirauá Project–Sociedade Civil Mamirauá, Brazil

Jose Marcio Ayres, Dierctor Instituto, de Sesenvolviimento Sustentável Mamiraua

The Mamirauá Sustainable Development Reserve (MSDR) is the largest (1,124,000 ha.) Brazilian protected area conserving flooded forests of any kind, and the only functional area (as opposed to "paper parks") conserving the Várzea flooded forests present along many Amazon white water rivers. Due to its importance, the area has been listed under the Ramsar Convention as an Internationally Important Wetland and is among a number of Amazon Conservation Areas under consideration for nomination as Biosphere Reserves. This area is unique not only because of its high level of biodiversity but also because it is the first Brazilian conservation area designed to integrate the preservation of fragile habitats with the sustainable development of local resident communities.

The need to integrate natural resource conservation and social development goals has been widely discussed in environmental and sustainable development circles over the past few years but, at Mamirauá, this approach has been under development in the field since 1992. Through a participatory system, approximately sixty small várzea communities have participated in research, monitoring, extension, and protection efforts. During the first four years (Mamirauá Project Phase I, 1992–1996) research has been the number one priority, with the aim of generating sufficient scientific knowledge for effective management. At the same time, monitoring, extension, and protection programs were also initiated. Logistical and financial constraints limited the work to a "Focal Area" 260,000 hectares wide. Plans to expand the work to the remainder of the Reserve ("Subsidiary Area") depend on securing funds (figure 1).

The Mamirauá Management Plan for the Focal Area produced by the end of Phase I represented the effort of some eighty researchers and extension workers from various Brazilian and foreign institutions. Development of the Mamirauá Management Plan was a pioneering effort and required considerably more input than more typical plans. To carry out this task, it was necessary to mobilize interinstitutional and multidisciplinary resources to study the principal resources and economic activities in Mamirauá and at the same time develop participatory and social extension programs to facilitate the integration of local communities in the planning and management process. The management of natural resources in areas of tropical rain forest, such as the Brazilian Amazon, is considerably more complex than resource management in developed countries, where much more is known about the ecology of the resources involved, as well as the social and environmental context.

The exploitation of natural resources, such as fish, timber, game, and other forest products, is of great importance for the economy and subsistence of the Amazonian rural population. In general, the principal markets for these resources are local and regional urban centers. Goods are exchanged through an extensive network of intermediaries who control the local and regional political power. Altering relations in the system of resource access, production, and exchange provokes both direct and indirect impacts on the economy and the subsistence of a population that is much larger than that which lives within the borders of the conservation area. It also leads to involvement in policy development for natural resource management and conservation.

The human population is a fundamental part of the proposed sustainable development/ biodiversity conservation model proposed in the Management Plan. Most Amazon protected areas, as in many other parts of Brazil and the underdeveloped world, are not achieving the effective levels of protection, often due to the neglect of the social aspects in the analysis of the situation. Public institutions have not been able to bear the full costs of compensation for removal of the inhabitants from protected areas, nor do they manage effectively the areas within those conservation areas. The large size, and consequently the large perimeter of these areas, together with a regional economy that is heavily based on extraction of forest resources and unaccustomed to legal restraints, are additional factors constraining protected area management in the Amazon.

The binomial of poverty and resource degradation in the region appears to exercise its effect in a quite clear manner. Underdevelopment leads to greater pressures on the natural resources, especially when markets and inadequate policies foster resource exploitation rather than sustainable management. The Management Plan clearly recognizes the presence and the way of life of local populations and other regional stakeholders, such as commercial fishermen from nearby towns, and identifies them as partners in the management of the MSDR. This is fundamental for the management of the MSDR because the continued presence of local communities inhabiting the area reduces the expense of expropriations and relocation and makes possible the development of a surveillance system that is cheaper and certainly more effective than contracting outside personnel as guards. At the same time, it makes it possible to forestall the collapse of the local economy by encouraging the sustained use of natural resources.

The continued presence of the local population requires an adequate zoning system together with management rules to reduce conflicts between productive activities and biodiversity conservation, thereby increasing sustainability. The Mamirauá model proposes effective, low-cost environmental protection and reduction of the demand for natural resources by

Figure 1. The MSDR is located between the Amazon and the Japurá river floodplain in the center of the Amazonas State, Brazil. The Focal Area, the center of activities carried out by the Mamirauá Project has approximately 260,000 hectares. The Subsidiary Area, which contains four fifths of the total area, is practically unknown.

means of the development of new economic alternatives (Fisheries Post-Harvest Project, Community Forestry Management, introduction of new crops adapted to várzea environmental conditions, etc). Although the model has been developed specifically for Mamirauá, we believe that the approach could be applied successfully in other Amazonian areas as well.

Phase II of the Mamirauá Project started in September 1997 with the aim of implementing the Management Plan for the Focal Area of the MSDR. Monitoring systems were put in place to follow the expected changes in socioeconomic, key economic, and biological indicators following the establishment of the Zoning System along with management regulations and economic alternatives interventions carried out to compensate for income losses resulting from restricted use of natural resources. For example, in the case of fisheries, a program was initiated to organize the fishermen from an area of the reserve (Jarauá Sector, figure 2). Through this program fishermen received training to improve their fish processing techniques (salting, drying, icing, etc.), training in management, and support for marketing their products—all requirements to introduce a new fisheries postharvest system. Investments were also made to purchase a boat to transport fish and ice and to build a fish processing and storage floating base. The idea was to improve the quality of fisheries products, to remove the middleman, and to sell the products in more valuable markets. This system was devised to compensate fishermen for their losses due to restrictions imposed by the Management Plan rules, which limited their access to fishery resources. A decrease in production (and income) is balanced by an increase in marketed prices and by the diversification of the exploited fish species. All fish marketed through the Fisheries Post-Harvest Program are recorded to provide biological and fishing effort information.

Stocks of the two major fish species (the pirarucu, *Arapaima gigas* and the tambaqui. *Colossoma macropomum*) are also monitored through mark and recapture experiments and other methods (visual census in the case of pirarucu, an air breathing species that comes regularly to the surface to breathe) to follow the expected changes in abundance. It is paramount that all information produced by the monitoring systems is readily available to back management decisions, and GIS is an extremely useful tool for achieving this.

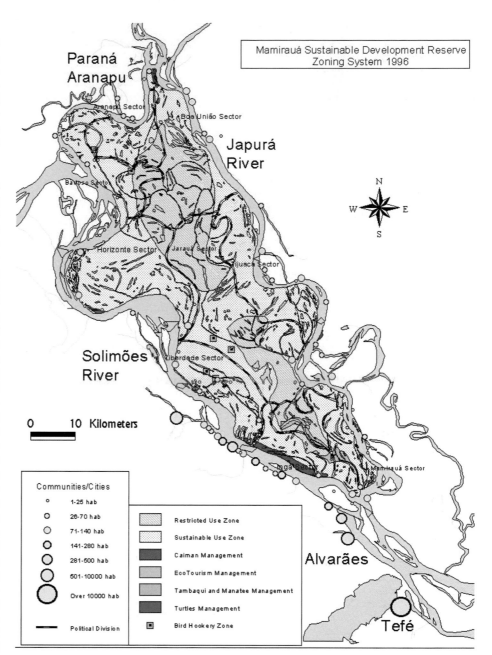

Figure 2. Zoning System of the MSDR as of 1996. Zoning was established considering inputs from residents (people living within the reserve area) and users (people living in the river banks outside the reserve who traditionally used the area), and inputs from researchers working with key economic and endangered species such as manatees, caimans, turtles, and so forth. The Focal Area of the Reserve is divided into nine political units (or sectors) grouping more than sixty communities and over six thousand residents and users.

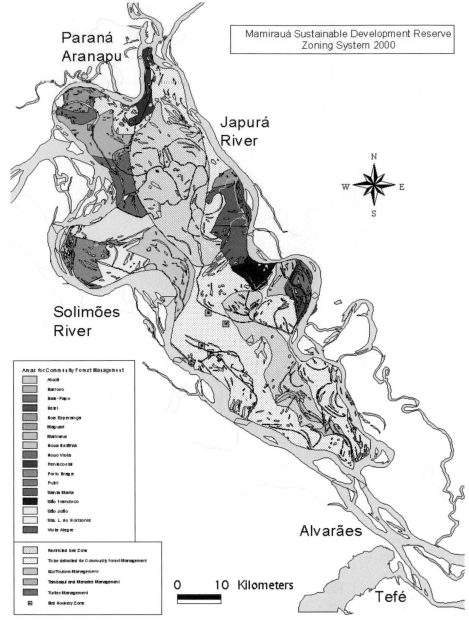

Figure 3. Zoning System of the MSDR as of 2000. As the work for the implementation of the Management Plan progresses, new economic alternatives are being introduced in the Mamirauá Reserve. In the case of forestry, for example, areas in the Sustainable Use Zone were identified and delimited for the management of wood species. So far seventeen communities are participating, but this initiative will be extended to other areas in response to demands by other communities. People from these communities are being trained in stock assessment techniques and receiving support for the elaboration of management plans and also for the establishment of associations through which the wood will be marketed. This is a requirement of the state and federal environmental authorities to grant permits for community forest management.

GIS Work in the Mamirauá Project

GIS work started in Phase I using a MS-DOS®-based program. Although useful at that time, this software did not provide a user-friendly interface and was limited to mapping applications. The need to spread the use of GIS among Mamirauá Project staff, coupled with the requirement to continuously update the information available for the management of the reserve, required the migration of all information to a modern GIS software. The grant provided by ECP to the Mamirauá Project-WCS allowed us to start the process of implementing an effective GIS for the management of the MSDR.

During the first year of the grant, the work focused on the migration of more than one hundred fifty layers of information into ArcView GIS. Unfortunately, the application used for the conversion did not handle "label" archives, and this required tremendous work to edit the imported layers and several maps. This process is largely completed, and we are already working toward updating the layers to account for the changes in the Zoning System and location of human settlements (the updated version of the MSDR Zoning System is shown in figure 3).

Future GIS Work

The basemap currently in use was digitized using a 1:250,000 map produced by the RAD-AMBRASIL Project that surveyed the area in the early 1970s using aerial radar sensors. Due to the high dynamics of floodplains (the MSDR is located between two large floodplain rivers, the Amazon and the Japurá—figure 1), there have been significant changes in the channels over the past thirty years. In addition, some important features for the management of the MSDR, such as internal floodplain lakes and the complex network of channels that connect these bodies of water, need to be better represented. It is also a priority to produce a vegetation map for the area that will be used to estimate forestry species stocks and to monitor land clearings by subsistence agriculture. Because this information is not readily available, the Mamirauá Project is investing on the application of satellite imagery to extract this information and to make it available for management purposes. This, together with the integration of the monitoring databases, will provide a fundamental tool for the management of the MSDR.

Mesoamerican and Caribbean Program, Guatemala

Robin Bjork, Research Fellow

Introduction

I am a research fellow with the Wildlife Conservation Society currently conducting my dissertation research (Department of Fisheries and Wildlife, Oregon State University) on design and management of protected area networks in lowland tropical forests of Central America. My investigation is regional in scale with the goal of documenting the annual habitat and spatial requirements of a highly mobile parrot species dependent on a threatened mature forest habitat and using that information to evaluate the distribution and management of a regional system of protected areas. As part of that study, I am also comparing the habitat and spatial use characteristics of parrots living in intact forest with those in a fragmented, degraded forest landscape. The nature of my investigation requires data analysis across multiple geographic layers including habitat maps, bird locations and home ranges, and land management boundaries.

Prior to learning ArcView GIS, I was competent with another GIS, CAMRIS[1]. However, because ArcView GIS has certain features that provide greater GIS capability (specifically in working with satellite images) than CAMRIS and because of its extensive use around the world, I felt it was a critical skill for me to develop to aid in my line of research and to expand my professional abilities. Because of ArcView GIS software's wide-

spread use, I have been able to obtain numerous ArcView GIS files from other sources, including the Guatemalan government, to use in my analyses. I have used ArcView GIS graphics in various reports to granting organizations, Guatemalan land management agencies, and in presentations at international meetings (please see Documents and Presentations at end of this report). I have just completed a preliminary assessment of the area requirements of the Mealy parrot population for the World Wildlife Fund that will be used in the development of its conservation landscape initiatives in the neotropics.

My research is in progress; analyses are preliminary and many have not been completed. Below, I provide a brief description of the project and some of the ArcView GIS analyses that I have been able to execute as a result of my training.

Summary of Research

Title: *Reserve Network Design and Management in Moist Lowland Tropical Forests: Habitat Site, and Spatial Requirements of the Mealy Parrot as a Guide*

As native tropical forest habitats rapidly shrink, we want to know that the areas protected in reserves and biological corridors are doing their job—namely, being large enough and configured in such a way as to maintain regional biological diversity and ecosystem functions. My research addresses questions of reserve network design in a case study of a lowland frugivorous bird species, the Mealy parrot (*Amazona farinosa*), a species that I predicted would exhibit characteristics valuable in conservation planning of regional reserve networks and corridors. The study site spans large intact and highly altered tropical lowland forests in northern Guatemala including the World Heritage site, Tikal National Park, and

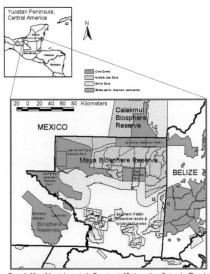

Figure 1. Map of the study area in the Department of Petén, northern Guatemala. The enlarged box displays the distribution of land management categories in the region. The red box shows the area across which the radio-collared birds' nesting territories were distributed.

the Maya Biosphere Reserve (MBR), which forms part of the largest contiguous area of lowland forest remaining in Mesoamerica (figure 1, above). The Mealy parrot belongs to a highly endangered family of birds, many of whose members are suspected of engaging in regional movements although little empirical data exists on this aspect of parrot life histories.

Primary research objectives of the project

1 Document annual movement patterns, home ranges, habitat affinities, and diet of Mealy parrots throughout the annual cycle.

2 Document spatial and temporal patterns of abundance of Mealy parrots' food resources as an explanatory variable for their habitat and spatial use patterns.

3 Evaluate the distribution and management of protected areas with respect to annual spatial, habitat, and resource requirements of the Mealy parrot population.

4 Provide recommendations for design and management of northern Guatemala's reserve/corridor network and for protection of the Mealy parrot population.

In two and a half years of research using radio-telemetry, I have documented the Mealy parrots' dependence on mature moist lowland forest and shown that a large portion of the adult population breeding in northeastern Guatemala engages in predictable regional migrations during the nonbreeding season, covering up to a few hundred kilometers within forested tropical lowlands. After the breeding season

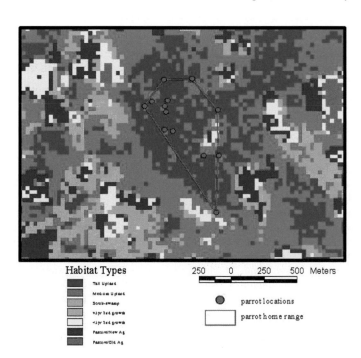

Habitat Types

Tall Upland
Medium Upland
Scrub-swamp
+3yr 3ed growth
+3yr 3ed growth
Pasture/New Ag
Pasture/Old Ag

250 0 250 500 Meters

● parrot locations

☐ parrot home range

Figure 2. Example of Mealy Parrot breeding home range as calculated from its locations during the breeding season. The underlying map shows the habitat types that occur within its home range.

in June, birds traveled to sites northeast up to 84 kilometers from their breeding territories where they remained for three to five weeks. In late August, they returned to breeding territories for a few weeks and then traveled southwest up to 190 kilometers where they remained until November or December, at which point they again returned to their breeding territories. Population surveys of the parrot community reveal that some other species exhibit seasonal abundance patterns similar to that of the Mealy parrot, suggesting that parrot migration may not be uncommon.

I have begun to utilize the results of my Mealy parrot research in conservation planning for this globally important and highly threatened region by providing a stronger scientific basis for biodiversity impact assessments for regional biological corridor strategies and forest management plans. I am providing conservation recommendations for the reserve system and the Mealy parrot population to appropriate government agencies and nongovernmental organizations. These entities include CONAP (Guatemalan Council for National Protected Areas), PROSELVA (German NGO responsible for conservation initiatives in southwestern Petén), Arbol Verde Forestry Concession (community forestry concession containing critical seasonal Mealy parrot habitat in the MBR), and The Nature Conservancy (U.S. NGO coordinating review of the MBR Master Plan).

My results point to the need to expand the scope of conservation strategies for effectively conserving neotropical ecosystems. This research is among the first studies to provide baseline data on spatial needs of wild parrots and to elucidate the necessity of regional habitat heterogeneity and connectivity for an animal population in lowland tropical forest.

GIS Analyses

ArcView GIS has been used to (1) estimate the size of individual Mealy parrot's breeding season home ranges, (2) gauge the area of forest within those home ranges, (3) display Mealy parrot nonbreeding season home ranges in relation to protected areas, and (4) calculate the population's annual area and site requirements as follows:

1 Geographic locations (latitude/longitude) for each radio-collared parrot were imported into ArcView GIS tables from Excel files that had been saved as .txt files. They were then saved as event themes that were then projected into UTM format (for use with other UTM files). With that file of all locations for an individual, breeding season locations were selected based on date (i.e., locations from 15 December through 14 June) by editing the attribute table and making a new shapefile. Home ranges were

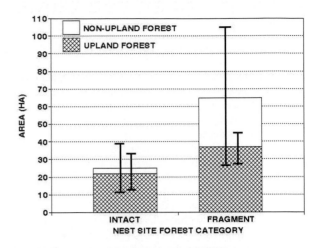

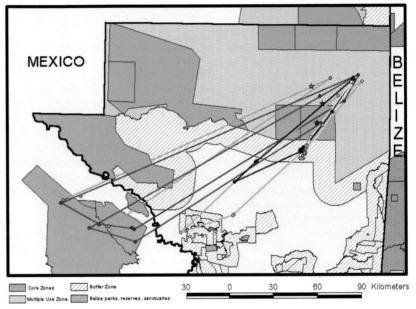

Figure 4. Nonbreeding season home ranges of six Mealy Parrots during the nonbreeding season, July–December 2000.

calculated using the Animal Movement Analysis extension and selecting "minimum convex polygon" around those locations for each individual. Areas of breeding ranges were calculated using the ReturnArea script.

2 Because a number of habitat types not used by Mealy parrots (e.g., scrub-swamp forest and agricultural land) also occur within their home ranges, especially for those birds nesting in the fragmented zone, I also calculated the area of "Mealy parrot habitat," or upland forest in each polygon. I used a classified Thematic Mapper image of the region containing 14 habitat classes[2] and converted it to a grid. The minimum convex polygon home ranges (number 1 above) were converted into grids as well. I used "Analysis–Summarize Zones" to calculate the area of each habitat type within each breeding home range. From the 14 habitat classes, I selected classes that I considered to be habitat of Mealy parrots based on habitat types used by the radio-collared birds and

summed the areas to derive the area of Mealy parrot habitat. Figure 2 (previous page) provides an example of the layers used in ArcView GIS for this analysis including Mealy parrot locations, the minimum convex polygon that was calculated from those locations, and the habitat map used for analyzing the area of each habitat type within a home range. Figure 3 (this page, top) displays a summary of those results for all the parrots in the analysis: Mealy parrots nesting in forest fragments in the buffer zone (n=5) had breeding season home ranges more than twice the size of those nesting in intact forest (n=3). The difference between the two categories of intactness was reduced when the area of upland forest (Mealy parrot habitat) was compared. This data suggests that forest in the fragments is of lower quality than in intact forest relative to the birds' needs.

3 Similar to the steps carried out for breeding season home ranges (see GIS analysis section 1, above), nonbreeding season locations

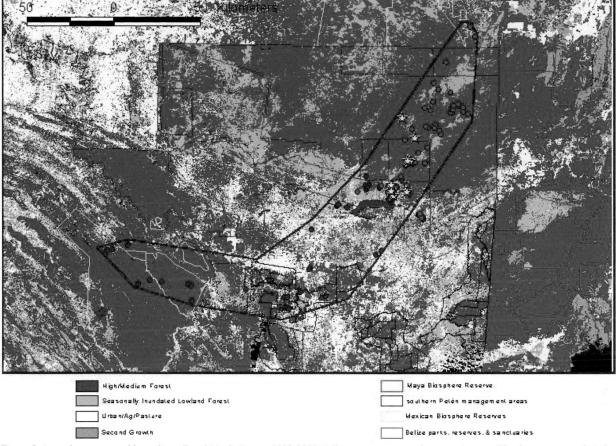

Figure 5. Annual area covered by radio-collared Mealy Parrot, 1998-2000. Yellow stars represent the parrot's nests. The dots represent their locations during the nonbreeding season (pink=June–August, red = September–November) covering an area of approximately ten thousand square kilometers (blue outline). Vegetation map of the Maya Forest (Cons. Intl. 2000) and distribution of protected and management area show the striking contrast in forest cover between lands inside and outside protected areas. The forest of the Maya Biosphere Reserve's buffer zone (southern pink band) and the southern Petén management areas are highly altered and continue to disappear rapidly.

of each Mealy parrot were selected from the file containing all locations, and the associated minimum convex polygon was calculated. These results were overlaid with protected area boundaries to illustrate the connectivity of the various protected areas for the Mealy parrots and the extent of overlap of nonbreeding season home ranges (figure 4).

4 Finally, I compiled several data layers including a habitat basemap for the trinational region of Guatemala, Belize, and Mexico[3]; protected area boundaries; and a polygon that incorporates all Mealy parrot locations (figure 5, above). Total area encompassing the parrots' locations equals approximately 10,000 square kilometers, a value that I cautiously provide as an estimate of the minimum area required to support a population of possibly 900 pairs of Mealy parrots nesting in northeastern Guatemala (the estimated size of the population represented by my radio-collared parrots). Figure 5 illustrates that although the overall area covered by the parrots is considerable, they concentrated in a few regions for most of the nonbreeding season. From mid-June through August, parrots concentrated in the multiple-use zone of the Maya Biosphere Reserve. From October through November, they concentrated in the southwestern Petén management zone and the Lacantún and Montes Azules Biosphere Reserves in southeastern Mexico.

Documents and Presentations Using ArcView GIS Analyses and Graphics

Bjork, R., J. López, O. Aguirre, M. Córdova, and J. Madrid. 2000. "Migraciones Locales del Loro Real (*Amazona farinosa*) en El Petén, Guatemala: Consecuencias Para la Planificación de la Conservación Regional." Proceedings of the conference Nuevas Perspectivas de Desarrollo Sostenible en Petén, December 1999. Flores, Petén, Guatemala.

Bjork, R. 2000. *Intratropical Migration of a Lowland Parrot: Implications for Conservation.* Society for Conservation Biology, 9–12 June. Missoula, Montana.

Bjork, R. 1999. "Habitat Use Patterns of Mealy Parrots (*Amazona farinosa*) in Guatemala: A Landscape Perservice." Neotropical Ornithological Congress, 4–10 October. Monterrey, Mexico.

Bjork, R., J. Lopez, and J. Madrid. 1999. "Loros en El Petén: Una Perspectiva del Paisaje." Presented in Spanish. Mesoamerican Society of Biology and Conservation, June. Guatemala City.

References

1 Ford, R. G. 1989. "CAMRIS, Computer Aided Mapping and Resource Inventory System." Unpublished document, 124 pp.

2 Sader, S. A., T. Sever, C. Soza, and N. B. Schwartz. 2000. "Time-Series Forest Change, Land Cover/Land Use Conversion, and Socioeconomic Driving Forces in the Petén District, Guatemala." Progress report.

3 Bjork, R. 2000. "Vegetation of the Maya Forest." Digital map. Collaborators: USMAB, NASA, Center for Conservation Biology, Clark University, Comisión Centroamericana de Ambiente y Desarrollo, CONAP, ECOSUR, Land Information Centre, MIAI, Programme for Belize, and Proselva. Conservation International, Washington, D.C.

China Program

By Yongpei Wu, WCS–China Program/East China
Normal University, Shanghai

A copy of ArcView GIS 3.1 from an ESRI grant has been used in the evaluation of the habitat quality of the Chinese water deer (*Hydropotes inermis*) in Yancheng Reserve, Jiangsu Province, China, since November 1999, in the hope of getting overall suitability of the habitat quality of the water deer in the Reserve by integrating the suitability of various ecological factors (i.e., plants, land-use types, water, roads, and other human activities). A basemap of the research site is constructed using the software and is enclosed. Use of the software in the project goes smoothly.

However, because the GIS program is in its early stages and due to limited manpower and other resources, we are unable to complete all GIS activities on our own (e.g., remote sensing, digitalization, and plotting). As a result, we have made an extra effort to seek help from other resources. Therefore, only the analysis function of the software is most important to us to date. Funding for the GIS program has not expanded this year. In the future we expect the whole GIS work to be simpler and more convenient for field ecologists. We also hope to have the information on what kind of baseline data is available and how can we obtain it. We would like to see the following improvements in the coming version of the software:

- The software will become more user-friendly if it can integrate more functional modules that we can develop by ourselves for different objectives.

- Unnecessary data transition can be avoided if we know what data format ArcView GIS can access before we collect data in the field. This also helps avoid unexpected data translation errors because of small differences in coordinates.

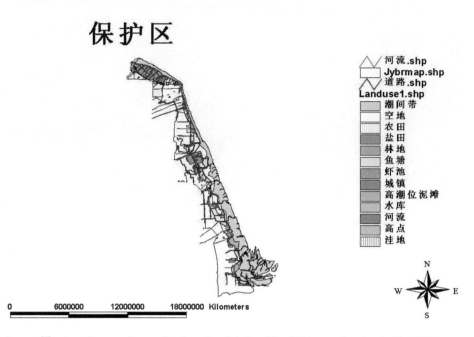

A map of Yancheng Reserve, Jiangsu Province, People's Republic of China, made using ArcView GIS.

Myanmar Program

GIS Progress

The ArcView GIS copy obtained by the Asia Program was used for the following activities in 2000:

The first activity included learning the application of ArcView GIS through attending an intensive two-week workshop on GIS and its applications. During this workshop, ArcView GIS was used in a project titled "A Preliminary Gap Analysis of Myanmar's Protected Areas."

The following question was addressed: What is the degree of representation (percent area) of forest cover and ecoregions within the existing protected area network in Myanmar?

Burma's protected area coverage amounts to almost 1.2 percent of the total land area. The long-term goal is to expand the coverage to about 5 percent of the total land area. In the absence of data on biodiversity either inside or outside existing protected areas, it becomes necessary to prioritize expansion using criteria such as forest cover, land use, ecoregions, and threats.

Methods

Three data layers were used in the analysis. They included:

1 Forest cover data (based on 1:1,000,000 Landsat MSS and RBV Images [1979–1981])

2 Protected area coverage (digitized from 1:1,000,000 topographical map [1943])

3 WWF Ecoregion data

All three of the above data layers were first converted to shapefiles (Azimuthal Equal Area Projection). Forest cover and terrestrial protected areas were overlaid using the UNION function to determine forest cover within protected areas. Ecoregion and protected area data was overlaid using the INTERSECT function to generate percent of ecoregion covered within the protected area network.

Results

The preliminary analyses determined that protected areas represented approximately 3.29 percent of Myanmar's forest cover. Large areas of forest north of 21 degrees latitude could be incorporated into protected areas.

Less than 1 percent of the total area of 12 ecoregions was found to be included in the existing protected area network.

The next step of the project is to conduct a landscape-level GIS analysis using the best available data to help identify potential expansion areas as target sites for wildlife surveys. This would constitute a significant step in the expansion of the protected area network to represent 5 percent of the total land area in Burma.

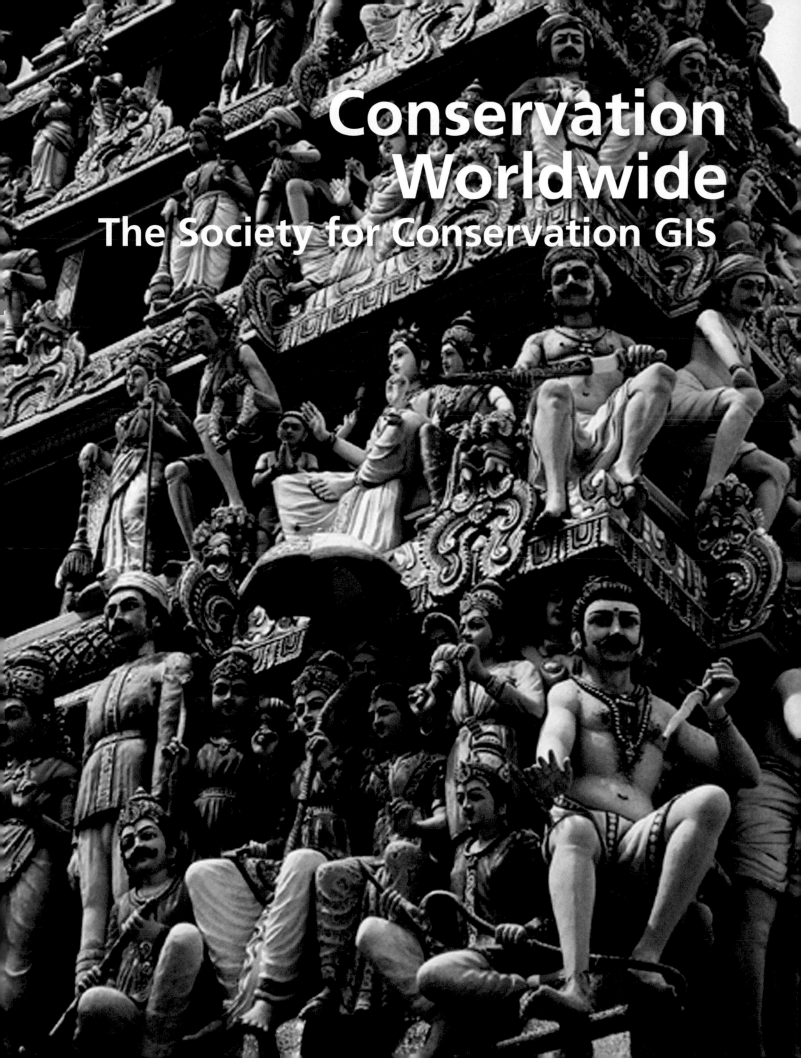

Conservation Worldwide
The Society for Conservation GIS

Introduction

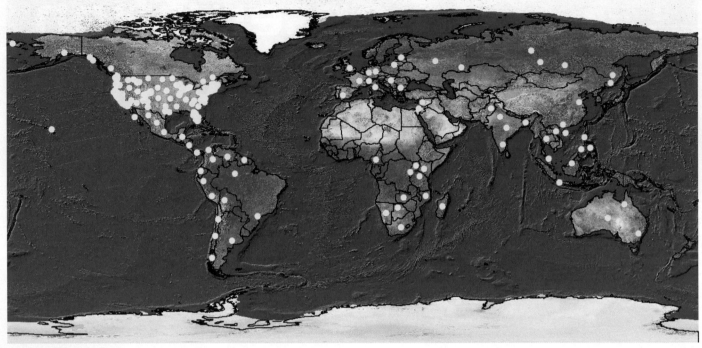

World map of SCGIS members, projects, and partners.

The final section of *Conservation Geography* is devoted to the work of the all-volunteer Society for Conservation GIS. Founded in 1995, it is the primary source of support, help, guidance, and mentorship for hundreds of struggling nonprofit and nongovernmental organizations around the world. The map (above) shows the location of the current SCGIS members and projects, active in dozens of countries, with new chapters forming every year. Also featured here is a message from the SCGIS persident and an account of the scholarship grants program. Featured too is a list of the recipients in 2001, when the program doubled in size thanks to help from two new donors, the Infodev Program and the Ecology Center, Montana.

Message from the SCGIS President, 2001
Eric Treml, National Oceanic and Atmospheric Administration   eric.treml@noaa.gov

Over the past three years, I have had the opportunity to experience the Society from several different levels: general member, member of the board of directors, and now as president of SCGIS. I am constantly and consistently reminded of the dedication to conservation, the enthusiasm in helping our global community, and the high quality conservation GIS work characteristic of our members. This book is a product of such efforts and I'd like to thank all of the contributors for making this,

and future editions, a reality. Briefly, I'd like to share with you my thoughts on where SCGIS is headed this year. I believe this is a pivotal year for the Society, one which will be both challenging and rewarding. As the Society continues to grow, we need to remain focused on our mission and dedicated to accomplishing our goals as stated in the SCGIS Strategic Plan. Within the next several months, some of our highest priorities include the annual conference, development of the executive director position, completing a more robust long-term strategic plan, building a solid Society infrastructure, securing additional funding, and improving communication between the board, staff, and general membership. Undoubtedly, there will be plenty of opportunities for more members to join in and help shape SCGIS throughout this next year.

The SCGIS 2001 International Program
Roberta Pickert, SCGIS International Committee   RPickert@archbold-station.org

Through our International Scholarship Program we attract professionals involved in conservation around the world using GIS and digital mapping technology. The SCGIS training courses we offer are scheduled around the SCGIS and ESRI annual conferences. These activities have been carried out successfully over the past three years, training nearly a hundred students to date. Specialized training

is conducted in collaboration with the ESRI Conservation Program, University of California James San Jacinto Mountains Biological Reserve, and University of Redlands, California. These organizations have supported us by providing the infrastructure including computer lab space and equipment, ESRI Users Conference registration fee waivers, by hosting the training programs, granting training fee waivers, and providing software and hardware to organizations that display exceptional capacity and have financial need. This year, thanks to new funding sources, we have expanded our scholarship program to include nearly thirty grantees from seventeen countries.

The following list includes the scholars accepted into our 2001 program:

Students sponsored directly by the SCGIS:
Lizz Wandag
Palawan NGO Network, Inc.
Zanzibar Building
Rizal Avenue, Puerto Princesa City
5300, Palawan

Peter Potapov
e-mail: p_potapov@rambler.ru
Greenpeace Russia
Novaya Bashilovka Str., 6
Moscow 101428, GSP-4
Russia

Olga Hernandez
e-mail: olhernandez@wwf.org.co
WWF–Colombia
Carrera 35 #4A-25
San Fernando, Cali
Valle del Cauca
Colombia

Lawrence Luhanga
e-mail: nature50@hotmail.com
Malawi Ornithological Society
Malawi

Alexander Yumakaev
e-mail: yumakaev@ab.ru
Fund for 21st Century Altai
P.O. Box 845
Barnaul, 656015
Russia

Yolanda Wiersma
e-mail: ywiersma@uoguelph.ca
University of Guelph
Department of Zoology
50 Stone Rd. East
Guelph, ON N1G 2W1
Canada

Luis Fernando Gomez-Navia
e-mail: lfgomez@wwf.org.co
WWF–Colombia Program Office
Cr 35 # 4A-25
San Fernando
Cali, Valle
Colombia

Merin Anak Rayong
e-mail: rinely20@yahoo.com
Sahabat Alam Malaysia
P.O. Box 216
Marudi, Sarawak
98058
Malaysia

Ventzeslav Dimitrov
e-mail: vdimitrov@space.bas.bg
CLEAN WATER
Sofia-1592
P.O. Box #31
Sofia 1592
Bulgaria

Pascal Andriamanambina
e-mail: rubis3@yahoo.com
Centre Ecologique Libanona
Box 255
Fort-dauphin
Fort-Dauphin 614
Madagascar

Daniel Arancibia
e-mail: daniel@wwfperu.org.pe
WWF–Peru
P.O. Box 11-0205
Av. San Felipe 720
Jesús María
Lima 11
Perú

Susanna Paisley
e-mail: susy.paisley@ioz.ac.uk
Durrel Institute of Conservation and Ecology
University of Kent
Canterbury, Kent
CT2 7NS
UK

Andres Moreira-Muñoz
e-mail: mmunoz@mnhn.cl
Taller La Era
Montana 7516, Vitacura
Santiago
Chile

Liliana Díaz Riveros
e-mail: sdiazriv@banrep.gov.co
Fundación PUIQUI
Calle 107A #13-75 Barrio Santa Paula
Bogotá
Bogotá, D.C.
Colombia

Trina Galido-Isorena
e-mail: tgisorena@edsamail.com.ph
Anthropological Watch
46-c Mahusay Street
UP Village, Diliman
Quezon City
Metro Manila 1101
Philippines

Students sponsored by the Infodev Program (www.infodev.org)
Herbert Tushabe
e-mail: htushabe@hotmail.com
Makerere University Institute of
Environment & Natural Resources,
PO Box 7298, Kampala - UGANDA
Job Title: Data Bank Manager

Jayanta Ganguly
email: joi@auroville.org.in
Pitchandikulam Forest
Auroville, TN, 605101, India
Job Title: Systems Analyst

Sreekesh Sreedharan Nair
e-mail: sreekesh@teri.res.in
TERI
Darbari Seth Block, Habitat Place
Lodhi Road, New Delhi 110003, INDIA
Job Title: Research Associate

Marius Paul Hubert Rakotondratsima
e-mail: wcsmad@dts.mg
c/o The Wildlife Conservation Society.
BP 8500 Antananarivo 101, Madagascar
Job Title: Conservation Biologist

Lovasoalalaina Razafimbololona
email: lalaina_ra@hotmail.com
CARE International - Madagascar
BP 1677 , Antananarivo 101, Madagascar
Job Title: : Design project, monitoring and
evaluation in using the GIS.

Mr. Andry Razanajatovo
e-mail: andry_raza @hotmail.com
CAVEPI - Madagascar
BP 816, Antananarivo 101, Madagascar
Job Title: : GIS Analyst

Wycliffe Mutero
e-mail: : muterow@kws.org
Kenya Wildlife Service (KWS)
P.O. Box 40241 , Nairobi , KENYA
Job Title: : GIS Officer

Leonard K. Mubalama
e-mail: radiorm.bukavu@wfp.or.ug
CITES/MIKE (Monitoring of Illegal Killing
of Elephants)
C/o PNKB/GTZ project
BP 86 Cyangugu, Rwanda
Job Title: : National Elephant Officer

Lucy Chege
e-mail: lwaruingi@acc.or.ke
Country: Kenya

Students sponsored by the Ecology Center, Missoula, Montana (www.wildrockies.org/teci)
Valery Sinukov
e-mail: valery@as.khb.ru
Russia

Denissov Pavel
e-mail: siblarus@online.ru
Russia

Trigoubovitch Alexei
e-mail: siblarus@online.ru
Russia

Anna Grocholevich
e-mail: astranna@mail.ru
Russia

GIS Activities at World Wildlife Fund Canada for 2000

Hussein M. Alidina, Manager, Geographic Information and Conservation Analysis

The wilderness campaign that changed the Canadian landscape • 1989-2000

The year 2000 was an exciting one for GIS at WWF. It saw the completion of the ten-year Endangered Spaces Campaign and the birth of a new restructured Conservation Program. GIS was identified as being critical to WWF Canada's future conservation activities and past successes. Highlights of the year that relate to GIS are noted here. Endangered Spaces comes to an end. WWF Canada's Endangered Spaces Campaign officially ended on July 1, 2000. This ten-year campaign to establish protected areas across Canada's natural regions was launched in 1989. In 1992, the campaign won federal, provincial, and territorial governments' commitments to complete the protected areas system of Canada. GIS played a crucial role in the development of conservation science applications for the campaign and was effectively used to conduct and maintain a countrywide gap analysis, identify candidate sites for protection, and report annually on the progress made to establish new protected areas. Almost 96 million acres were added to Canada's parks and protected areas network. This increase more than doubled the land area protected in Canada. GIS activities relating to the Endangered Spaces Campaign in 2000 involved mapping and analytical support for conservation initiatives across the country including protected area campaigns in Alberta, Manitoba, New Brunswick, Labrador, and Quebec. A detailed report examining the life of the campaign was published and is included.

WWF also developed a leading-edge conservation suitability analysis to help define candidate-protected areas. The analysis supported efforts in Ontario to secure 2.4 million hectares of new protected areas as part of the Lands for Life initiative. This effort was documented in the summer 1999 issue of ArcNews (Vol. 21, No. 2). Organizational Restructuring and New Conservation Programs: July 2000 saw WWF Canada move to a new conservation program structure. The new structure consists of five programs:

International

Our international team is responsible primarily for undertaking conservation activities in Cuba and for leading our international toxicity efforts.

National

The national program is responsible for granting monies under the Conservation Science and Solutions Fund to field-based projects to protect and restore habitat, recover species at risk, and reduce pesticide use. The national program will also be producing a national report on biodiversity trends in Canada.

North American

The North American program is responsible for conservation initiatives across North America with an emphasis on forest certification and trade issues. GIS decision support tools will be a key component to ensure that landscape planning is undertaken as part of the certification process.

Marine

Our marine program is responsible for undertaking activities related to the establishment of marine protected areas and the protection of marine biodiversity.

Arctic

Our arctic team is primarily responsible for conservation activities in the northern territories with an emphasis on balancing resource extraction with protection or natural capital.

A subsequent process of assessing GIS needs with these new programs was initiated and is currently underway. The outcome of this process will be a longer term vision for GIS conservation support at WWF and the resources required for implementation.

Marine Related Activities
In early 2000, WWF Canada released a scientific document entitled *Planning for Representative Marine Protected Areas—A Framework for Canada's Oceans.* This document laid out the principles for a Marine Ecosystem Classification System for Canada and highlighted the approach by which Marine Protected Area planning will be undertaken. The report illustrated these principles in an ecosystem mapping exercise in the Bay of Fundy and Scotian Shelf region in Canada (www.wwf.ca/en/res_links/pdf/marineappendixandmaps.pdf). The mapping will form the basis on which a regional marine gap analysis will be conducted. A national-scale exercise mapping

marine natural regions was also completed. It is anticipated that finer mapping nested within the natural regions will continue for the foreseeable future. GIS activities planned for the marine program in the coming year include the identification of sites of high conservation priority and the development of a national marine protected areas database.

World Wildlife Fund Canada's Endangered Spaces Campaign
(From: *Endangered Spaces,* published October 2000 by World Wildlife Fund Canada)

World Wildlife Fund Canada's Endangered Spaces Campaign has literally changed the Canadian landscape. Through a ten-year effort supported by hundreds of thousands of Canadians, the campaign put the need to protect our fast-disappearing natural heritage front and center across Canada and produced significant, measurable, on-the-ground results. Between 1989 and Canada Day 2000, almost 96 million acres were added to Canada's parks and protected areas network through the designation of more than 1,000 new parks and reserves. This increase more than doubled the land area protected from coast to coast. The spaces campaign set a significant challenge for Canadians and their leaders—to protect a representative example of each of the country's 486 terrestrial natural regions by the year 2000 and its marine regions by 2010. At the heart of the campaign was the idea of protecting a complete template of Canada's natural diversity, from the coastal rain forests of British Columbia and the almost-vanished grasslands of Saskatchewan and Manitoba to the rich deciduous forests ringing the Great Lakes and the Acadian forests of the Maritimes and on up to the vast boreal expanses of Canada's northern regions and beyond. For the first time, Canadians were presented with a national action agenda that addressed such significant issues as our ever-expanding list of endangered species, the near disappearance of certain ecosystem types such as grasslands and old-growth forests, and the unseen but potentially disastrous loss of genetic diversity in wild systems. The campaign embraced the idea that by protecting representative examples of Canada's habitat types, we could safeguard our natural heritage and create future opportunities—opportunities for everything from the chance to gain knowledge and understanding by observing natural processes to the sense of personal renewal that can come from spending time in untamed places. Before the spaces campaign, efforts to secure a healthy

Level of Representation of Canada's Terrestrial Natural Regions at the Beginning of the Endangered Spaces Campaign (1989)

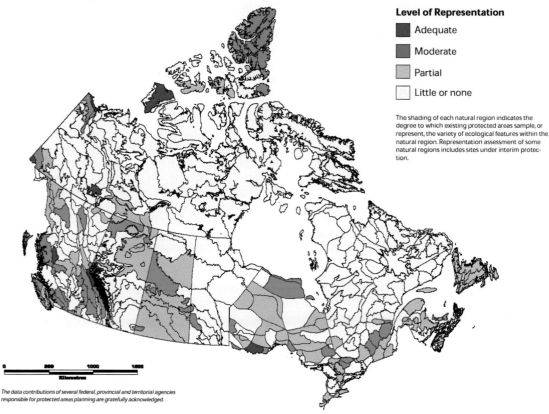

Level of Representation

- Adequate
- Moderate
- Partial
- Little or none

The shading of each natural region indicates the degree to which existing protected areas sample, or represent, the variety of ecological features within the natural region. Representation assessment of some natural regions includes sites under interim protection.

The data contributions of several federal, provincial and territorial agencies responsible for protected areas planning are gratefully acknowledged.

World Wildlife Fund Canada / Endangered Spaces

ecological and economic future by protecting wild areas and wild species had often been sporadic and fragmented. Park establishment was, with a couple of exceptions, a process marked by short bursts of activity followed by long stretches of inertia. The trend line was clear—the task of protecting wilderness was proceeding at a snail's pace and progress in the 1980s was dangerously close to grinding to a halt, a fact that did not go unnoticed by the auditor general of Canada in 1989 when he urged the federal government to get on with the task of finishing the national parks system.

The spaces campaign helped to change that pattern in two ways. First, thanks to the political commitments to wilderness protection secured during the campaign, there was a dramatic increase in the pace of designation, leading to a trend of park and protected areas establishment that outpaced anything seen previously in Canada. Second, the campaign successfully introduced the concept of a consistent, systematic approach to protected areas designation. The Endangered Spaces Campaign gave us a national road map, destination, and deadline.

In trying to ensure that our protected areas system really did represent the natural diversity of the Canadian landscape, the description of natural regions would have to be based on more predictable long-term factors than forest

types or particular species populations—qualities that could be changed overnight by sweeping natural events like fire or flooding. There was strong agreement about the need to look at the underlying factors that shape ecosystems, such as geology, topography, and climate, and to use these to rationally subdivide the Canadian landscape into distinct "ecological" or "natural" regions. This method became known as the "enduring features approach." It was strongly endorsed by the CCEA, which recommended that natural regions be defined based on "enduring features of the environment . . . relatively stable landforms and seaforms and their accompanying plant and animal communities." The next step was judging how well protected areas represented the enduring features found within each of the natural regions. Again, it was clear that without sound guidelines, the approach could vary from jurisdiction to jurisdiction—some might look at how well the protected area encompassed the enduring features of the region, while others might judge a natural region represented if it simply contained a protected area. The approach developed by the spaces team was to first measure the extent to which each enduring feature in a natural region was protected, and second, to measure how well all the enduring features of the region were protected as a whole. Finally, an assessment of the suitability of the protected area for such things as wide-ranging

species or the area's ability to withstand large natural disturbances such as wildfires was also factored in. All of this led to the ultimate goal: the ability to map out natural regions and to indicate for each the level of ecological representation by protected areas—"adequate," "moderate," "partial," or "little or none." WWF could now show decision makers and Canadians at large where the holes in the representation template were and could help design new protected areas to fill them.

This technical work would prove to be vitally important in making it clear that the spaces campaign had an agenda that was science-based and achievable. In fact, the work led Reed Noss, editor of the prestigious science journal, Conservation Biology, to remark that the spaces campaign was unmatched worldwide "in terms of its vision, breadth, scientific defensibility, public acceptance, and possibility of ultimate success."

There were a handful of events that shaped the development of the spaces campaign. The first was the battle for South Moresby (Haida Gwaii) in the Queen Charlotte Islands of British Columbia. This campaign to protect the deep-green, almost mystical forests of these rain forest islands from clear-cut destruction had raged for years with road blockades, arrests of Haida elders, demonstrations of international outrage, and threats of economic collapse. In 1988, British Columbia and the federal government finally agreed to protect Haida Gwaii as a national park. But even then, new conflict sites loomed—the Stein Valley and Carmanah in British Columbia, Temagami in Ontario, and many others. To protect wilderness one site at a time, dedicating years and enormous resources to every single conflict, was clearly only going to result in the loss of more areas than it was going to save. A more systematic approach was needed. WWF and others had a taste of such an approach with Ontario's Strategic Land Use Planning (SLUP) process in the early 1980s. The inelegantly named SLUP process had resulted in more than 150 new sites being protected across the province and had broken new ground with its attempt to take a systematic, provincewide approach to protected areas establishment. The experience of the SLUP process also dovetailed with the recommendations of the federal Task Force on Park Establishment, which spelled out the need for a more coordinated and deadline-driven approach to finishing the national parks system in its 1986 report titled *Parks 2000: Vision for the 21st Century*. The final piece of the puzzle lay in the pages of a manuscript begun by Doug Pimlott, a leading Canadian conservationist. When he died in 1978, Pimlott left behind his warning about our fast disappearing wilderness heritage. That unfinished manuscript, passed to Monte Hummel, president of WWF, gradually evolved into an assessment of the state of wilderness across Canada—an assessment that clearly called for a strong response.

Hummel took up the challenge by pulling together *Endangered Spaces: The Future for Canada's Wilderness,* a national bestseller containing contributions by knowledgeable conservationists from across the country. The book was the launch vehicle for the spaces campaign's clarion call to action—the Canadian Wilderness Charter, which invited Canadians to endorse a new protected areas vision. That vision was spelled out in the final chapter of the book, which called for the establishment of a representative system of protected areas from coast to coast by 2000. The Endangered Spaces Campaign was born. Wrote Hummel:

Endpoints are a time of reflection. Beginnings fuel anticipation. As I reflect on our 10-year Endangered Spaces Campaign to protect Canadian wilderness, I find myself revisiting what we anticipated 10 years ago. In our 1989 book Endangered Spaces, which launched the campaign, I said I thought it was 'going to be an interesting struggle.' More precisely, I predicted that the campaign should strive to be characterized not so much by '...a win/lose, extremist, fight-to-the-death over wilderness, but by a more professional competition as to

who can sound the most reasonable. Nevertheless, what is at stake will be either saved or not.' So, was it an interesting struggle? Many who worked tirelessly on the campaign would argue that was the understatement of the decade! Was wilderness saved? Yes, 96 million acres of it. Was enough wilderness saved? Unfortunately, no. Today, a big job still remains to be done in about one-half of Canada's natural regions. Aside from acres on the ground, I believe that Endangered Spaces left other important, if less tangible, legacies. It made people and institutions think differently about the business of protecting wilderness. And as advocates for nature, we sharpened our existing tools, developed new lobbying techniques, extended our networking capacity, and proved that we could sustain all of this over the long haul. We set the conservation bar high, and we pursued our goal

through actions that were backed by the best science in the world. In short, the campaign was eminently reasonable, we engaged new players from all sectors of Canadian society, and we achieved not just commitments but actions that made a difference in every province and territory of our Country. Clearly, much was achieved. But the bottom line remains: Wilderness protection in Canada has not yet advanced even to the minimum goal set out by the Endangered Spaces Campaign. The cold lesson for me has been how incredibly difficult modern society finds it to simply leave some things alone. In the end, the question I posed in the beginning is still worth asking: 'How important is it to you that wilderness plays a role in the future of our country?'

Conservation GIS in Altai Mountains, Russia

Misha Paltsyn, Science Director   katunsky@mail.gorny.ru

The Altai Mountains, the highest mountain region in Siberia, are located at the nearly exact geographic center of the Eurasian continent, where such countries as Russia, Kazakhstan, China, and Mongolia meet. The Altai Mountains are one of the most interesting places of biodiversity in the temperate zone of the world. About 2,200 plant species grow in the region, among them are 212 rare species, occurring only in this part of the world. Eighty mammal, about 260 bird, and 11 amphibian and reptilian species inhabit Altai. Some of them (snow leopard, argali sheep, manul cat, Altaian snowcock, otter, and others) are endangered and listed in the Red Data Books of Russia and Altai Republic. A variety of Altai biodiversity, rare species habitats, and last stands of pristine forests are protected in nature reserves and refuges.

A herd of argali sheep females in the Chikhacheva range.

"Argali Sheep Database and Census System" (1997–1999) containing all data on argali sheep (the largest ram in the world and an endangered species) in Altaisky Nature Reserve and the surrounding area. Our lab successfully provided GIS support for the argali research expeditions; we made special work maps of the study area, worked with portable GPS receivers, and conducted all mapping in the field. PC ARC/INFO, ArcView GIS, and ArcView Spatial Analyst were intensively used for making all necessary DEMs and maps, calculating the observed area and argali population density, getting statistics, conducting spatial analysis of the data, and modeling. A special census system was developed using GIS. We have found GIS technology by ESRI to be a very valuable tool for the research and planning of argali sheep habitat protection in Southern Altai Mountains.

At the beginning of 1998 we started to collect information on distribution and population density of ungulates and some carnivorous mammals in Altaiski Nature Reserve. The data was used to create a GIS project to assess and compare different parts of the reserve for their ability to protect large mammal species. We tried to demonstrate high conservation significance and vulnerability of the area where space junk falls. As the result of our work, the paper and poster titled "Booster Fallout Problem and Conservation of Sustainable Populations of Ungulates and Some Carnivorous Mammals in Altaisky Nature Reserve" (1998) were made and presented at the 1998 ESRI International User Conference.

In 1998, in cooperation with wildlife biologists of Altaisky Nature Reserve we made the "GIS Database on Amphibians and Reptilians of Altai Republic." The database includes all known data on animal distribution in Altai Mountains from the 1890s to 1998 and is used for the Altai Republic Red Data Book publication. One more GIS project of Argali GIS Lab concerned snow leopard distribution in Altai. The data is useful for planning snow leopard research and assessment of contemporary status of the species in Altai Mountains.

In the framework of the Altai Mountains Conservation Initiative Project (supported by the Turner Foundation), starting in 2001, the Argali GIS Lab will help rangers of Altaisky Reserve prevent poaching. We will create a GIS database and map all cabins, trails, mountain passes, river crosses, possible poacher routes, usual poaching sites, and rare species habitats in Altaisky. The database will be used to discuss and plan optimal routes for the rangers as well as sites for new cabin construction. It will also include suggestions for extending the Reserve area for better protection of endangered species such as the snow leopard, argali sheep, reindeer, and manul wild cat. Our last ESRI Conservation Program (ECP) grant is perfect support for the future work.

Two years later, in 1999, another GIS Lab with a focus on conservation issues of Altai Mountains was established at Katunsky Biosphere Reserve (KBR). Sasha Yumakaev, GIS specialist of the Trust Altai for the 21st Century, helped us get an ECP grant (PC ARC/INFO, ArcView GIS, and ArcView Spatial Analyst), which provided us with our first GIS training. In the same year, our lab infrastructure benefitted from a scanner, CD writer, and a good printer courtesy of the World Wildlife Fund (WWF). One of the primary objectives of the lab is the creation of large-scale digital data for the reserves area and surrounding territories to support general activities of KBR on biodiversity inventory and ecological monitoring. So we constantly scan and digitize large-scale paper maps creating good quality digital data for field research and spatial databases.

Katunsky Range—the highest mountain chain of Altai and Siberia (up to 4,506 meters above sea level).

There are two nature reserves (Altaisky and Katunsky), five refuges, and about 100 nature memorials in Altai Mountains. They comprise an area of 20,000 sq. km, about 22 percent of the Altai Republic territory. Altaisky and Katunsky Nature Reserves and Ukok Plateau Nature Refuge were listed as the UNESCO World Nature Heritage Sites along with other valuable natural features such as Teletskoe Lake and Belukha Mountain. Recently Katunsky Nature Reserve was awarded status as a UNESCO Biosphere Reserve.

The Argali GIS Lab of Altaiski Nature Reserve was created in 1997 when we received our first ESRI Conservation Program donation—PC ARC/INFO and ArcView GIS—and a digitizer with the help of The Ecology Center, Inc., Montana. The first and most interesting GIS project of Argali GIS Lab was devoted to the

Along with data building, we are trying to find necessary resources for development and training of our staff. Thus in November 1999, in cooperation with Mike Beltz, GIS director of the Ecology Center, Inc., we applied for the Conservation Technology Support Program (CTSP) grant and last summer received an excellent set of CTSP computer equipment and software that greatly enhanced KBR's GIS lab. Our GIS specialist, Andrey Klepikov, participated in the Ecology Center's International Internship Program last January and learned valuable skills in ArcView GIS and PC ARC/INFO. At the same time, with the help of Alex Philp, assistant director of EOS Education Project at the University of Montana, we obtained two perfect Landsat 7 images for KBR's area.

Due to WWF support last winter and in cooperation with KBR's field researchers, we created a conservation spatial database for the whole Ust–Koksa region, Altai Republic. The database includes different thematic layers: protected areas, distribution of endangered species, vegetation cover and landscape classifications, soil map, land-use map, digital elevation model, and others. These data sets are intensively used by KBR's staff for planning of the protected areas system in South Altai (nature reserves, parks, and monuments). This map was printed for the presentation of the conservation system for local authorities and the government of Altai Republic. We believe this project to be the first step in creating

a transboundary biosphere reserve in the junction of Russia, Kazakhstan, China, and Mongolia.

We have paid more and more attention to the social and sustainable economic development in Altai to save biodiversity and natural resources in the region. We cooperate with the Land Management Committee of Altai Republic in its research on deer farming, one of the sustainable ways of land use in Altai. GIS helps us to facilitate the deer farming system and save endangered species habitats and migration routes of wild ungulates around the deer farms.

Our staff also uses GIS software and powerful CTSP computer equipment to support activities of KBR to organize a protected area for the snow leopard in Yungur Valley. Yungur is one of the most untouched and remote parts of the Altai Mountains and is also perfect shelter for snow leopards and other wild animals. We utilize our equipment to print good maps to present the project to local community and authorities and get their support. The new reserve will result in additional jobs for locals and will save not only snow leopards

Winter expedition for argali sheep research in the Chikhacheva range.

but historical and cultural monuments, sacred areas, and traditional ways of land use.

Support from the Conservation GIS community and GIS by ESRI helps us in nearly all our conservation activities in the Altai Mountains. We hope to enhance our GIS capabilities every year and help other nature reserves in South Siberia to get and use this useful tool to protect Russia's natural features.

⊕

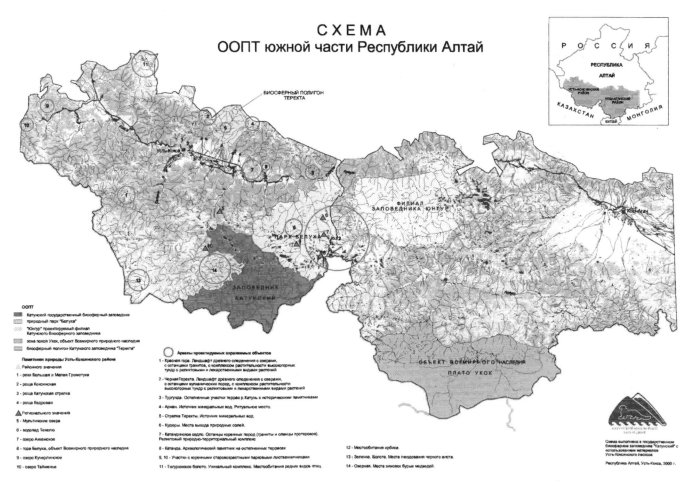

"Protected Areas of the Southern Part of Altai Republic" was made to show present and proposed protected areas in the Ust–Koksa and Kosh–Agach regions of Altai. Here you can see Katunsky Biosphere Reserve (green area), Belukha Nature Park (dotted green area), Ukok Plateau (dark green striped [or crosshatched] area), and proposed Yungur Reserve for snow leopard protection (light green striped area with dashed boundaries). Green triangles are existing nature memorials in the Ust–Koksa region. Green circles are proposed nature memorials and refuges in the region.

A Trip Around the World with the SCGIS International Community

Roberta Pickert, SCGIS International Committee   RPickert@archbold-station.org

On Thanksgiving eve, SCGIS members Kai Snyder, Prashant Hedao, Peter Schlesinger and Roberta Pickert returned from a most successful trip to Khabarovsk, Russia where we attended and spoke at the first Russian Far East SCGIS conference. We were able to witness the birth of the formal Russian Chapter of SCGIS. It truly was a return to the same excitement, intensity, and clear focus of purpose we all experienced during the early 1990s when the SCGIS was first forming. Many hours of discussion and debate ensued while details were worked out, people were cajoled, begged, or bludgeoned into volunteering for the many roles necessary to get this embryonic Russian SCGIS birthed. Sasha Yumakaev, the bright shining star who pulled the far flung (literally from all over Siberia and the Russian Far East) Russians together, patiently kept the debates focused and the spirits high. At the end, thanks to all the participants' (about sixty conservationists) intense efforts the rules were made and the gauntlet was thrown down for them to make this new Russian SCGIS work. The following points detail progress that was made at the conference itself:

- The SCGIS membership form has been translated, thanks to Sasha, into Russian. I am now receiving membership forms from a number of the Russian participants which have been filled out and re-translated into English. As decided by the board we are allowing the Russians free membership this first year and then they will pay if they can. If not they will, of course, still have membership in SCGIS.

- The 2001 scholarship announcement and form has been translated into Russian by Sasha and is being distributed among the conference participants. My talk was about SCGIS and about the scholarship program. So I hope we get a number of applications as a result.

- RSCGIS has begun a listserv of its own. I am on it though I have to go through this translation program, which is fine. This is also acting as a sort of online newsletter too.

- This year's conference was put together by Mike Beltz and company at The Ecology Center, in Missoula, Montana, with a grant from the Trust for Mutual Understanding. RSCGIS is now looking for a similar grant to front next year's conference which is already being planned for Kamchatka, Russia.

- Several of this year's participants have already written their papers, had them translated to English, and have sent them to me. I will be forwarding these on to our communications chair for inclusion in a special section of this year's newsmagazine.

- I was also able to bring home all the posters that were presented at the conference. Charles Convis and I will be creating a special section at the ESRI conservation arena for their display and I will be working with Steve Beckwitt to get many of them displayed at the ESRI poster session.

The SCGIS is very proud of this new Russian group. They have worked so hard to pull their first conference together and the coming years will see a lot of hard work and rewarding exchanges as we work together in this new collaboration.

The Last Frontier Forests of the Russian European North: Current State, Problems, and Perspectives

Greenpeace Russia

The remaining frontier forests in the European North of Russia represent unique natural habitats with key significance in conservation terms. Most of these areas, in the hands of the Russian Federal Forest Service, are currently threatened by extensive timber exploitation by a number of logging companies. In response to these threats, Greenpeace Russia, in collaboration with the Biodiversity Conservation Center and with assistance of a number of other research organizations and experts, has produced a map of the area's remaining frontier forests and also has made a list of major organizations whose activities led to the rapid destruction of these forests.

Frontier taiga is a forest area that is composed of forest and associated nonforest ecosystems (e.g., marshy, river, lake, rocky, mountain) that formed and has since developed without intense human influence and that complies with the following criteria:

- It has no permanent settlements or transport infrastructure.

- It exhibits no evidence of intense human activity (such as intensive cutting, mining, land clearance, industrial pollution, or plantation development).

- Its forest habitats are predominantly ancient forest as defined below.

- It maintains fundamental ecological processes that are manifest in terms of both forest dynamics and spatial structure. Such processes include those associated with natural forces, such as gap dynamic, as well as those associated with ancient, seminatural forces such as fires.

- It encompasses natural hydrological networks of minor and medium rivers or lakes and their watersheds.

- The area and configuration of the forest range are sufficient to conserve the biodiversity present including populations of large vertebrates that require a considerable area for their survival. The area and configuration should also provide for a high level of protection of central parts of the forest from human impact. At present, there is no single and sufficiently grounded expert opinion concerning the minimum size of a taiga range meeting all the requirements stated above. The overwhelming majority of data suggests that the minimum size ranges from 100 hectares up to several million hectares. However, it is clear that guaranteed conservation of natural taiga in a state that is as close as possible to the natural state is feasible only within very large areas of one million hectares and over.

Unfortunately, the last frontier taiga areas in European Russia are today at the edge of extinction. To prevent total extinction of plain European taiga it is necessary to map all of the frontier taiga areas and to protect them immediately. In response to these threats, Greenpeace Russia, in collaboration with the Biodiversity Conservation Center and with the assistance of some other environmental and scientific organizations and experts, has produced a map of the remaining frontier forests of Russian European North. Mapping of remnant frontier taiga has been carried out using satellite images (Resours-O-3, resolution 35 and 150m, 1997–1998 years and Resours-O-4, resolution 150m, 1998 year) and data from expeditions of Greenpeace Russia and other environmental and scientific organizations.

Another step necessary for creating an ecological frame of northern Russian regions is revealing and mapping small-sized fragments of ancient forests (from two to twenty thousand hectares). This work was executed by Greenpeace Russia and the Biodiversity Conservation Center for the Republic of Karelia, where the fastest destruction of ancient forests happened, between 1997 and 1998. On the following page is the scheme of all Karelian ancient forests areas exceeding 2,500 hectares (this map is accessible also in scale 1:1,000,000 as paper or electronic map).

The map that was prepared by Greenpeace Russia and the Biodiversity Conservation Center (next page) shows the state of almost undisturbed natural forests of Karelia proposed for conservation by January 1, 1998.

The map of "potential areas of almost-undisturbed natural forests of Karelia" reflects the present distribution of forest areas in Karelia, which provides current information on the condition of the forest cover in the Republic, and can refer to the natural almost-undisturbed forests with the highest degree of correlation. Potential massifs of almost undisturbed ancient forests of Karelia included the following categories of natural ecosystems:

- Natural forest ecosystems (growth of several generations of trees almost untouched by man).

- Ancient seminatural forest ecosystems (growth of several generations of trees indirectly affected by human influence, typical of the postindustrial period; e.g., regular forest fires).

- Sustainable disturbed forest ecosystems, which structure is close to the structure of natural or ancient seminatural forest ecosystems and ensures their stability in time.

- Areas of other mature forests, spotted in massifs of forest ecosystems of the three categories given above.

- Nonforest natural ecosystems (swamps, lakes, rivers, brooks, rock outcrops, etc.) with a low extent of disturbance, located inside massifs of forest ecosystems of the categories given above.

Potential massifs of almost undisturbed natural forests of Karelia include compact areas of natural ecosystems if their total area exceeded 2,500 hectares and if natural and ancient seminatural forest ecosystems prevailed on their territory in a rectilinear fashion. This map contains both areas of the researched potential massifs of almost undisturbed natural forests, which was confirmed during the field research, and areas documented by remote sensing and land capacity checks.

Russian World Heritage Protection

In 1994, the State Committee for the Environment of the Russian Federation and Greenpeace Russia concluded an agreement to nominate several Russian territories for inclusion in the World Natural Heritage List.

In 1994, Greenpeace prepared the documents necessary to nominate the "Virgin Komi Forests" for the World Heritage List. In December 1995, it became the first-ever Russian World Natural Heritage Area. The Virgin Komi Forests is the largest primary, or fully natural, forest of its size (3.2 million hectares) remaining in Europe. Due to its newly acquired international status, logging of the unique forests by the French company HUET Holding was prevented. A gold mining project in the northern part of the park was also suspended.

The Virgin Komi Forests is the largest primary forest of its kind remaining in Europe (3.2 million hectares), represented by almost undisturbed ecosystems that have not been affected by intensive human activities. The area counts more than 40 species of mammals, including the brown bear, the sable, and the highly migratory elk; 204 species of birds, including the white eagle and osprey, recorded in the Red Book; and 16 species of fish, the most valuable of which are glacial relics such as the lake char and arctic grayling. The forests consist of age-old fir, cedar, and spruce, with underwood of unique and rare slow-growth flora species. Mid and northern taiga change over to forest tundra.

The Golden Mountains in the Altai Mountains, at the juncture of Central Asia and Siberia, is a varied region of unbroken forests, steppes, glaciers, and mountain tundra. The site is world-renowned for its geological formations; swift, beautiful rivers cut from the rock; vast expanses of natural forest; and the many rare and endangered species that are protected across this Republic that is itself a natural wonder.

The genuine masterpiece of Altai is Mount Belukha, which, at 4,506 meters above sea level, is the highest point in Siberia, towering nearly 1,000 meters higher than the closest ranges. A remarkable part of the Altai are the river valleys, the most important of which are the Katun and Chulyshman. Flowing in deep, thin canyons, they are comparable in inspirational beauty to the famous Grand Canyon in America. The pearl of Altai is the Teletskoye Lake, often referred to as "Little Baikal," with pure water, rich animal life, and a backdrop

of mountain chains and thick taiga forest. The high density of endemic species in Altai exemplifies the diversity in landscapes and their characteristic types. A significant variation in elevations has created five distinct altitudinal zones, inhabited by 60 species of mammals, 11 species of amphibians and reptiles, and 20 species of fish. From the animal kingdom, it is necessary to single out the snow leopard as a reason for permanently protecting this territory. This is one of the world's rarest animals, with only a few left in Altai and the surrounding regions.

The Central Sikhote–Alin range is located in an ecologically critical area at the juncture of Central and Southeastern Asia. This, combined with the relative intact nature of the large wilderness areas, has created a unique center for ancient relic floral and faunal species. The area is characterized by a deep penetration of both animal and plant species of different natural zones. The Sikhote–Alin protects more rare and endangered species than any other

region of Russia. There are approximately 1,200 species of higher vascular plant species alone. The region also counts approximately 65 species of mammals and 342 species of birds. The Sikhote–Alin Nature Reserve and adjacent areas is one of the last refuges for the Siberian, or Amur, tiger. Population estimates range from 200 to more than 400 while human populations remain low. Many other rare, endangered, or otherwise valuable species inhabit the territory, such as the Himalayan bear, musk deer, Japanese and hooded cranes, Far Eastern and black storks, merganser, ginseng, and rosy rhodiola.

The nature of the nominated World Heritage area is highly valued for its aesthetic qualities. The area includes many natural phenomena of high aesthetic and recreational values: numerous mountains and rocks intermittent among the taiga; waterfalls and rapids such as the Kema rapids and the Big Amga Falls; as well as the peculiar rocks, rock formations, and sandy bays along the Sea of Japan.

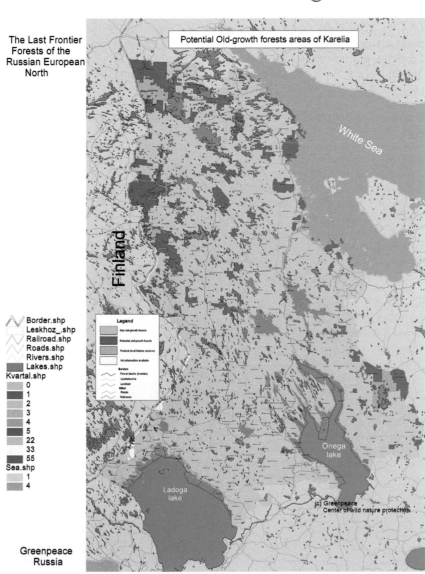

E-misszió Environmental Association, Hungary

Csaba Botos

I think it is important that the leadership of an organization gives full support to a challenging technology such as GIS. On January 27, 2001, the E-misszió Association's General Assembly was held. This assembly was organized to correctly inform our members and important individuals (e.g., the director of Environmental Authority, representatives from Hortobágyi National Park) about our work in an informal atmosphere. During this event I introduced our GIS activity that we had accomplished in the last few months since we received the ECP grant.

Besides the projects mentioned below, we held presentations about mapping water pollution sources and took part in the "Light and Dark Side of Nyíregyháza" photo exhibition that gave the opportunity to introduce our computer-based information system showing illegal deposits and natural areas with linked multimedia applications.

GIS Projects

Surveying of *Orchis laxiflora* ssp. *elegans* (see map, next page)

On May 11, 2000, we started our survey with the Young Botanists' Society beginning to monitor the above-mentioned orchid species. This is the most popular protected orchid in Szabolcs–Szatmár–Bereg region. Using topographic maps we selected wet grassy areas that became the objects of our work (likely to be orchid habitats) and planned the order of fieldwork.

We have used presence–absence examination and estimation of individuals. Up to this point we have already found forty orchid habitats, some of them in bloom.

Results
- Examined habitats: 151

- Found orchid habitats: 40

- Total stocks (estimated): 21,700

- Largest population (estimated): 8,000 stocks

Additional Achievements
- By handing over the results to the Hortobágy National Park they can make more well-informed decisions in their work.

- We have specified the extension of the examined wet habitats caused by land-use changes.

- As a result of the survey, at least fifteen valuable areas, among them a bog, were discovered near Nagykallo (these are "ex lege," protected in Hungary).

- Highly protected species were found such as the following: *Angelica palustris* declared by the Berne Convention; *Iris spuria* (first found in the Nyirseg, near Bokony); and *Dactylorhisa incarnata*.

Survey and Qualification of Illegal Waste Deposits Near Nyíregyháza
We started this project in 1999. Our aim was to find and analyze the deposits near Nyíregyháza and afterward in the whole county (see map below). We have analyzed the deposits using twenty-five factors such as size and contents. We have already found and analyzed eighty-three deposits. Their numbers were changed as a result of our work because there was some TV and radio talk about this project and four deposits became recultivated partly or in full by the local government.

In addition there was our initiative to recultivate five deposits with local people and students. It was in the scope of a nationwide "Landscape surgery" project organized by Hungarian environmental NGOs.

We collected information of the proposed location of deposits from local authorities and governments and by analyzing aerial photos. This way we selected the suspicious patches, identified them on maps, and started the survey. We digitized the location of the deposits as points and linked pictures and data to these features. It became an efficient representation of our work on the exhibition mentioned in the introduction. We also printed maps for re-cultivation plans and brochures for authorities and local leaders and representatives. We have compiled the work in Nyíregyháza, and we are going to do the same work for the whole county.

Tisza River Project

Pollution Sources
We digitized the officially recorded water, soil, and air pollution sources as points with linked images, data, and descriptions.

This information system was presented at the conference on water quality organized by E-misszió Association in close collaboration with the Regional Development Agency and Carpathians Euroregio in the end of 2000.

We have published a bilateral A2-sized brochure based on the collected data. The first part contains descriptions about our work, and

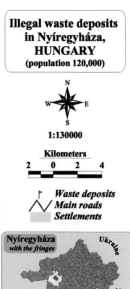

Illegal waste deposits in Nyíregyháza, HUNGARY (population 120,000)

1:130000

Kilometers
2 0 2 4

ᛕ *Waste deposits*
/\\/ *Main roads*
▢ *Settlements*

Nyíregyháza with the fringes

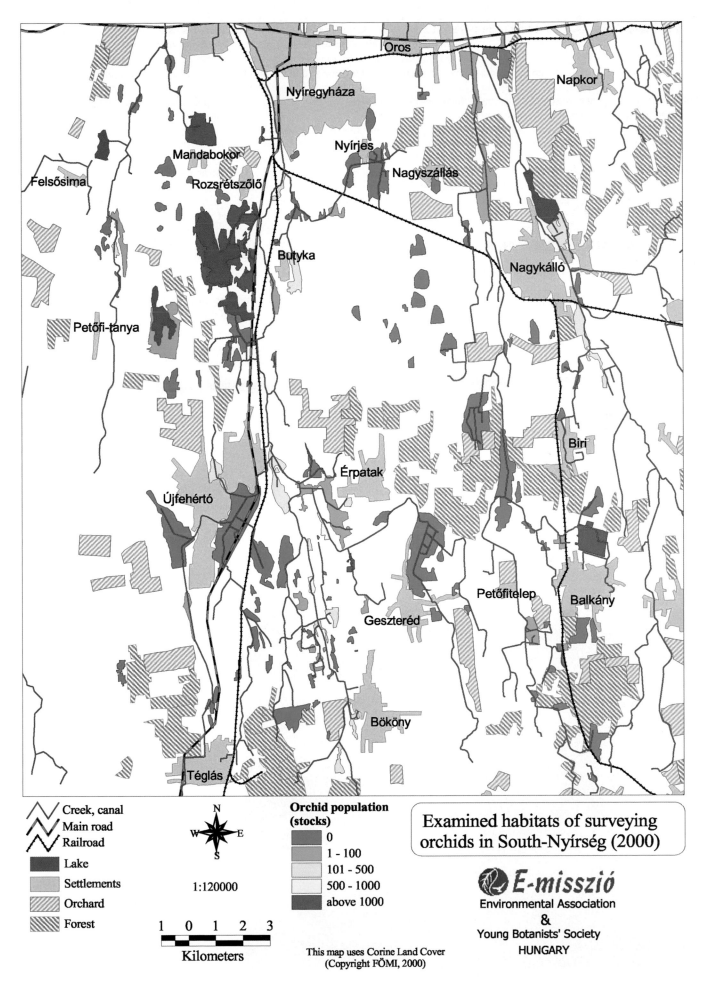

Creek, canal
Main road
Railroad
Lake
Settlements
Orchard
Forest

N
W   E
S

1:120000

1 0 1 2 3
Kilometers

Orchid population
(stocks)
0
1 - 100
101 - 500
500 - 1000
above 1000

Examined habitats of surveying
orchids in South-Nyírség (2000)

E-misszió
Environmental Association
&
Young Botanists' Society
HUNGARY

This map uses Corine Land Cover
(Copyright FÖMI, 2000)

the map contains symbols representing pollution emission.

We have sent these to each of the governments in Szabolcs–Szatmár–Bereg regions with an enclosed letter asking them to send us information about the environmental problems they have, especially regarding pollution sources in living water so we can update our database.

Flood Situation in the Bereg Area

The work we did after the huge flood on Tisza River in 2000 may not seem to be a typical environmental problem, but we rescued or fed many of the inhabitants (we did this only on the first night because the local older people did not want to leave their homes), as well as domestic and wild animals. During this time we noticed a lot of pollution sources (sewage cleaners, illegal deposits, illegal fuel deposits, chemicals). This experience gave a new perspective to our surveying of pollution sources.

We worked in close collaboration with the Animal Friend Foundation (Nyíregyhaza, Hungary), a local NGO, and other Hungarian organizations for animals: Rex Foundation, White Cross Foundation, and international organizations such as WSPA and RSPCA.

There were eleven settlements under water, and nineteen settlements (30,000 people) were evacuated. We tried to contact the authorities to cooperate in the rescue of animals, but they were too busy rescuing people and other settlements that we did not succeed in getting a single motorboat. There were sufficient resources to rescue people, but the authorities did not deal with animals. One activist in the

region who had a kayak reported that lots of domestic animals were left behind in the flooded areas. Many of them were still in their sheds because the water came so quickly. Wild animals were also in danger, and there were huge losses among them.

We completed seven days of rescue operations at the flooded area. We covered most of the flooded villages, approximately 1,800 homes. We had special water-resistant garments complete with rubber boots. We found and fed cats, sheep, pigs, chicken, and so forth, in almost all the yards we entered. The army and government units were making efforts to rescue animals, but they dealt mostly with large livestock (cattle, horses). Special rescue teams also helped us in difficult areas where we could not manage with canoes. During the field operation we found out that the villages were quite accessible by canoe and motorboat. The motorboat became very useful in rescuing animals (wild animals, too) stranded on small dry plots of land where the water was deeper. The canoes were useful in tackling the "waterways" within villages for relatively small distances, and the special garments and boots were useful when our volunteers had to walk on flooded streets or courtyards where boats or canoes could not enter.

We later purchased a motorboat with the financial support of the World Society for Protecting Animals, England (WSPA) and the Royal Society for Protecting and Care of Animals, England (RSPCA).

These floods come every year, and they become more and more severe. This means that any preparation and purchase we make now will serve our activities also in the future.

We plan to analyze the changing forests, which we suspect to be the cause of frequent floods in the basin area of Tisza River, by using satellite images.

River Watching Network

This autumn the E-misszió Association built a water quality controlling network with groups of students from Ukraine, Romania, and of course, Hungary. The groups used equipment and chemicals to examine water content from time to time. They will share their results with us.

We also took part in training for planning sustainable development because we have a project to make development plans for a microregion near the Tisza River. We worked together with numerous NGOs (e.g., WWF Hungary, Environmental Partnership Foundation, Green Action, Göncöl Foundation, and Ipoly Union) and heard about GIS-based methods and applications to make better decisions.

We have also had discussions with the Ministry of Environment about rebuilding the Bereg area and the environmental problems of floods, but they seem to be fairly powerless and are under the influence of political interests.

Clean Water Club, Bulgaria

Ventzeslav Dimitrov   vdimitrov@space.bas.bg
Research Fellow, Space Research Institute

First of all I would like to express my gratitude to SCGIS for awarding me an international scholarship and giving me the chance to participate in such significant events as SCGIS and ESRI conferences as well as GIS training.

I am a research fellow in the Space Research Institute, Department of Remote Sensing (RS) and GIS. My professional interests are in the field of satellite image analysis, RS, and GIS integration and applications development. Applications in ecology are of great interest to me. I have participated in projects concerning mapping of forest ecosystems, erosion risk, and monitoring of land use, as well as designing GIS databases of a national park and reserve. I have always considered that protecting the environment should be everyone's responsibility, and I admire the efforts and enthusiasm of people who volunteer in conservation organizations.

My country, Bulgaria, is a relatively small territory, but it is rich in natural resources and biodiversity, and in my opinion considerable effort is being applied to conservation at different levels in the society. Aside from the well-structured governmental institutions, there are nongovernmental organizations (NGO), ecological clubs, and groups involved in conservation. Unfortunately, GIS is implemented mainly in national parks and a few other protected areas. I think there is a need to educate and train conservationists so they understand the potential and specific applications of GIS in their work. About ten NGOs have shown good work during the last year, carrying out different projects. Their main achievements are in biodiversity conservation, public awareness, bird protection, and developing management plans of protected areas. Soon they are going to have a workshop on coordinating and incorporating their activities. One of the issues to be discussed is building and maintaining a common information pool from the considerable data volumes they have acquired.

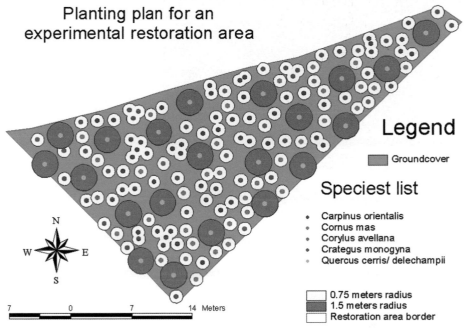

Planting plan for an experimental restoration area

Legend

■ Groundcover

Speciest list

- Carpinus orientalis
- Cornus mas
- Corylus avellana
- Crategus monogyna
- Quercus cerris/ delechampii

□ 0.75 meters radius
■ 1.5 meters radius
□ Restoration area border

7   0   7   14 Meters

Conservation organizations and especially NGOs maintain contact with the scientific community. Scientists like me take part in many of the projects and many of us are members of these nonprofit organizations. This facilitates our knowledge and applications transfer to NGOs.

I joined the environmental Clean Water Club about two years ago. They wanted to use satellite images of one of the study areas and approached me for consultation, and that was the beginning of our collaboration. I was attracted by their ideas and enthusiasm and decided to become a part of this community of young people.

Clean Water started as a high school youth environmental club including students from the National High School of Mathematics and

Landuse/ Landcover Map of Kutina

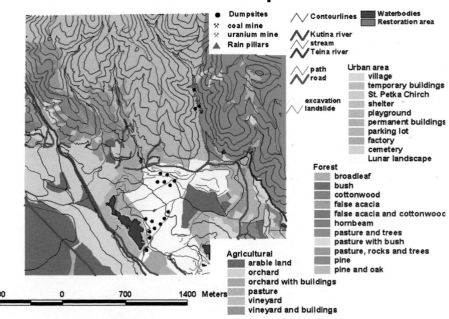

● Dumpsites
※ coal mine
※ uranium mine
▲ Rain pillars

/\/ Contourlines
/\/ Kutina river
/\/ stream
/\/ Teina river

/\/ path
/\/ road

/\/ excavation
/\/ landslide

■ Waterbodies
■ Restoration area

Urban area
village
temporary buildings
St. Petka Chirch
shelter
playground
permanent buildings
parking lot
factory
cemetery
Lunar landscape

Forest
broadleaf
bush
cottonwood
false acacia
false acacia and cottonwood
hornbeam
pasture and trees
pasture with bush
pasture, rocks and trees
pine
pine and oak

Agricultural
arable land
orchard
orchard with buildings
pasture
vineyard
vineyard and buildings

700   0   700   1400 Meters

Distribution of pH value over Sofia city, 14-16.02.1997

Analyze and processing - Geoinformation centre of Sofia University-fil@gea.uni-sofia.bg

The spatial mapschemes show the development of the rainfall process for the period of two days.

The data were recorded for the snow that fell after a two week long period without precipitation, followed by a slight warming.

Well distinguished areas with high acidity are observed on the map:

✓ In a local industrialised zone in the northern – northeastern part of the city

✓ In the neighbourhoods of the main pollutant in the eastern part of Sofia - Metallurgical Plants Kremikovtsy.

✓ In the central parts of the city, caused by the concentration of vehicles.

The southern part of the city is comparatively cleaner. It could be explained by the lack of local pollutants and the circulation of the atmospheric air.

The direction of the izolines shows that, in this case, the pollution is mainly result of local pollutants and not of the transfer of polluting substances

Rain pillar phenomenon in Bulgaria (Source: Lubka Roumenina).

We believe that how old you are and what your profession is are not as important as what your attitude is to the problems of nature conservation. So we try to help everyone who cares about nature and wants to put some effort to its preservation and protection!

From Lubka Roumenina, SCGIS 2000 Scholarship winner:

"The entire area of Kutina village is highly impacted by the coal and uranium mining. A large area is planted with Pinus sylvestris, which is pollution resistant, but it is not native for this area and is not common for this elevation. This GIS allowed us to create a viable reforestation plan and will enable us to determine the success of our restoration plan over time. Similar projects, which are common in the USA, give positive results and the designed plant communities perform very well. Cascadia Quest and the King County World Conservation Corps in Seattle carry out many projects like this with great success. That is why we hope that the restoration attempts in Bulgaria will be successful and will contribute to establishing principles for environmental restoration on the local level. I'd like to thank my teachers in environmental restoration—Cascadia Quest and the King County World Conservation Corps and my GIS teachers—SCGIS and ESRI for the knowledge and experience I acquired during my training with them."

Science and some teachers and scientists as scientific advisors. Last year we were granted two ArcView GIS software licenses that let us start informal classes on the basics of GIS and remote sensing. Our goal was to prepare the young people studying geography, geology, biology, ecology, and computer science for the world of new technologies. Most of them will work in the field of environmental science and in conservation, and this knowledge will give them a good background and a better understanding of how beneficial GIS could be toward their efforts.

Clean Water members carry out research projects related to monitoring acid rainfall—a unique natural phenomenon in the area of Sofia, our capital. GIS software was used to integrate field measurement and other relevant data, and GIS analyses were performed. We reported the results of our work at several environmental forums: competitions for young scientists, conferences, and symposia.

Another interesting event that should be mentioned was our participation in GIS Day 2000. It showed us that young people at this school are very interested in GIS and are ready to learn this new technology.

Last year we maintained our contact and cooperation with scientific organizations. This let us increase the number of older people in the club. Now two Ph.D. students have joined us, and this not only helps a lot in the education and training of our younger members but also broadens their possibilities to participate in research projects.

Disturbing the area – changing the landscape with:

❖ **Coal extraction sites**
❖ **Earthworks**
❖ **Mine dump sites**

This leads to distorting native ecosystems and lowers the biodiversity.

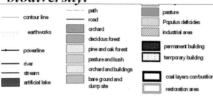

contour line	path	pasture
	road	Populus deltoides
earthworks	orchard	industrial area
	decidous forest	
powerline	pine and oak forest	permanent building
river	pasture and bush	temporary building
stream	orchard and buildings	
artificial lake	bare ground and dump site	coal layers combustion
		restoration area

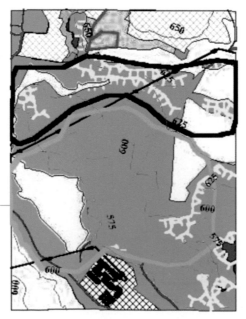

GIS at the Alexander von Humboldt Institute, Colombia

Dolors Armenteras, Principal Researcher, GIS Unit, Instituto de Investigación de Recursos Biológicos Alexander von Humboldt, Colombia

Colombia is considered one of the world's most diverse regions for flora and fauna and has been identified as a megadiverse country. The loss of biodiversity and landscape transformation is occurring at such a rate today that entire ecosystems are under threat of being lost forever. The mission of the Humboldt Institute is to promote, coordinate, and carry out research that contributes to the conservation and sustainable use of the biological diversity in Colombia. The Institute is in charge of the realization of basic and applied research on the genetic resources of national fauna and flora and of creating and arranging the scientific biodiversity inventory throughout the country.

The GIS Unit was formed in 1998. In 1999 and 2000 the unit consolidated as a support unit for the rest of the research programs within the Institute. The main objective is to help with the storage, management, analysis, modeling, and display of geographic data required by the Institute's programs, projects, and researches and develop GIS-based decision support systems for decision makers in conservation and biodiversity matters. 2001 is the year in which the unit has began to undertake high-quality research on GIS, and it is expected to have at least two international publications from recent projects submitted for approval before the end of the year. It is seeking to strengthen the unit and to realize more specialized spatial analysis. In particular, we are focusing on geostatistical analysis, biogeography, and biodiversity modeling as well as adapting gap analysis to local conditions.

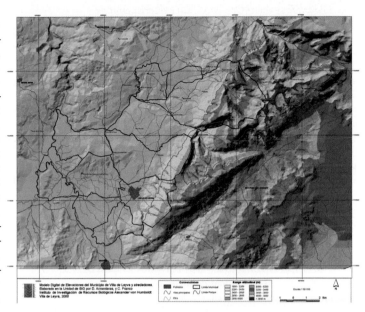

Examples of our recent work include research at different levels—local, regional, and national. At the local level, we helped the local authorities with management plans for the Municipio of Villa de Leyva, where Humboldt has its headquarters. We provided support and basic GIS information and generated a digital elevation model of the area that includes a flora and fauna sanctuary in the vicinity of Iguaque (see map above).

At a regional level we have done some research on the Eastern Andes, and we submitted a paper of our research on ecosystem fragmentation and the representatives of protected natural areas in the eastern Andes. This work suggests areas of high priority for conservation actions. We hope that this paper will soon be accepted for publication. We also had the Ecosystems of the Eastern Andes Mountain Range in Colombia map accepted for the ESRI Map Book, Volume 16, thanks to the conference pass given by the ESRI Conservation Program in 2000. We expect to keep contributing as much as possible to the knowledge and conservation of biological diversity. At this level, we also did some analysis of the distribution, degradation of habitat, and threat to 88 bird species dependent upon montane forest in the Eastern Andes.

At a national level, we tried to give some indications of possible botanical gaps (see map at left), analyzing sites where botanical inventories have been carried out and areas where there are a clear lack of them, for future field expeditions.

Finally, we hope to continue on with the work undertaken so far in a biologically rich country such as Colombia where the role and importance of new technologies such as remote sensing and geographic information systems are rapidly growing. These technologies are a research tool in the investigation of biological diversity and can make data readily available for scientists and decision makers to execute their biodiversity planning and lead to better conservation and management strategies for Colombia.

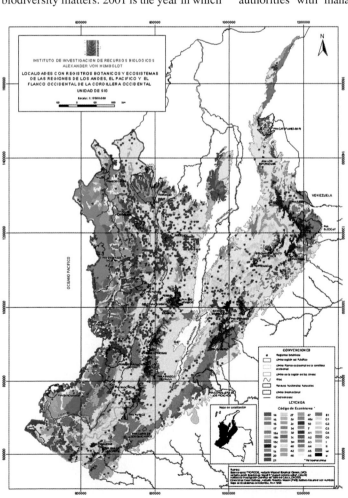

A Biodiversity Vision for the Northern Andes Ecoregional Complex

Daniel Arancibia Vega–Centeno   davc@terra.com.pe
World Wildlife Fund, Peru

The biological diversity of the Northern Andes Ecoregional Complex (CEAN in Spanish), located in South America and ranging from western Venezuela to northern Peru, is considered by many scientific and conservation institutions to be both globally outstanding and significantly at risk. World Wildlife Fund–U.S. (WWF–US) has classified this ecoregional complex as one of the organization's twenty-five priority ecoregions based on its unique biodiversity and threatened status. Conservation International (CI) has identified the Northern Andes as one of the world's twenty-five biodiversity "hot spots" (Mittermeier, et al. 1999). Bird Life International also classifies virtually the entire ecoregional complex as an Endemic Bird Area (EBA) of global importance (Statersfield, et al. 1998; Wege and Long 1995), while the International Union for the Conservation of Nature (IUCN) and World Wide Fund for Nature (WWF–I) recognize the uniqueness of the flora of this region, identifying at least nine Centers of Plant Diversity (Davis, et al. 1997) in the tropical Andes.

The CEAN comprises an area of approximately 490,000 km², covering montane forests and paramos of western Venezuela, Colombia, Ecuador, and northern Peru.

In recognition of this, collaborators from five countries—Ecuador, Colombia, Peru, Venezuela, and the United States of America—have joined forces with WWF to analyze the patterns of biodiversity in the region and establish minimum spatial and distributional requirements for its long-term conservation. The analysis process followed the recently established approach of ecoregional conservation planning (ERC, Dinerstein, et al. 2000), a procedure that scientifically establishes conservation priorities and identifies opportunities for achieving them.

This paper summarizes the framework used to develop the biodiversity vision for the CEAN, where ERC aims to identify the conservation actions required to safeguard, over the long term, natural communities such as the paramo communities of Espeletia spp. viable populations (of the highly endangered Andean Bear [*Tremarctos ornatus*] and Mountain tapir [*Tapirus pinchaque*]), ecological and evolutionary processes, large blocks of natural habitat, and connectivity among blocks of natural habitat.

To address the conservation goals for the CEAN and to make use of available biological and socioeconomic information, the development of this biodiversity vision combined several methodological techniques adapted from Dinerstein, et al. (2000). This method defines and prioritizes areas where conservation action is needed to preserve the ecoregional biodiversity and maintain natural processes. This method also defines a framework for actions and conservation opportunities in the ecoregional complex to a broader audience, which includes governments, local NGOs, and other environmental and civil organizations.

This priority-setting exercise, which is based on extensive GIS work, involved assessing the biological diversity through countrywide analysis, vegetation analysis (a vegetation cover map was elaborated, using satellite imaginary), experts workshops (which helped establish the most biological outstanding areas), and conservation gap analysis. Once a clear understanding of the area's conservation status was reached, the identification and ranking of priority areas for conservation followed (based on biological importance, ecological processes, and intactness).

Daniel Arancibia, Northern Andes coordinator at WWF Peru Programme Office, received an ECP grant last year and is a member of the Northern Andes Ecoregional team. This paper was made based on the biodiversity vision document (under preparation).

These priority areas are also analyzed in terms of their socioeconomic condition to determine threats and opportunities and prepare effective strategies. Approximately 60 percent of the Colombia and Ecuador population lives in the CEAN, putting pressure on the natural resources and demanding a sound development model, which should balance protection (through protected areas) and sustainable resources use. This analysis shows how ERC has become an effective and integral conservation tool.

The biodiversity vision for the CEAN is on the last stage of review and consulting, and its publication and dissemination to the stakeholders involved will be in three months. WWF and its partner organizations who have been working for years in the area, hope that this joint effort will be useful as a scientific working tool for all people who want to support the conservation of one of the most important and threatened areas in the planet.

NORTHERN ANDEAN ECOREGIONS

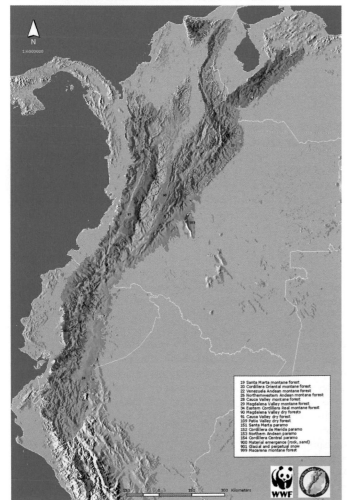

19 Santa Marta montane forest
20 Cordillera Oriental montane forest
22 Venezuela Andean montane forest
26 Northernwestern Andean montane forest
28 Cauca Valley montane forest
29 Magdalena Valley montane forest
34 Eastern Cordillera Real montane forest
90 Magdalena Valley dry forests
91 Cauca Valley dry forest
109 Patia Valley dry forest
151 Santa Marta paramo
152 Cordillera de Merida paramo
153 Northern Andean paramo
154 Cordillera Central paramo
900 Material emergence (rock, sand)
901 Glacial and perpetual snow
999 Macarena montane forest

Ecoregional Planning and Monitoring for the Chocó

Gómez–Navia, Luis Fernando, Junior Program Officer   lfgomez@wwf.org.co

Abstract

This project presents the partial results of the process carried out by World Wildlife Fund Colombia Program Office (WWF CPO) between February 2000 and June 2001 to define the most important areas for conservation in the Chocó Ecoregion based on the available information about biodiversity and socioeconomic areas and the respective analysis to define an ecoregional long-term conservation vision.

Background

The word Chocó comes from the tribes Kuna and Embera, or Chocó, the first inhabitants of the west coast of Colombia (WWF 2000). The biogeographic region of the Neotropical Pacific (Chocó) constitutes an extensive forest zone that extends from the north of Ecuador, including all the Colombian Pacific coast, to the south of the Republic of Panama. The climatic, ecological, and geomorphologic characteristics of the region have allowed the sprouting of an extraordinary biological diversity including numerous species of typical fauna and flora of South America (IberoMaB, 2000; conservation.org).

For several years, WWF has developed projects for the conservation and sustainable use of natural resources in Chocó biogeográfico. The geographic priorities were defined in an analysis process carried out between 1985 and 1986 and updated in 1990. Based on these analyses, and the present processes in the Colombian Pacific and northwestern Ecuador, WWF Colombia formulated a project: "Conservation and Sustainable Management for Chocó Biogeográfico Ecoregional Complex." This project is executed jointly with diverse partner institutions that are distributed throughout the Ecuadorian and Colombian Pacific corridor. As a result of this work, different cartographic products have been processed that have served as support for the decision making on use and management of natural resources at the local scale.

The present proposal tries to update and develop an integral long-term vision of the ecoregional complex that will optimize the decision making related to the conservation activities at the ecoregional level based on the information produced by the GEF Biopacífico project (1993–1997) and Zonificación Ecológica del Pacifico Colombiano (IGAC) (1995–1998).

The objective of the project was to define the elements for a conservation strategy for the Chocó Ecoregional Complex with a basis in a biogeographic analysis of distribution patterns of Chocó's biodiversity and understanding the social, cultural, and economic dynamics at an ecoregional level.

Methods

The analysis was carried out using the following GIS tools: ARC/INFO 7.2.1 and ArcView GIS 3.2 with ArcView Spatial Analyst extension for GPS information support and satellite images analysis. Information was analyzed at 1:250,000 scale. We compiled, edited, homogenized, and analyzed information on biodiversity and socioeconomic aspects including protected areas, indigenous territories, forest reserves, infrastructure, base information, and original and actual ecosystems maps (shown at left).

The process has included the attainment of the necessary cartographic sources, digitalization and editing, database joints, maps production, support from the biodiversity and

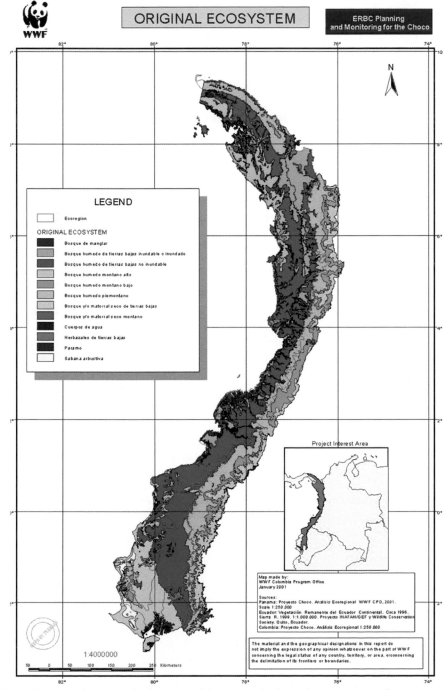

Respecting the project areas, these are some of the zones and cultures that we support in the "Conservation for Sustainable Development in the Chocó" project.

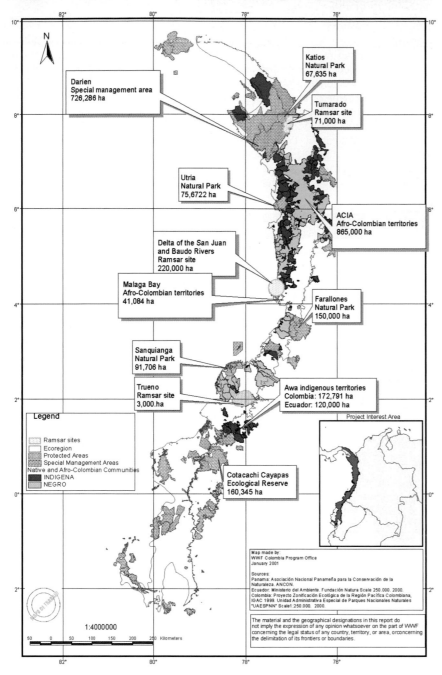

LEGAL STATUS Vs PROTECTED AREAS

ERBC Planning
and Monitoring for the Choco

Katios
Natural Park
67,635 ha

Darien
Special management area
726,286 ha

Tumarado
Ramsar site
71,000 ha

Utria
Natural Park
75,6722 ha

ACIA
Afro-Colombian territories
865,000 ha

Delta of the San Juan
and Baudo Rivers
Ramsar site
220,000 ha

Malaga Bay
Afro-Colombian territories
41,084 ha

Farallones
Natural Park
150,000 ha

Sanquianga
Natural Park
91,706 ha

Trueno
Ramsar site
3,000.ha

Awa indigenous territories
Colombia: 172,791 ha
Ecuador: 120,000 ha

Project Interest Area

Legend

Ramsar sites
Ecoregion
Protected Areas
Special Management Areas
Native and Afro-Colombian Communities
INDIGENA
NEGRO

Cotacachi Cayapas
Ecological Reserve
160,345 ha

Map made by:
WWF Colombia Program Office
January 2001

Sources:
Panama: Asociación Nacional Panameña para la Conservación de la
Naturaleza. ANCON.
Ecuador: Ministerio del Ambiente. Fundación Natura Scale 250.000. 2000.
Colombia: Proyecto Zonificación Ecológica de la Región Pacífica Colombiana,
IGAC 1999. Unidad Administrativa Especial de Parques Nacionales Naturales
"UAESPNN" Scale1:250.000. 2000.

The material and the geographical designations in this report do
not imply the expression of any opinion whatsoever on the part of WWF
concerning the legal status of any country, territory, or area, or concerning
the delimitation of its frontiers or boundaries.

1:4000000

50   0   50   100   150   200   250 Kilometers

socioeconomic groups, preliminary analysis executed with grid covers and data models, satellite images analysis, supervised classification, field trips, and methodological coordination with project collaborators.

Selection of Areas
The priority areas will be selected by overlaying a series of variables from biodiversity and socioeconomic criteria such as the following:

- Strategic unique ecosystems.

- Diversity of species and ecosystems.

- Vulnerable ecosystems by deforestation.

- Restricted species.

- Status of conservation.

- Fragmentation analysis.

Socioeconomic Area
The development of socioeconomic criteria for the Chocó has required the identification of conflict areas for conservation based on the analysis of cultural space patterns, predominant productive systems, legal state of land tenure, and transformation of original ecosystems.

The socioeconomic analyses were carried out using the following information:

- Cultural landscapes analysis (geographic characterization of the region population such as African–American, indigenous, and peasant's groups).

- Productive agro-ecosystems analysis (categorization and analysis of predominant productive systems for ecoregions and their impact on the natural ecosystems).

- Land tenure and population density analysis

- Anthropogenic pressure analysis (criteria definition to evaluate land use impact over original ecosystems).

- Census information analysis for each of the countries and quality of life indicators analysis.

- Norms and policies that affect the state of conservation in the ecoregion.

Partial Results
The GIS tools have been used to produce a homogeneous geographic database at the ecoregional level; it has required very intensive work because of the different parameters and quality of information from each country (Panama, Colombia, and Ecuador).

The consolidated information allows the generation of partial spatial analysis with the biodiversity and socioeconomic information and their respective planned overlaying, as support for the analysis and definition of the conservation priorities at the ecoregional level.

Biodiversity Area
A new database system for biological information was designed and implemented. This system includes standardization and homogenization processes, which are important for importing the information to the new database system. This information homogenization was made in approximately thirty-six tables of databases and included the following steps:

- Elimination of duplicated codes (in the different dictionaries and data cards).

- Elimination of duplicated localities, authors, taxonomies, and sources (by spelling, syntax, abbreviations, types, and others).

- Overhaul of official taxonomies to make synonymous their sources, authors, and spelling to complement dictionary definitions (collections and herbariums, bibliographical sources, localities).

- Overhaul and corrections of the precise registries with the previous corrections.

- Overhaul and fit of connections when configuring the database to the new database system administrator.

Also, programs to detect and correct errors were developed to consolidate the information. Without these programs the work would be more expensive and delayed.

Another activity carried out was the georeferencing of the specific localities for each registry included. This process included determination of the geographical coordinates and evaluation of the data precision of each registry.

With the information systematized into the database, preliminary analyses to identify biodiversity distribution patterns at the ecoregional level were made, and now the process has advanced toward defining biogeographic zonification for the Chocó Ecoregion that permits identification of high-priority areas for conservation based on the results of the analysis of priorities from the biological and socioeconomic approaches.

Socioeconomic Area

Using the census information of each country supported by additional secondary information, the geographic characterization and analysis of population was made including the analysis of some quality of life indicators and population density.

Parallel to this analysis, the land tenure analysis was made where the protected areas and the territories titled to the communities are considered (see map on previous page).

Land Tenure vs. Protected Areas

Overlaying
The map of productive systems attempts to categorize the different forms of natural resources use in the region, associating these to the type of land tenure conditions and to the corresponding ethnic group. As a result, a matrix of impact valuation on ecosystems is generated from the type of predominant productive system and the related population group.

Predominant Productive Agro-Ecosystems
With the information of planned infrastructure megaprojects for the region and the analysis about norms and regional policies, we will be

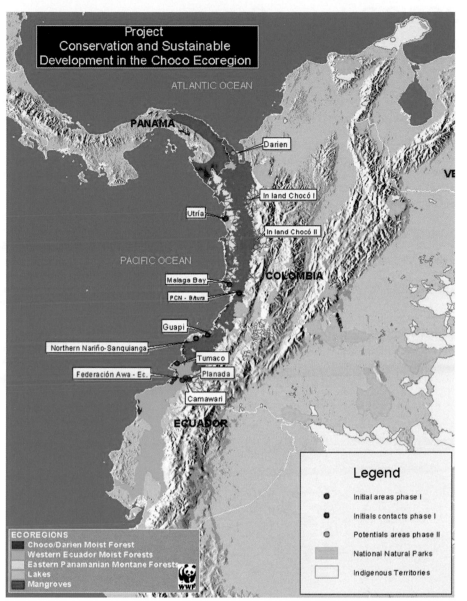
Chocó Ecoregion Map with the WWF Project Influence Areas.

developing a map of threats and possibilities for conservation initiatives that will also have an impact valuation on ecosystems, according to the type of project and ecosystem.

With these elements a human pressures map is constructed as one of the main inputs for overlaying biological and socioeconomic information for the definition of priority areas for the conservation strategy. The high-importance areas from the biologic vision related with the areas that, from the socioeconomic point of view, make possible the work of conservation, and also those that require immediate attention would be defined as high-priority areas. At the moment, we are beginning to make tests for running the analysis, with the necessary adjustments. Also, we are planning an international workshop to present the results to experts from biological and socioeconomic areas of each country to validate and adjust the criteria to identify and define a conservation strategy for the Chocó.

Project collaborators
WWF: Mary Lou Higgins, Carmen Candelo, Mauricio Castro, Cesar Suarez, Silvio Sanchez, and Giovanni Tovar.

Socioeconomic Area: Enrique Sanchez and CECOIN staff.

Biodiversity Area: Thomas Walshburger, Adriana Hurtado, Milton Romero, Maria Lucia Rosas, and Andres Rubio Trogler.

Institutions
Fundación Natura Ecuador

Centro de Datos para la Conservación–Ecuador

Asociación Nacional para la Conservacion–ANCON (Panama)

SCGIS Scholars Report

Luis Fernando Gómez Navia, 2000 SCGIS Scholar
World Wildlife Fund, Colombia Program Office

I am an agronomist, with a touch of poet and genes of pedagogy. My experience in GIS was born with the opportunity to work with WWF Colombia through an internship program. For about four years I have been enjoying the opportunity to work with new communities that include farming, indigenous, and African–American peoples.

In this effort important interinstitutional and interdisciplinary alliances have been established at regional, national, and international levels to promote and enrich the work of conservation that WWF develops.

I share my experience with a great staff at the WWF Colombia Program Office with excellent professional performance. We are working to promote alternatives for a better future; we have had good experiences applying GIS tools. How we have done so is the subject of my presentation at the conference.

I extend to all of SCGIS members a warm and tropical greeting from the corner of South America, the beautiful Colombia.

The following is the introduction to the conservation effort of the WWF that is referred to on our Web page at www.wwfcolombia.org.

WWF in Colombia has focused its work on the ecoregions of the Northern Andes and Chocó with very positive results. It has progressed with the establishment of new collective territories in the Pacific, which is an important step in the long term to assure the sustainability of these areas. In the Andes and other regions of the country, the establishment of private reserves is increasing, and the lease and strategic creation of networks of reserves promises to increase its impact on conservation. We have also started voluntary forest practices certification programs in neighboring communities.

WWF in Colombia has made important gains and had good opportunities as far as the creation of more favorable environmental policies related to incentives, establishment of protected areas, environmental infrastructure impact, and voluntary forest certification. All these elements have been supported through the mandate, directed to strengthen the role of the civil society in conservation (citizen formation and environmental education). The impact of the training and technical attendance and the priorities in the development of tools and abilities have demonstrated that the qualification helps people to find their own solutions.

La Campana–Peñuelas Biosphere Reserve, Central Chile

Andrés Moriera–Muñoz, Director, Taller La Era, Santiago de Chile

La Campana–Peñuelas Biosphere Reserve is located in the geographic and political center of Chile, between the country's two most populated regions (Valparaíso and Metropolitan) where half of the country's population lives (about 7 million people). The reserve protects mainly the central Chilean matorral, one of the twenty-five most important ecoregions for biodiversity conservation worldwide (Myers, et al 2000; Dinerstein, et al 1995). About five hundred fifty plant species can be found in the reserve, and among them are a large number of endemics. Also many endemic and endangered animal species inhabit the reserve. The principal protected ecosystems are the Chilean palm tree formation (one of the three important existing "palmares" in the country); the "roble" formation, in fact the northern limit for the South American distribution of Nothofagus, the most important genus for the subantarctic forests; important remnants of sclerophyllous woodlands, the original most extended formation before the human settlement in the central valley; hidrophyllous forest in the deepest valleys, including important trees like "patagua" (*Crinodendron patagua*) and "belloto" (*Beilschmiedia miersii*) declared as a national monument due to the decrease of its populations; and xeric formations composed basically of endemic Bromeliaceae.

The biosphere reserve comprises an area of about 17,000 hectares, actually composed of two separate cores, La Campana National Park and Peñuelas National Reserve. For the moment we concentrate our work in the national park and its surrounding area because of its higher biological value. The biosphere reserve was declared in 1984, but the matrix that surrounds the core has never been managed as a buffer zone, and the historical threats that affect the reserve have not been solved and have increased in the last decades. These threats have a long history due to human activities that affected central Chilean matorral for more than ten thousand years. The reserve has been affected directly by mining and road construction, woodcutting for construction and charcoal, goat and cattle pasture, and the introduction of exotic fauna. In the last years, unsolved legal problems for the land property have increased the threats and generated a bad relationship with the surrounding rural communities.

Thanks to national and international founding agencies, like Fondo de Las Américas, and also a donation from the ESRI Conservation Program (PC ARC/INFO and ArcView GIS), we can begin working on a project for managing the national park at its biosphere reserve status.

The first goal of the project was to present prior research on the reserve in a more simple format for the benefit of the general public, visitors, and surrounding communities. We are developing a multilevel education program for the park guards, teachers, students, and the general public, which includes the main species and ecological relations, the high biodiversity of the area, and also the main historical and traditional issues of the local communities related. Some important research themes were underdeveloped like detailed vegetation cartography. We use Landsat images, aerial photos, and field research integrated through ArcView GIS 3.2 software, making maps that help both objectives of management and education in the reserve.

The most important objective is to work on a participatory cartographic approach for the design of the buffer and transition zones around the core, developing a "Cooperation Strategy for the Participatory Management of the Biosphere Reserve" with the surrounding rural communities. People realize that a better knowledge of their territory can be a good chance for the design of sustainable development strategies, improving their quality of life. Therefore, the traditional position against conservation objectives is now changing, and communities are beginning to act as partners on a common strategy for the conservation and sustainable use of this unique ecosystem.

Some of the emerging challenges are to:

- Build a Visitor and Education Center to continue the education program (we are already working with ten schools, and thirty more schools in the area are anxious to join the program).

- Translate GIS knowledge and technology directly to the communities around the reserve and to the schools.

- Create a conservation spatial database for the maximum potential extent of the biosphere reserve including land use, forest fragments, endemic and endangered species concentration areas, and land tenure.

- Explore and design sustainable land uses (like certified organic agriculture) and ecotourism that will give additional jobs to locals and will save not only the ecosystem but also historical and cultural monuments and traditional ways of land use.

- Build a Biological Research Station, which will help the work of many ecologists who are constantly studying ecological processes in the reserve.

Davis, S. D., V. H. Heywood, O. Herrera–Macbryde, J. Villa–Lobos, A. C. Hamilton, eds. 1997. *Centres of Plant Diversity: A Guide and Strategy for Their Conservation*. Vol. 3: The Americas. WWF, IUCN.

Myers, N., R. A. Mittermeier, C.G. Mittermeier, G. A. B. da Fonseca, J. Kent. 2000. *Biodiversity Hot Spots for Conservation Priorities*. Nature 403:853–858.

Fundacion Conservacion Internacional, Peru

Carol L. Mitchell, Director   ci-peru@conservation.org; carolm@wayna.rcp.net.pe
Luis Espinel, Project Director   luisec@terra.com.pe

We use GIS technology to support sustainable development in one of the most diverse, remote, and least technologically developed areas of the globe, the rain forests of Tambopata, Peru. Our program produces maps that rural communities use in sustainable development projects and provides geographic information to local government offices involved in land and resource use planning. We need this grant to improve our efficiency and the visual impact of our work, allowing us to reach more communities in our area and to produce maps and other information that serve in local and national decision making procedures concerning conservation issues.

Our primary goal in Tambopata is the conservation of the astounding biodiversity of Bahuaja–Sonene National Park, a one million hectare area of pristine rain forest harboring one of the most diverse communities of flora and fauna in the world. This corner of Peru is the least altered area of the entire country and forms part of the southwestern Amazon Tropical Wilderness Area, where contiguous large tracts of forest are still present in Peru, Brazil, and Bolivia. Our mission is to ensure that Bahuaja–Sonene remains intact and that human activities in the surrounding Tambopata National Reserve and buffer zone are developed in a sustainable manner that preserves the health of ecosystems inside and outside the national park.

We combine basic research on biodiversity and rain forest ecosystems with work on sustainable development and appropriate use of natural resources in the Tambopata area. We have current projects on wildlife and fisheries management in native communities and sustainable use and economic development of forest products including timber species, palm species, and Brazil nuts. All of these projects require GIS services, but the project that uses most of our GIS capacity at this time is the Land Capability and Appropriate Use (LCAU) Project.

The LCAU project works to produce maps of theoretically appropriate land use systems for the rural agricultural communities living in the Buffer Zone of Bahuaja–Sonene National Park. Using eight different variables based on vegetation and edaphic factors, we create maps that indicate the areas of community lands appropriate for agriculture, two different levels of agroforestry systems, and forestry systems. For each community, each of the eight variables as well as the final integrated appropriate land use map are mapped and plots produced. We also are creating maps of current use to be able to monitor the changes that occur and track the approach to (or retreat from!) the optimum. In addition to being a tool for use within the community, these theoretical maps are a necessary part of the legal paperwork required for the development of community projects with local government

institutions. At the request of farmers, we are beginning to use this technique at the level of individual agricultural plots within communities as well.

Part of our conservation strategy in Tambopata includes support to local government institutions, especially the National Institute for Natural Resources. In 1999–2000 we were invited to participate in government land planning procedures, producing maps used for the zoning of the Tambopata area into national park, national reserve, and buffer zones. We have also provided geographic information to the local Ministry of Agriculture and the local police force. We are currently the only local office with GIS capacity, and as such, are able to disseminate a conservation message with each map and service we provide.

Block 78 EISA Project Map

The map on this page shows petroleum exploration Block 78 in Peru, where Fundacion Conservacion Internacional carried out a project (EISA) evaluating the environmental impacts of seismic exploration and exploratory drilling for hydrocarbons. Shown are protected areas, borders of the hydrocarbon exploration block, seismic lines, and our field study sites. Much of this geographical information was acquired from Mobil Exploration Producing, Peru, the company in charge of hydrocarbon exploration in Block 78.

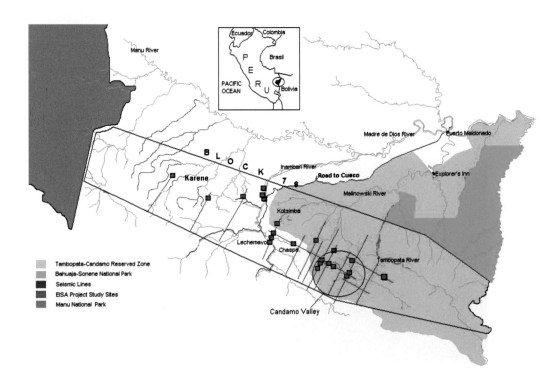

Optimum Paths Map

Another map (this page, top) was created during a project to improve the harvesting procedures for Brazil nuts in concessions in the Tambopata area. Brazil nuts are harvested year after year, and the huge trees are accessed by permanent trails cut through the forest. Nuts are collected at the base of the tree and carried out in gunnysacks on worker's backs along these same trails to an outlet port. These trails are created initially by "rumberos," who specialize in finding the Brazil nut trees simply by walking through the forest. The trails begin as slightly meandering, but over the years become more roundabout due to tree falls, regrowth, and recutting.

Our project used GPS units to locate all the Brazil nut trees in a particular concession and then created a program to find the minimum path length to reach all the trees and access the outlet port. These trails were then cut through the forest; and these optimum paths reduced the length of trail by about 30 percent, saving time, effort, and money for the Brazil nut harvesters.

Tambopata Rural Axes Map

This map at the bottom was created for our Tambopata Strategic Conservation Plan and shows native community lands, protected areas, and areas where different cultural groups live in Tambopata. This kind of visual aid is key to the success of planning workshops.

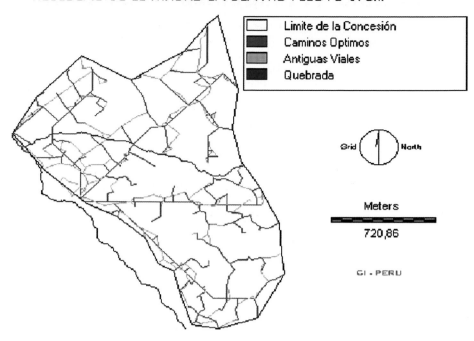

REDISEÑO DE ESTRADAS EN CENTRO PILOTO J. Ch.

Límite de la Concesión
Caminos Optimos
Antiguas Viales
Quebrada

Grid | North

Meters
720,86

GI - PERU

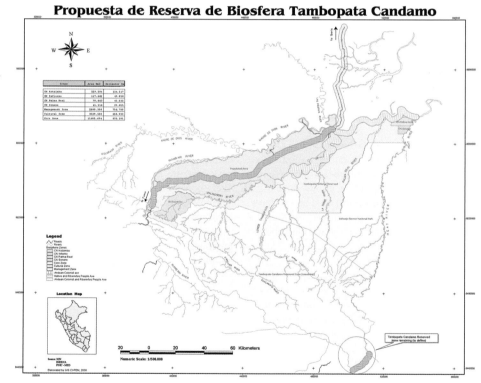

Propuesta de Reserva de Biosfera Tambopata Candamo

Anthrowatch, Philippines

Trina Galido–Isorena anthrow@psdn.com.ph
www.anthrowatch.org

I am Trina Galido–Isorena from Manila, Philippines. I work with a nonprofit organization called Anthropology Watch or Anthrowatch. Our organization is composed of researchers with different expertise and training but bonded by concern and development for the work done with indigenous people and other marginalized populations.

I came to know of the organization in 1998 when I was able to join one of its activities in the Sierra Madre Mountains—the longest mountain range in the Philippines. It was part of the ethnographic survey of the Agta, a seminomadic tribe who resides in mountains and river valleys of the Sierra Madre range.

Drawing from members with various expertise, Anthrowatch aims to facilitate and advocate a holistic and integrative approach to the development process. We support a process that is sensitive to the nuances of cultural diversity, humanization, and ecological principles.

Research is the core component of Anthrowatch's involvement in the development work. The research it conducts is aimed at understanding the living circumstances of tribal groups and the problems they face as part of the less privileged of society. Through this, the organization is better equipped in implementing the Indigenous Peoples' Rights Act (IPRACT) that legally acknowledges their right to claim their ancestral land and to have sole decision in the management, sustainable utilization, and preservation of the land and its resources.

Anthrowatch also provides other organizations working with indigenous people with training on cultural sensitivity because such training is important in understanding and working with marginalized groups.

The organization is active in the advocacy of issues affecting indigenous people (e.g., lobbying for the passage of the IPRACT). Recently it was invited to be part of the Presidential Task Force on the Ancestral Domain Claim to help in the review of pending applications by different tribal groups.

Mapping and GIS was incorporated in the studies and projects that Anthrowatch was involved with in the past. Today, there is a specific GIS section that aims to assist communities in the delineation of their ancestral domain and in its sustainable development and protection plan. In the future Anthrowatch intends to build a spatial database of information related to indigenous people that could help guide government as well as nongovernment programs in interventions for these communities.

Most of the teaching or training that I have conducted involved government staff rather than nongovernment groups. This is because in my country nongovernment groups usually have more capability and are more progressive in their approach to their work. That is why training and capability building activities are mostly geared toward government staff.

My experience included training the Department of Environment and Natural Resources technical staff as well as local government staff. Training ranged from simple lectures on the basic concepts of cartography that help them understand GIS processes more easily to the actual process of spatial data building and using such data for planning. Because most of the training also included people from the planning and development divisions of local governments, the training has been fruitful. They now believe in the usefulness of GIS and are more confident that they can use the latest technology in their planning process. As a result, most of them have pushed for the purchase of computerized mapping and GIS equipment in their respective municipalities.

It is common for the people I work with to initially see GIS as a "pie in the sky." Most of them think it is too high-tech for their understanding as well as too expensive. I am happy and proud to say that I have helped in changing this perception (although not the price tag) and convincing them that GIS is worth having, especially in times when there are a lot of factors to consider in trying to deal with environmental problems in their areas.

I believe what is most challenging is to be able to bridge the gap between the community and the government in the decision process in any endeavor—whether it is for planning infrastructure or developmental activities or in managing protected areas and ancestral domains. My work as a geographer has helped support development decisions in local governments that are more collaborative and less top-down.

"Survivors": For Borneo's People, It's Not Just a TV Show

Dan Scollon
From the Earth Island Journal, *Winter 2000*

The journey from San Francisco to the village of Keluan in northeastern Borneo was an arduous one with more than twenty hours of flying time, and stops in Tokyo, Bangkok, and Brunei. Then four hours upriver by boat, followed by four hours waiting in the oppressive heat of a logging camp and, finally, an hour crashing down a logging road in a four-wheel-drive vehicle.

It was all worthwhile, though, when my wife and I reached the top steps of the longhouse and were greeted by the smiling faces of my Keluan friends, who are ethnic Dayaks. I'd last seen them five years ago when I'd helped train the villagers in basic mapping techniques.

The fruits of that work can now be seen everywhere: dozens of villages mapped, an experienced team of local mappers, and exchanges with other mappers in Indonesia. The most exciting news is that the maps have been used to support land claims in the courtroom. Cases filed several years ago are finally being decided, and we've already won four major legal victories in the last eighteen months.

Over the ensuing days, we took part in the daily life of the longhouse—communal meals on the floor of the verandah, hunting and gathering trips in the forest, and singing and dancing in the evenings. Amidst the joyful reunion came the sad realization that things had changed for the worse since my last visit.

Traveling upriver, one encounters a seemingly endless procession of barges carrying logs downstream. Logging camps, often staffed by young Dayak men in need of jobs, roar furiously day and night. The labyrinth of logging roads makes travel between villages easier, but further degrades stream and forest habitat.

The most shocking site was the dramatic expansion of oil palm plantations carving deep swaths through the forest. The native communities of Sarawak province continue to show resilience in their struggle to protect their rights and culture in the face of these threats.

Oil Palms Threaten Forests

I had returned to Keluan to help the local mapping team increase the accuracy of their maps by using geographic information systems (GIS), sophisticated computer software that turns huge amounts of data into multi-layered maps.

While the technical complexity of GIS usually makes it both unnecessary and difficult to implement at the village level, it is particularly useful for combining village maps with regional data to address large-scale landscape changes, in this case, massive oil palm plantations.

The GIS map will allow opponents of these plantations, including one that will destroy 30,000 hectares of rain forest inhabited by seven villages, to compare the village boundaries with the plantation boundaries and force the government to recognize that the latter violate traditional landownership laws.

In the future, GIS will help communities develop sustainable resource management plans and provide the Borneo Project with an opportunity to expand coordination with local NGOs focused on addressing Borneo's native rights and conservation concerns.

Diversity vs. the Global Economy

After a week of training mappers, I traveled with two community activists to a conference sponsored by the Borneo Research Council in Kuching, capital of Malaysia's Sarawak province. Academics, government officials, and a few activists gathered to discuss issues facing Borneo's immersion in the global economy.

Much of the research focused on the island's biological and cultural richness. A World Wide Fund for Nature study, for example, revealed the presence of five hundred fourteen plant species (including twenty-two previously unknown to science) in the Ulu Padas region of neighboring Sabah province. Despite its tremendous floral diversity, this region is currently included in a government-granted logging concession.

Indigenous land rights were also a hot topic at the conference. Borneo's indigenous peoples live in a symbiotic relationship with the plants and animals of the forest. Although many researchers presented evidence that indigenous land use practices are consistent with conservation objectives, the government has amended the constitution to make it easier to extinguish native customary rights (NCRs).

An amendment passed in spring 2000 limits NCR land to that cleared for cultivation, disregarding communal forests, hunting and fishing grounds, and areas of forest cultivation. The amendment also allows individual community members to sell farms traditionally held as communal property, a tactic aimed at undermining community solidarity.

With forest resources dwindling, the Sarawak government is aggressively pushing to convert NCR lands to oil palm plantations. Oil palms have recently replaced logging as Sarawak's main source of export revenue. Many villagers I spoke to expressed particular concern about these plantations. At least with logging, one man told me, the forests can eventually rebound. Plantations, however, require clearcutting and are leased out for a period of sixty years.

Although communities do receive some monetary compensation from oil palm plantation developers, they must first surrender their land, the foundation of their culture and livelihood. This will force many forest dwellers to either migrate to coastal cities in search of work or become indentured laborers on what used to be their own land.

The forces of global economics and local politics are formidable, but not insurmountable. Native communities believe they can adapt to these changing realities without losing the essence of their cultures. The Borneo Project's efforts are helping to ensure that Dayaks can remain in their forest homeland in a way that is both ecologically and economically sustainable.

As I left Borneo, I peered out of the airplane window at the verdant expanse below and wondered what I would find on my next visit to this mystical island. Will my friends still have a forest to sustain their livelihood or will they have become oil palm wage earners in the global economy? My hope is that, through willpower and effort and with a just cause on their side, they will find a way to retain their vital culture as they transition into the new century.

Sahabat Alam Malaysia (Friends of the Earth Malaysia)

smidris@tm.net.my
Jok Jau Evong   jvon_ei@yahoo.com
Harlan Thompson   borneo@earthisland.org

Sahabat Alam Malaysia (SAM or Friends of the Earth Malaysia) works to change national and state policies regarding destructive development projects and commercial logging, as well as discrimination against Malaysia's indigenous groups. Since its founding in 1977, SAM has been one of the most established and well-respected environmental and human rights organizations in Southeast Asia. For its efforts, SAM and SAM staffers were awarded the Alternative Nobel Prize in 1988, the Goldman Award in 1991, and the Conde Nast Traveler Environmental Award in 1998.

Our activist programs include ecologically sustainable development, income-generating programs, and protection of legal rights. In Sarawak, community-based mapping has become the primary program supporting legal land claims.

SAM has offices in Kota Kinabalu, Penang, and Marudi, with our mapping program centralized in Marudi. SAM Marudi has been on the forefront of indigenous struggles in the Malaysian province of Sarawak since the mid-1980s. For the last six years, SAM Marudi has been closely based with the U.S.-based Borneo Project. Both groups share an exclusive focus on the Baram River Basin of Northern Sarawak.

SAM Marudi provides community organization and training to a vast network of villages in the Baram watershed. All too often, indigenous people living in these areas are overwhelmed by the extensive resources of outside forces that seek to profit from the natural resources there. Lacking legal representation, technical skills, and often even basic literacy, historically these people have been pushed

SAM community mappers.

aside and left behind while large companies from far away exploited their traditional lands. Typically, as the forests on which they base their livelihoods are diminished, people are forced to give up their traditional ways of life and culture and move to large coastal towns to face an uncertain future (see map below).

PETA MENUNJUKKAN KAWASAN SEMPADAN TANAH NCR DAN TATA-LAHAN GUNA TANAH SERTA SUMBER HUTAN KAMPUNG PENAN LONG SAYAN, SUNGAI APOH, TUTOH, BARAM.

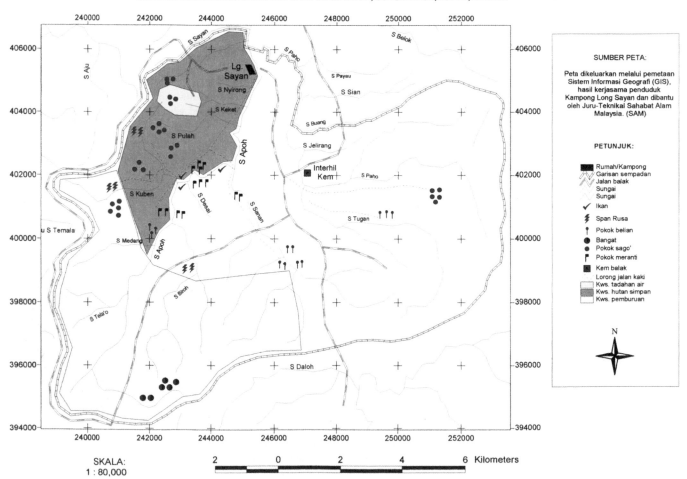

SAM promotes alternative models of development that aim to benefit these people directly, giving them an opportunity to adjust to the modern world on their terms. The key to this effort is securing land ownership. Thus, the major focus of SAM Marudi is mapping to document ancestral land claims. These maps are then presented directly to encroaching logging or plantation companies who arrive expecting little opposition to their industrial scaled efforts. If these initial demonstrations of local organization and land claims are not enough, the maps are taken to the courts to bring about an injunction, and in many cases, eventual legal ruling in favor of the indigenous inhabitants.

However, defending land tenure is not the end of SAM's work. A viable economic model to logging and plantations needs to be pursued to ensure long-term economic sustainability and happiness. SAM provides paralegal training, education, support, and documentation to indigenous communities facing injustice. SAM has looked for and built economic generating possibilities such as fish farms, the marketing of traditional crafts in the First World, communally owned rice huskers, village rice banks, and other projects. One new example is a microhydro power generation project that provides power to villages that previously relied on polluting diesel generators. Another example is a reforestation project to promote the growth of tree species with particular uses for food and materials.

The GIS project to be undertaken by SAM will encompass two broad objectives:

1 Legal defense of aboriginal land rights.

2 Land and resource management.

By developing a GIS in the SAM office in Marudi, Sarawak, there will be an opportunity to produce a database of geographic information to facilitate the production of maps for use in legal land claims. A wealth of information has already been collected showing the present and historic presence of tribal groups in the region including invaluable knowledge of places and resources from village elders (see map, previous page). This data includes compass and tape surveys, GPS points, interviews, and hand-drawn and to-scale maps. The improved presentability and completeness and accuracy of GIS-based, computer-designed maps compared to the current hand-drawn maps will greatly improve the negotiating position toward governmental officials, timber companies, and oil palm plantations.

Malaysian law protects indigenous land ownership, if it can be proved that a community has continuously lived in and used the land of a specified area since 1958. Making these claims not only requires the mapping of current land boundaries and prominent features but also requires evidence that proves long-term occupancy. This can include sacred sites, graveyards, previous habitation sites, locations of previously tilled fields, and the tying in of oral testimony to specific geographic facts. The need to show how information has changed through time would seem a natural fit with GIS. An especially exciting opportunity GIS could facilitate would be the comparison of land use data from 1940s, 1950s, and 1960s in British and Japanese colonial maps and aerial photos. Nothing could better document 1958 land tenure for a village than aerial photo information overlaid onto current land boundaries.

Furthermore, the GIS will provide a valuable tool for resource and land use planning within individual villages. Despite extensive resource destruction in their midst, indigenous communities are asserting their rightful ownership to land and resources. They are working to develop plans for how to best use their land. This includes designating forest preserves for hunting and gathering, identifying farming history to better plan for future crops, and designating areas for special projects. For instance, the village of Uma Bawang is currently adding rattan, a widely used and versatile building material, to their already extensive Forest Restoration Project area. Sites for fruit tree cultivation, fishing, and water resources are also among the resources for which GIS would provide a valuable planning tool.

Access to maps and photos within Sarawak has proven to be problematic. However, as a former British Colony, historic maps and photos of the area are in existence in England. Acquisition of these documents will allow for an evaluation of land cover change over time. This is important information given to dramatic destruction of Sarawak's forests over the past thirty years. Satellite imagery is also a source of information that will be useful for demonstrating land cover change over time.

The world's oldest rain forests are on Borneo, and they house a wealth of biological diversity unparalleled in the world. Yet these magnificent forests have been destroyed at an alarming rate with some logging companies operating twenty-four hours a day. The more recent threat of oil palm conversion threatens to permanently take even degraded forest lands away from their rightful owners.

The native people of Borneo who have lived sustainably in the forests for thousands of years are nonetheless fighting back with an indomitable spirit. Gradually these communities have begun to assert their legal rights to the land, which they and their ancestors have occupied for centuries. Maps have proven to be their most effective tool in demonstrating their rightful ownership of the land and the resources they need to survive. The development of GIS presents the opportunity to use mapped information with even greater effectiveness. In addition to providing a central and standardized means of storing and updating geographic data and producing high-quality maps, GIS will be used to improve local decision making as a land use and resource use planning tool.

By building a geographic database, producing professional maps, and guiding rural development and resource planning efforts, SAM intends to support the indigenous communities of Sarawak to improve their way of life and assert a productive future for subsequent generations.

Madagascar Biodiversity Plateforme D'Analyses Project

Lantoniaina Andriamampianina
Wildlife Conservation Society Madagascar

The island of Madagascar is known as one of the so-called world biodiversity hot spots. Madagascar fauna and flora have interested numerous naturalists and biologists for a very long time. A large volume of biodiversity data has been collected in the field. However, many of these data sets are not available, and those that are available are scattered all over the island and outside the island. In addition to the difficulty of accessing biodiversity data, the insufficiency of technical capacities within the country constitutes a serious handicap to environmental management decision making at all levels. The Madagascar Biodiversity Plateforme D'Analyses Project, locally known as PDA, aims to put in place a structure that will help organize and analyze biodiversity data in Madagascar. This structure will include a network of biodiversity databases through which data can be accessed/shared, permanent analysis tools that will enable users to conduct advanced spatial analyses, and an institutional framework for the management of the whole network system. This structure will hopefully improve conservation planning, natural resource management, and environmental decision making in Madagascar. Specifically, the objectives of the projects are to add value to existing biodiversity data, to favorize biodiversity data repatriation, to facilitate biodiversity data access and analyses for environmental management and conservation planning, and most important, to strengthen national capacity.

Initiated by the Wildlife Conservation Society (WCS) in collaboration with the Center for Conservation Biology (CCB) at the University of Stanford and the USAID-funded project, Projet d'Appui à la Gestion de l'Environnement (PAGE), the PDA seeks the collaboration of all individuals and institutions working with Madagascar biodiversity to reach its goal. Potential partners for this project would include data providers as well as data users and institutions working in the public and private sectors. In fact, much more than a simple network of databases, the final output of this project is more characterized by its analytical tools designed to facilitate/support decision making.

Combining Microsoft Access and ArcView GIS (with several of its extensions), different routines have been created and incorporated in the data analysis tools using a very easy to use interactive dialog box to facilitate users. A fully georeferenced and taxonomically classified ArcView GIS biodiversity database for Madagascar has been compiled. Routines that are already incorporated in the tools include plotting observation points of a given species or group of species on the map, generating a list of species for a given site, and calculating patterns of biodiversity richness or endemism. A mechanism of prediction of the range distribution of a given species is also incorporated based on real data points where the species was observed and its ecological requirements. The tools will have a wide range of users, from

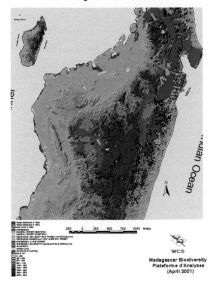

The distribution of the indri (Indri indri)*, the largest of all lemurs in Madagascar in 1972 and 1992. The light blue in the map represents the loss of habitats for indri between these two periods.*

simple researchers to environmental planners and decision makers. The application of the tool will cover various domains including the protected area network management, individual species monitoring through time and space in relation to habitat changes, priority settings for conservation and research, conservation planning, environmental impact assessment, and so on.

During the first year of the project (2000), a prototype of this conservation analysis tool has been created using more than fifteen-thousand records for 1,228 species from ten different taxonomic groups. Those are mainly WCS Madagascar data completed with some other published data. Several institutional partners in and outside Madagascar that have substantial databases on Madagascar biodiversity have already agreed to participate in this project.

There is no doubt that the setting up of this structure, once it is in place and adopted by all its target users, will help to better manage Madagascar's unique natural resources and improve the Malagasy environment management system.

Training and transfer of responsibility for database maintenance to Malagasy technical personnel is scheduled at the end of the project.

Lemurs distribution range

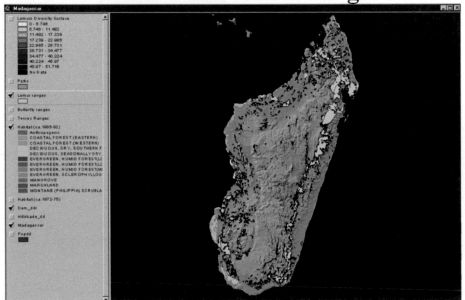

Care Madagascar: Safe Water Project

Lalaina Razafimbololona   *<lalaina_ra@hotmail.com>*

Care Madagascar received a Society for Conservation GIS scholarship in 1999 and 2000. The skills we obtained from this support and training allowed us to undertake the Safe Water Project.

The Safe Water Project enables the community to take charge of all the activities of development and conservation. We did a collaborative analysis of water-quality problems and the constraints they put on the community. Then we elaborated on a strategy and a detailed implementation plan that included monitoring and evaluation. GIS will be used for monitoring, the first component of this project. Other components are planning, research, launch, penetration survey, and final evaluation. We will adjust our activities according to our ongoing analysis and survey result. GIS is useful in making our project more efficient.

Lalaina Razafimbololona, right, SCGI 1999 scholar, discusses a Madagascar conservation map with a Malagasy colleague.

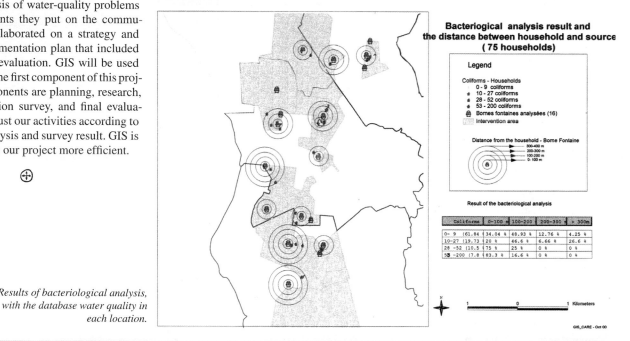

Bacteriogical analysis result and the distance between household and source (75 households)

Legend

Coliforms - Households
- 0 - 9 coliforms
- 10 - 27 coliforms
- 28 - 52 coliforms
- 53 - 200 coliforms
- Bornes fontaines analysées (16)
- Intervention area

Distance from the household - Borne Fontaine
- 300-400 m
- 200-300 m
- 100-200 m
- 0 - 100 m

Result of the bacteriological analysis

Coliforms	0-100 m	100-200	200-300	> 300m
0- 9 (61.84	34.04 %	48.93 %	12.76 %	4.25 %
10-27 (19.73	20 %	46.6 %	6.66 %	26.6 %
28 -52 (10.5	75 %	25 %	0 %	0 %
53 -200 (7.8	83.3 %	16.6 %	0 %	0 %

Results of bacteriological analysis, with the database water quality in each location.

N

1   0   1 Kilometers

GIS_CARE - Oct 00

Cheetah Conservation Fund

Laurie Marker, Executive Director   cheeta@iafrica.com.na
Dr. Richard M. Jeo   rjeo@earthlink.net
www.cheetah.org

The Cheetah Conservation Fund (CCF) is dedicated to saving wild cheetahs, through research, education, and community-based conservation efforts. The mission of the Cheetah Conservation Fund is to be an internationally recognized center of excellence in research and education on cheetahs and their ecosystems, working with all stakeholders to achieve best-practice in the conservation and management of the world's cheetahs.

We recently completed a new research and education center based in central Namibia amidst the largest population of wild cheetahs remaining in the world. From our Namibian and United States offices, CCF conducts conservation activities including the following:

- Development of science-based conservation strategies for cheetahs. We have conducted scientific research in cheetah genetics, morphology, demographics, and habitat use in collaboration with the National Zoo, the National Cancer Center, Oxford University, Round River Conservation, the University of Namibia, and many others. We believe that understanding the basic biology of the Namibian cheetah is critical to the success of conservation efforts in Namibia, and many of the principles learned apply throughout the cheetah's range. We are also developing conservation projects in other countries including South Africa, Botswana, Zimbabwe, Niger, and Iran.

- Reducing conflict between cheetahs and livestock ranchers through the development and implementation of sound livestock management practices. The most immediate threat to cheetahs in Namibia is thought to be indiscriminate killing by ranchers, and our conservation programs have focused on reducing human-caused mortality through research, education, and community outreach. We encourage ranching techniques that reduce conflicts, and we donate Anatolian Shepherd livestock guard dogs to farmers experiencing predator problems. These, and similar measures, have substantially reduced the numbers of cheetahs that have been killed by ranchers over the last ten years and have halted precipitous decline in Namibian cheetah populations.

- Conduct conservation education programs for local villagers, ranchers, and school children. We have developed partnerships with several organizations to implement conservation education programs. We have a United States-based, university accredited field school in cooperation with Round River Conservation Studies and the University of Namibia. We also visit schools throughout Namibia and host local school groups at our Wilderness Camp and more than 30,000 Namibian students have viewed or participated in our environmental education program over the past three years.

Our new GIS program will be expanding into completion of GIS maps of an intensive cheetah study area. We are near completion of small-scale habitat mapping of ranches owned by CCF (about 15,000 hectares). This project will greatly enhance our ongoing habitat and ecosystem surveys.

Matti Nghikembua, Fund Education officer and 2000 SCGIS scholar and friend.

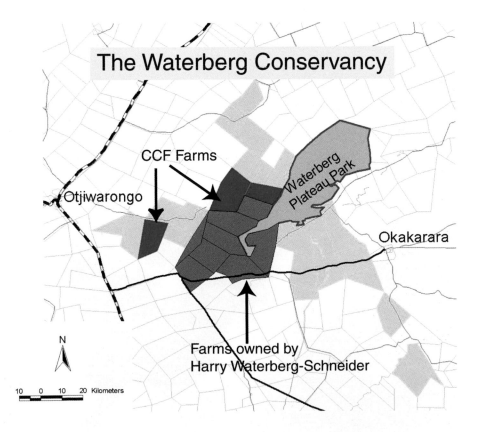

Innovative Resources Management, Inc.

Michael Brown, President

IRM is using the power of maps and mapping technology to help communities realize a sustainable future. We are using these technologies in an innovative way, helping local communities in Africa conserve their natural resources, promoting the sustainability necessary to ensure their livelihoods. The mapping activities have mobilized and energized entire communities, along with local NGOs, toward assuming a greater degree of self-determination and ensuring the long-term sustainability of their resource base. The communities have seized the power and potential of this new tool and are extremely enthusiastic about using the maps to develop land-use plans and even to negotiate with the government to acquire "community forest" status for their lands. Information is power, and the communities of Mokoko, Djoum, and Tikar have proven that they are ready and able to use these maps and the information they contain to better their lives and ensure a sustainable future. One of our roles in helping the communities achieve their goals is to assist them in using available technologies to their best advantage. By translating the raw data into a digital format and using it in conjunction with other data sets, we can provide them with extremely valuable and powerful information that they have not had access to before. Participatory mapping (PM) is a tool of tremendous potential. IRM is committed to using PM to help local communities ensure a sustainable future and to achieve lasting conservation and development goals in the region.

IRM believes that biodiversity conservation can be achieved if the development aspirations of key stakeholders are responsibly addressed. The challenge of achieving sustainable development is daunting. IRM recognizes that sustainable development starts with forging sustainable relationships among partners and that pathways to sustainable development require skillfully managed solutions that match the complexity of the problems. IRM is working to develop and test innovative tools that can build successful teams and multistakeholder coalitions, enhance stakeholder capacities for sustainable development and resource conservation, transform conflicts into broad-based consensus, and enable local communities to assess and negotiate their own conservation and development options. We work closely with communities in Africa and around the world to identify equitable conservation and development options in decentralized, as well as highly structured, governance environments.

In 1998, IRM began an ambitious project to introduce PM to communities in Africa.

Adapting a methodology used in Central America and South America, IRM began by educating and training communities in map technologies and map production. Using GPS and basemaps and assisted by U.S. experts and Cameroonian cartographers from the government of Cameroon's Ministry of Environment, these communities set out to produce land use maps that reflected their vision of the environment and their own resource use patterns. These maps serve as the basis for improved land/forest use agreements that integrate biodiversity conservation and sustainable development.

PM (as practiced by IRM) is an iterative process, involving whole communities and requiring several months of data collection and transcription to complete. The maps are georeferenced and hand-drawn to scale. In their final form they are very large format maps, each map in a set often measuring in excess of six square feet. The communities of Mokoko, Djoum, and Tikar produced map sets of four, six, and forty-two maps respectively, with the latter covering an area of several thousand square kilometers. While it would be technically possible to produce these maps in the field directly by computer, use of this technology in the initial stages

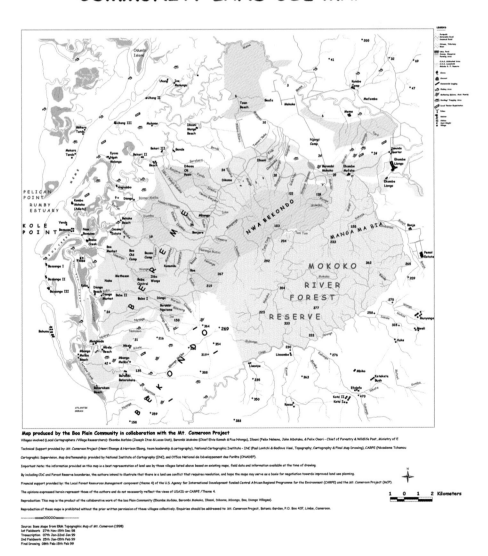

BOA PLAIN
COMMUNITY LAND USE MAP

Map produced by the Boa Plain Community in collaboration with the Mt. Cameroon Project effectively eliminates the community from any degree of tangible participation. We believe that the mapping approach we have opted for is "appropriate," in that with Phase 1 we merged the best of technically sophisticated GPS-based mapping together with participatory, community-driven methods. We believe that it is critical in Phase 1 that these maps be produced using technologies that are both appropriate and feasible for use by local communities. After the hand-drawn maps are produced, IRM can digitize them and create different data layers that can be easily manipulated to produce different maps as needed. This "two-phase approach" is powerful, as it brings the best of technology and participatory methods together into one package.

IRM believes, based on experience, in the validity and power of PM as a new and powerful tool to mobilize communities to take more effective steps in conservation of their natural resources. We have incorporated Phase 1 mapping into a comprehensive five-step management model for decentralized natural resource management (http://ag.arizona.edu/OALS/ALN/aln48/brown&hutchinson.html). Soon, after introducing the mapping as part of this methodology, we began to see our efforts pay off as the communities of Mokoko, Djoum, and Tikar produced detailed maps outlining areas important for fishing, hunting, and

gathering; land used for cultivation was defined as sacred sites and village shrines. Communities also indicated areas of (potentially illegal) timber extraction and areas where their own subsistence livelihoods were in conflict with land appropriated for government-owned industrial farms.

The results of this first phase of activities (which terminated in 2000) were extremely positive. Based on the results of the mapping, one community formally protested the government zoning of the forest in which they live (a bold step in a political context where the government has legal tenure over virtually all the forest resources), and another is using the maps to develop a management plan for the wildlife resources in their region. In both cases, the maps are enabling communities to negotiate new management arrangements. We, and others, feel this level of community involvement is crucial for conservation of the region's biodiversity.

The level of enthusiasm among project participants led us to design the next phase of activities for FY 2001–2003. In an effort to help define land use options and develop negotiated plans for resource conservation and sustainable natural resource use, IRM plans to digitize the maps from Djoum and Tikar in Phase 2 (Mokoko is currently being digitized).

While communities can and have been using the hand-drawn maps effectively, digitization allows for the incorporation of additional data set layers. We believe this is critical for a more comprehensive representation of the resources in the area. It also will permit integration of other data sets, along with facilitated interfacing with other partners that must be involved to achieve conservation objectives.

As part of our five-step management model, IRM introduced a complementary resource inventory methodology that permits the communities to assess distribution, abundance, maturity, condition, and so forth, of several nontimber forest resources that are critical in terms of livelihood security. The communities established transects and conducted inventories of important food and medicinal species as well as several species of commercial value. This data will be used to create map layers that can be overlaid onto the digitized participatory maps, providing a picture of current and emerging resource use trends. Digitization allows us to continually update the maps as new data becomes available. Changes in agricultural patterns, forest cover, and species distribution can quickly be translated onto the maps. New data can also be used to produce models depicting the impacts of different management actions communities may take.

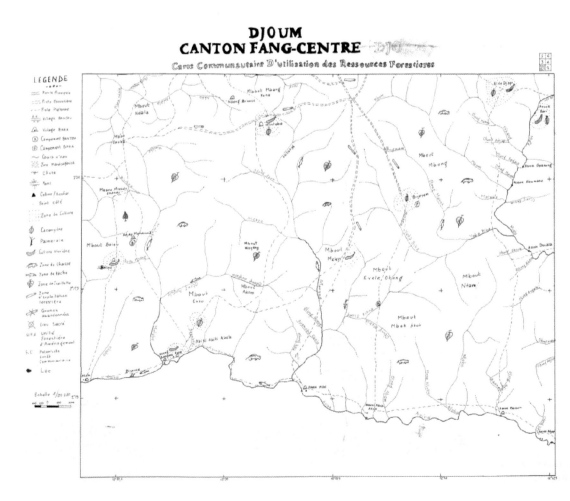

Malawi Ornithological Society

Lawrence Luhanga, Dip. Wild. Cons., AMCSSA, CTIA, Cert. (ESRI)

There are more than six hundred species roaming the mountainous greens and the rolling hills of Malawi. Those that scout the savanna land, such as the ground hornbill, do so majestically. Those that take refuge in the lake do it with glamour. But there are also those that are rare and must painstakingly survive the disparaging forces of the human world.

Malawi ornithology goes back many years. Sir Charles Frederic Belcher is probably well known for his publication of what might be the first literature on the birds of Malawi, the Nyasaland birds, in 1930. Since then, occasional surveys or expeditions have been conducted, some of which are well documented and some not known. An important publication came out in 1966 by the Bensons, whose checklist is still used today as a major reference. A few ornithologists have published short or noncomprehensive publications on birds of Malawi including Stewart–Johnston and Bob Medland. Presently, Robert Dowsette and F. Dowsette–Lemaire are working on what will probably be the most updated reference guide to Malawi birds.

But then this brings us to one important question: where does all this leave the country's ornithology? We all know that for conservation, research, and ornithology as a whole to step forward, we can never be successful without precise/accurate or cross-referenced data. Malawi may have had expeditions, it may have had a couple of publications on her birds, but if this information is not easily accessible or not available, then we are regarded as being unprogressive.

The Malawi Ornithological Society (MOS) formed in 1996 with the following objectives: to promote Malawi ornithology in a general scope, to promote avian recreational activities (such as bird walks, etc.), to promote conservation and preservation awareness, and to promote ornithological research and development of a database bank. MOS believes that by developing an ornithological database bank where all available and known ornithological data and literature will be centralized would help achieve not only its goals and objectives but also help the world of ornithology and avian conservation globally.

It is this activity that profoundly concerns MOS today. How will we do this? MOS's preliminary goal is to act as a communication and networking facility mediating between the members (local birders network) and MOS's central ornithological data (the National Ornithological Database Bank of Malawi, dubbed Nodab). MOS is working on two powerful computers (to be housed at Museums of Malawi, as Nodab base). Since 1998, MOS has worked harder in an attempt to locate any known available information on Malawi birds, scientific and nonscientific. Some of this has been easily located, obtained, and entered but some has not been easily obtained or is not accessible. Many of these data owners want to be paid for the work they did in Malawi for which the Society has no funds. Some of the data requires a long process of legal jumble in order to turn the data over to Malawi. Whichever the case, sooner or later MOS believes that Nodab will succeed and everyone will have equal access to Malawi ornithology and its database for conservation and other related purposes.

There are a number of strategies used in achieving Nodab, included in which is applying the use of GIS. In 2002, MOS will start using GIS in mapping bird species in major IBAs in Malawi. Depending on volunteers and funds, MOS believes that the use of GIS not only gives us accurate data but may bring us in the vicinity of precise species status. The Nodab database will include the most detailed GIS-based birds' atlas of Malawi. By using Mosnet,

MOS will distribute GPS units to all local birders who will be taking coordinates of each and every observation they encounter. These coordinates will be used in GIS analyses together with the existing database to produce the atlas. It is estimated that it will take six months to complete a midsize IBA like Thyolo and approximately ten years to complete 60 percent of the known distribution nationally.

This is our story, our dream, and passion, and we believe and live to achieve it!

Save the Elephants

Iain Douglas–Hamilton

Save the Elephants (STE) has helped the Kenya Wildlife Service (KWS) conduct regular aerial and ground surveys of Kenya's elephant populations. Much of the information gathered in the 1999 surveys was used to support Kenya's position during the CITES meeting in April 2000 that any reopening of the ivory trade would be extremely risky in light of population numbers, security, and a huge influx of automatic weapons into elephant ranges.

In conjunction with KWS, STE is compiling a report on the Status of Kenya's elephants during the 1990s. This report summarizes all the information on elephant numbers, distribution, threats, and mortality gathered by various sources throughout the last ten years and will be a definitive reference document that will be available to the local and international wildlife community.

In Laikipia, Samburu, and Amboseli, data from Dr. Iain Douglas-Hamilton's GPS collars is still providing fascinating insights into the elephant mind. We have found elephants with ranges as wide as 120 kilometers across

and elephants that appear to "streak" across areas of danger leading their family to protected reserves. In addition, we have established corridors between areas that have long been suspected but never proved, until now. Research has shown that separate protected areas and the corridors between should be treated as a whole entity, as elephants link them together with their movements. STE and KWS have initiated a project to monitor the Meru elephant population. This population suffered from horrific poaching during the late 1970s and early 1980s and has shown little signs of recovery in the last decade, remaining at an estimated 250–300 individuals out of the original 2,300. In June 2000, ten elephants were radio collared in Meru as part of an exciting collaborative project between Save the Elephants and the Kenya Wildlife Service, which aimed to increase the ability to monitor and protect the elephant population in this wild and volatile part of Kenya. We know that the elephants from this area move over great distances outside the protected areas of the park,

however, their exact movement patterns are not known. The data gathered during this project will provide the KWS with the information required to decide on the best management options such as patrols to guard elephant routes, the need to keep certain corridors open, and the use of fences. In 2001 STE will have a full-time researcher in the field to gather demographic data on the elephant population through an individual elephant identification study. At last we can increase our knowledge of this little known elephant population.

Our hope is that with improved protection the population will begin to grow and its future will be secure.

Samburu/Laikipia Elephant Movements 1998 - 2000

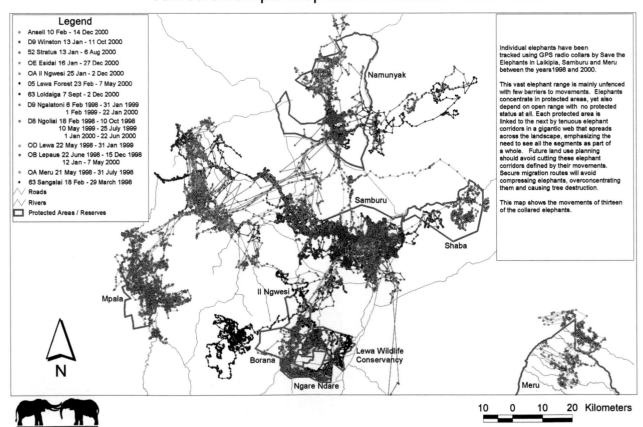

Durrell Institute of Conservation and Ecology

Annette Huggins

Heightening worldwide concern about the earth's marine ecosystems is directed toward the problems of habitat degradation, marine pollution, and heavy exploitation of living marine resources. The Dover Straits is an area of the English Channel that is difficult to monitor due to the interesting dynamics of its position between two seas. It is also heavily utilized by industries such as shipping and fishing and is controlled by the differing cultures and legislation of the UK and France, but falls under the common framework of the European Union.

Information concerning the biodiversity of the marine ecosystem in the Dover Straits is needed in order to develop strategies required by the Convention on Biological Diversity for the conservation and sustainable use of biological diversity.

Traditional single species management strategies have failed to support the protection of marine ecosystems and have allowed marine resources such as fish stocks to decline. It has been widely recognized that management

Annette Huggins, SCGIS 2000 scholar, on a marine research project n the Dower Straits, England.

strategies need to be implemented at the ecosystem level. Information regarding regions of importance for different species could be used to protect the whole ecosystem by providing marine refuges or protected areas. This research aims to identify regions of habitat that are important to plankton communities, providing information that could help to minimize the impact of anthropogenic activities on the marine ecosystem in this area.

Plankton species abundances are being mapped using various modeling techniques that include the aid of satellite imagery. Modeling will assess the extent to which each species is influenced by various environmental factors and how these factors could be used to predict abundance.

Habitat suitability models and maps are being produced for all species with a large spatial sampling extent from multivariate modeling of the environmental influences. The differing methods used will be compared. Where the data allows, an analysis of change of these models over time will be possible. Comparisons are also possible between the use of quantitative abundance data and the presence/absence of information, to highlight appropriate methodology for future monitoring.

Priority area analysis is to be used to attempt to pinpoint key areas within the study site that are particularly important to plankton communities. This analysis will also allow comparisons between quantitative abundance map use, presence/absence maps, and the use of higher taxa levels of classification, again highlighting differences in approaches to data collection and monitoring methods.

Factors common to these resulting areas will be identified and provide insight into environmental conditions needed by these larger communities and other possible regions that could also be important to plankton communities. This research aims to provide information concerning the biodiversity of the Dover Strait and the priorities for its future management, and has been facilitated by ESRI training given as part of a SCGIS international scholarship. A photo diary of the 2000 scholarship award training and conferences is available at www.conservation-gis.org/scgis. Details of this research and previous work are available at www.conservation-gis.org.

Our other research includes the Darwin Initiative Masai Mara Programme: Human–Wildlife Conflict and Solutions to It in and Around Masai Mara National Reserve.

GIS is being used as a key tool in four areas of study:

1 Spatial patterns of crop raiding by elephants in Trans Mara district.

2 Tourist impact in Masai Mara National Reserve, in particular predicting patterns of offroad driving and assessing the effect of vehicle density on wildlife spatial patterns.

3 Constructing a habitat map of Masai Mara National Reserve and investigating spatial patterns of browse density in relation to browser distribution and competition.

4 Investigating ecological and human correlates of rhino distribution in Masai Mara National Reserve with particular reference to the effect of pastoralism on rhinos, and predictive modeling of rhino distribution under different management regimes.

Kenyan M.Sc. and Ph.D. students are undertaking the work with a UK post doctorate; all are at DICE and analyses are currently underway after two years of fieldwork. Two papers have been published that deal in part with GIS issues, as follows:

Walpole, M. J. (2000) "GIS as a Tool for Rhino Conservation." *Pachyderm* 28:33–39.

Walpole, M. J., M. Morgan–Davies, S. Milledge, P. Bett, and N. Leader–Williams (2001) "Population Dynamics and Future Conservation of a Free-Ranging Black Rhinoceros Population in Kenya." *Biological Conservation* 99(2):237–243.

Home Range Position and Overlap

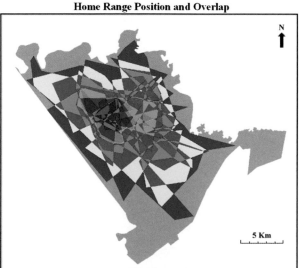

| Nº H.R. |
| 0 |
| 1 |
| 2 |
| 3 |
| 4 |
| 5 |
| 6 |
| 7 |
| 8 |
| 9 |
| 10 |
| 11 |
| 12 |
| 13 |

N

5 Km

Contact Information

Conservation Geography
Charles Convis, editor
ESRI Conservation Program coordinator
380 New York Street
Redlands, CA 92373-8100
Phone: (909) 793-2853 ext. 2488
E-mail: ecp@esri.com

Alexander von Humboldt Institute, Colombia
Instituto de Investigación de Recursos
Biológicos Alexander von Humboldt
Calle 37 # 8 - 40 Mezzanine
Santafé de Bogotá, Colombia A.A. 8693
Phone: +57-1-3406925 or 2877530
Fax: +57-1-2889564
Web: http://www.humboldt.org.co
Web: http://www.kcl.ac.uk/kis/schools/hums/
geog/phd.htm
Dolors Armenteras, principal researcher,
GIS Unit
E-mail: darmenteras@humboldt.org.co
E-mail: dolors.armenteras@kcl.ac.uk

American Wildlands
40 East Main #2
Bozeman, MT 59715
Phone: (406) 586-8175
Fax: (406) 586-8242
E-mail: amwild@wildlands.org
Web: www.wildlands.org
Lauren Oechsli, GIS specialist
E-mail: loechsli@wildlands.org
Linda Bowers Phillips, GIS lab manager

Anthropological Watch (Anthrowatch)
46-c Mahusay Street UP Village, Diliman,
Quezon City, 1101, Metro Manila,
Philippines
Phone: (063) 2 436-0992
Fax: (063) 2 436-0992
E-mail: anthrow@psdn.com.ph
Web: www.anthrowatch.org
Trina Galido–Isorena, GIS specialist
E-mail: tgisorena@edsamail.com.ph

Appalachian Mountain Club
5 Joy St.
Boston, MA 02108
Phone: (617) 523-0636
Fax: (617) 523-0722
E-mail: information@amcinfo.org
Web: http://www.outdoors.org
David A. Publicover, Senior Staff Scientist
E-mail: dpublicover@amcinfo.org

Bay of Fundy Marine Resource Centre
Martin Kaye
Phone: (902) 638-3044
Fax: (902) 638-3284
E-mail: martink@bfmrc.ns.ca

Bayou Preservation Association, Inc.
3201 Allen Parkway, Suite 200,
Houston, Texas 77019
Phone: (713) 529-6443
Fax: (713) 529-6481
E-mail: bpa@hic.net
Web: www.bayoupreservation.org
Mary Ellen Whitworth, P.E., executive
director

Biodiversity Associates, Inc.
P.O. Box 6032
Laramie, Wyoming 82073
Phone: (307) 742-7978
Web: www.biodiversityassociates.org
Jeff Kessler, GIS specialist
E-mail: jeff@biodiversityassociates.org

Bird Studies Canada / Etudes d'Oiseaux
Canada
P.O. Box 160, 115 Front Rd.
Port Rowan, Ontario
Canada N0E 1M0
Phone: (519) 586-3531, ext. 213
Fax: (519) 586-3532
Web: http://www.bsc-eoc.org/bscmain.html
Andrew Couturier, GIS analyst
E-mail: acouturier@bsc-eoc.org

The Borneo Project
1771 Alcatraz Avenue
Berkeley, CA 94703 USA
Phone: (510) 547-4258 v
Fax: (510) 547-4259 f
E-mail: borneo@earthisland.org
Web: www.earthisland.org/borneo
Dan Scollon, GIS specialist
Harlan Thompson, program manager

California Academy of Sciences
Department of Herpetology
Golden Gate Park
San Francisco, CA 94116-1363
Phone: (415)750-7031
Web: http://www.calacademy.org/
Michelle Koo, Dept. of Herpetology
E-mail: mkoo@calacademy.org

Cannon River Watershed Partnership
328 Central Avenue
Faribault, MN 55021
Phone: (507) 332-0488
Fax: (507) 332-0513
E-mail: crwp@means.net
Web: http://206.11.107.10/crwp/
Justin D. Watkins, watershed specialist

Care International Madagascar
BP 1677 Ambohitrarahaba
Antananarivo 101, Madagascar
Phone: 261 20 22 62142
Fax: 261 20 22 41174
E-mail: caremad@bow.dts.mg
Lalaina Razafimbololona, GIS staff
E-mail: lalaina_ra@hotmail.com

Center for Marine Geography
1725 DeSales St. NW, Suite #600
Washington, D.C. 20036
Phone: (202) 429-5609
Fax: (202) 872-0619
Jack A. Sobel
E-mail: jsobel@dccmc.org

Cheetah Conservation Fund
1012 West Ojai Ave
P.O. Box 1380
Ojai, CA 93023
Phone: (805) 640-0390
Fax: (805) 640-0230
E-mail: cheeta@iafrica.com.na
Web: www.cheetah.org
Laurie Marker, executive director
Dr. Richard M. Jeo (Round River Conserva-
tion Studies), GIS consultant
E-mail: rjeo@earthlink.net

Chesapeake Bay Foundation
Philip Merrill Environmental Center
6 Herndon Avenue
Annapolis, MD 21403
Phone: (410) 268-8816
Fax: (410) 268-6687
Web: www.savethebay.cbf.org
Steve Libbey, land planner, ext. 2154
E-mail: slibbey@savethebay.cbf.org

Clean Water Club of Bulgaria
Bigla Str. #54
P.O. Box #31
Sofia-1592. Bulgaria
Phone: 359-2-978-0782
Fax: 359-2-373-822
E-mail: cleanwater1996@hotmail.com
E-mail: nmpg@schools.acad.bg
Ventzeslav Dimitrov, Space Reseach Institute
E-mail: vdimitrov@space.bas.bg
Lubka Roumenina, GIS specialist
E-mail: roumenina@hotmail.com

Comanche Pool Prairie Resource Foundation
105 N. New York Ave.
P.O. Box 516
Coldwater, KS 67029
Phone: (316) 582-2211
Fax: (316) 582-2035
Loren Graff, GIS specialist
E-mail: lgraff@rh.net

The Community Builders, Inc.
95 Berkeley Street, Suite 500
Boston, MA 02116
Phone: (617) 695-9595
Fax: (617) 695-9205
Ronald Wong, planning project manager
E-mail: RonW@tcbinc.org

Connecticut Policy and Economic Council
179 Allyn Street, Suite 308
Hartford, CT 06103
Phone: (860) 722-2465
Fax: (860) 548-7363
Web: www.cpec.org
Richard Walker, research analyst
E-mail: rwalker@cpec.org

The Conservation Fund
10001 Main Street, Suite C
Chapel Hill, NC 27516
Phone: (919) 967-2223
Fax: (919) 967-9702
Web: http://www.conservationfund.org
William L. Allen, III, director of Geographic
Information Services
E-mail: will@tcf.arcana.com, or
will@willallen.com

Conservacion Internacional Peru
Jr. Chinchon 858-A San Isidro
Lima 27, Peru
Phone: 51-1-4408967
Fax: 51-1-4403665
E-mail: ci-peru@conservation.org
Dr. Carol L. Mitchell, director
carolm@wayna.rcp.net.pe
Luis Espinel, project director
luisec@terra.com.pe
Juana Silva, coordinador, Proyecto Uso
Mayor de Tierras
E-mail: jmsilva@terra.com.pe

Craighead Environmental Research Institute
201 South Wallace Avenue, Suite B2D
Bozeman, MT 59715
Phone: (406) 585-8705, 585-8220
E-mail: ceri@avicom.net
Web: http://www.grizzlybear.org
Lance Craighead, executive director
E-mail: lance@grizzlybear.org
Troy Merrill, GIS consultant
E-mail: troy1@moscow.com

Dolphin Ecology Project
Laura K. Engleby
Florida, USA
Phone: (305) 852-0649
E-mail: lengleby@aol.com
E-mail: Willaura@aol.com

Durrell Institute of Conservation and Ecology
University of Kent
Canterbury CT2 7NS, United Kingdom
Phone: 01227 823455
Fax: 01227 827839
E-mail: aeh6@ukc.ac.uk
Web: http://www.conservation-gis.org
Web: http://www.ukc.ac.uk/anthropology/dice/
dice.html
Annette E. Huggins, GIS specialist
E-mail: aeh6@ukc.ac.uk

E-misszió Environmental Association
H-4400 Nyíregyháza, Malom u.18/a
Hungary
Phone/Fax: +36-42-423-818
E-mail: emisszio@zpok.hu
Web: www.e-misszio.hu
Csaba Botos, GIS specialist

Environmental Defense Fund
Oakland, CA 94618
Phone: (510) 658-8008 ext 258
Peter Black, GIS specialist
E-mail:
Peter_Black@environmentaldefense.org

ESRI
308 New York St.
Redlands, CA 92373
Phone: (909) 793-2853 ext. 2146
Joe Breman
E-mail: jbreman@esri.com
Web: www.esri.com/oceans

Forest Community Research
4405 Main Street
P.O. Box 11
Taylorsville, CA 95983
Phone: (530) 284-1022
Fax: (530) 284-1023
E-mail: info@fcresearch.org
Web: http://www.fcresearch.org
Lee Williams, program manager
Will Kay, GIS specialist
E-mail: wkay@fcresearch.org
Jonathan Kusel, executive director

FGDC Bathymetric Subcommittee
David Stein
Phone: (843) 740-1310
E-mail: Dave.Stein@noaa.gov

Forest Issues Group
Steve Benner c/o Jeans 76, Highway 89,
Sierraville, CA 96126
Phone: (530) 994-3535
Fax: (530) 994-3535
Steve Benner, GIS specialist
E-mail: sbenner@jps.net

Gifford Pinchot Task Force
3932 Biscay St. NW
Olympia, WA 98502
Phone: (360) 753-4185
E-mail: gptf@olywa.net
Web: www.gptaskforce.org
David Jennings, chairman
E-mail: 71634.127@compuserve.com

Glacier Bay Field Station
Philip N. Hooge, Ph.D., research population ecologist
USGS–Alaska Biological Science Center
Glacier Bay Field Station
Gustavus, AK 99826
Web: http://www.absc.usgs.gov/glba/gistools/
index.htm

Greenpeace Russia
Novaya Bashilovka Str., 6
Moscow, 101428, GSP-4, Russia
Phone: 7 (095) 257-4116 or -4118 or -4124
Fax: 7 (095) 257-4110
E-mail:
greenpeace.russia@diala.greenpeace.org
Serguei Tsyplenkov, executive director
Dr. Alexei Iarochenko, head of GIS Department
Serguei Mikhailov, Petr Potapov, Nadia
Sudzilovskaja: GIS staff

Heal the Bay
3220 Nebraska Ave.
Santa Monica, CA 90404
Phone: (310) 453-0395
1-800-HEAL-BAY (in California only)
E-mail: Info@healthebay.org
Web: www.healthebay.org

Housatonic Valley Association
Lenox Station, P.O. Box 1885
Lenox, MA 01240
Phone: (860) 672-6678
Fax: (860) 672-0162
E-mail: housatonic@snet.net
Web: www.hvathewatershedgroup.org
Kirk Sinclair, GIS manager, HVA

Innovative Resources Management, Inc.
2421 Pennsylvania Avenue NW
Washington, D.C. 20037
Phone: (202) 293-8384
Fax: (202) 293-8386
Web: http://www.irmgt.com
Michael Brown, president
E-mail: brown1irm@aol.com

Institute for Wildlife Studies
P.O. Box 1104
Arcata, CA 95518
Phone: (707) 822-4258
Web: http://www.iws.org
Gregory Schmidt, wildlife biologist
(GIS coordinator)
E-mail: schmidt@iws.org

International Marine Life Alliance
2800 4th Street North Suite 123
St. Petersburg, FL 33704
Phone: (727) 896-8626
Peter Rubec
E-mail: prubec@compuserve.com

Katunsky Biosphere Reserve
Box 24, Ust-Koksa
Altai Republic, Russia
Phone: (38848)-229-46, (38848)-227-55
E-mail: katunsky@mail.gorny.ru
Misha Paltsyn, science director

LEGACY, The Landscape Connection
830 G Street, Suite 230
P.O. Box 59
Arcata, CA 95518
Phone: (707) 825-8582
E-mail: legacy@legacy-tlc.org
Web: www.legacy-tlc.org
Robert Brothers, Ph.D., project manager
• For Curtice Jacoby and Chris Trudel
  Curtice Jacoby
  Department of Forestry
  Humboldt State University
  Arcata, CA 95521
  E-mail: cej4@axe.humboldt.edu or
  jacoby@legacy-tlc.org
  Phone: (707) 826-9408
  Fax: same number, call first
  Chris Trudel: cpt@northcoast.com
• For Steven Day
  Leggett, CA 95585
  E-mail: maprap@zapcom.net

The Los Angeles & San Gabriel Rivers
Watershed Council
111 N. Hope Street, Suite 627
Los Angeles, CA 90012
Phone: (213) 367-4111
Fax: (213) 367-4138
Web: http://www.lasgRiversWatershed.org/
Rick Harter, executive director
E-mail: rickharter@verizonmail.com
Rumi Yanakiev, office manager

Marine Conservation Biology Institute
Lance Morgan
Noreen Parks
Redmond, WA
E-mail: Noreen@mcbi.org
Web: http://www.mcbi.org

Contact
Information

Mount Desert Island Water Quality Coalition
RR1 Box 2445
Bar Harbor, ME 04609
P.O. Box 911
Mount Desert, ME 04660
Phone: (207) 288-2598
Fax: (207) 288-2598
E-mail: lalaland@acadia.net
Lelania Avila, executive director
E-mail: lelania@mdiwqc.org

Museo Nacional de Historia Natural, Chile
Casilla 787, Interior Quinta Normal
Santiago de Chile, Chile
Phone: (56 2) 6804600
Fax: (56 2) 6804602
E-mail: mmunoz@mnhn.cl
Web: www.mnhn.cl
Andrés Moreira-Muñoz, director, Taller La
Era, Santiago de Chile
Patricio Pliscoff, geographer, Taller La Era
E-mail: artel@vtr.net

New England Aquarium
Central Wharf
Boston, MA 02110-3399
Phone: (617) 742-5198
Fax: (617) 723-9705
Rob Schick
E-mail: rschick@neaq.org
Web: http://www.marinegis.org

New Mexico Wilderness Alliance
202 Central SE, Suite 101
Albuquerque, NM 87102
Phone: (505) 843-8696
Fax: (505) 843-8697
E-mail: sully@nmwild.org
Web: http://www.nmwild.org/
Kurt A. Menke, GIS coordinator
Direct address: Earth Data Analysis Center
The University of New Mexico
Bandelier West, Room 111
Albuquerque, NM 87131-6031
Phone: (505) 277-3622 ext. 239
Fax: (505) 277-3614
E-mail: kmenke@unm.edu

New York Botanical Garden
Institute of Economic Botany
200th Street and Southern Boulevard
Bronx, NY 10458
Phone: 718-817-8727
Fax: 718-220-1029
E-mail: cpeters@nybg.org
Web: www.nybg.org
Dr. Charles M. Peters, Kate E. Tode, Curator
of Botany
Berry Brosi, MESc., Institute of Economic
Botany

New York City Environmental Justice
Alliance
115 West 30th Street, Room 709
New York, NY 10001
Phone: (212) 239-8882
Fax: (212) 239-2838
E-mail: mapping@nyceja.org
Emily Chan, community geographer
Leslie H. Lowe, executive director

NOAA Coastal Services Center
Tony Lavoi
2234 South Hobson Avenue
Charleston, SC 29405-2413
Phone: (843) 740-1200
E-mail: csc@csc.noaa.gov
Web: http://www.csc.noaa.gov

NOAA Coastal Services Center
2234 South Hobson Avenue
Charleston, SC 29405
Phone: (843) 740-1310
Fax: 843-740-1224
David Stein
E-mail: Dave.Stein@noaa.gov
Web: http://www.csc.noaa.gov

North Carolina Zoological Society,
Education Division
4403 Zoo Parkway
Asheboro, NC 27203
Phone: (336) 879-7718
Web: http://www.nczooeletrack.org
Web: http://www.nczooeletrack.org/elephants/
loomis_maps/index.html
Web: http://www.nczooredwolf.org
Mark MacAllister, coordinator, Online
Learning Projects
Phone: (919) 545-3068 (Home office,
Pittsboro, NC)
E-mail: markmacallister@mac.com

Oceanic Resource Foundation
Michelle Kinzel
Environmental Education Programs
1095 Calle Mesita
Bonita, CA 91902-2405
Phone (619) 251-5484
Fax: (415) 954-7199
Web: http://www.orf.org

Oregon Natural Resources Council
5825 North Greeley Avenue
Portland, OR 97217
Phone: (503) 283-6343
Fax: (503) 283-0756
E-mail: info@onrc.org
Web: www.onrc.org
Erik Fernandez, GIS coordinator
E-mail: ef@onrc.org

Pacific Biodiversity Institute
P.O. Box 298
Winthrop, WA 98862
Phone: (509) 996-2490
Fax: (509) 996-3778
E-mail: jason@pacificbio.org
Web: http://www.pacificbio.org
Peter Morrison, executive director
E-mail: peter@pacificbio.org
Jason Karl, senior GIS analyst
E-mail: jason@pacificbio.org

People for Puget Sound
Philip Bloch
Tom Dean
1402 Third Ave. Suite 1200
Seattle, WA 98101
Phone: (206) 382-7007
Fax: 206-382-7006
Web: http://www.PugetSound.org

Platte River Whooping Crane Maintenance
Trust, Inc.
6611 W. Whooping Crane Dr.
Wood River, NE 68883
Phone: (308) 384-4633
Fax: (308) 384-7209
Web site: www.whoopingcrane.org
Bob Henszey, GIS specialist
E-mail: henszey@hamilton.net

Pohatcong Creek Watershed Association
4 Old Canal Road
Washington, NJ 07882
Phone: (973) 426-7211
Fax: (973) 426-7115
Dave Dempski, GIS specialist
E-mail: ddempski@lucent.com
Alternate address: 450 Clark Drive, room
2B3I
Mount Olive, NJ 07828

Point Reyes Bird Observatory (PRBO)
4990 Shoreline Highway
Stinson Beach, CA 94970
Phone: (415) 868-1221 x52
Fax: (415) 868-1946
Web: http://www.prbo.org
Diana Stralberg, GIS specialist
E-mail: dstralberg@prbo.org

Predator Conservation Alliance (formerly
"Predator Project")
P.O. Box 6733
Bozeman, MT 59771
Phone: (406) 587-3389
Fax: (406) 587-3178
Web: http://www.predatorconservation.org
David Gaillard, GIS specialist
E-mail: gaillard@predatorconservation.org,
or gaillard@wildrockies.org

Sahabat Alam Malaysia
(Friends of the Earth Malaysia)
27 Lorong Maktab
Penang 10250, Malaysia
P.O. Box 216
Marudi, Sarawak, 98050, Baram Region,
Malaysia
Phone: 011 60 85 756 973
Fax: 011 03 2275705
E-mail: smidris@tm.net.my
Jok Jau Evong, program manager (SAM
Marudi)
E-mail: jvon_ei@yahoo.com
Louis Ngau, mapping coordinator
E-mail: lngau@yahoo.com
Alternate: Harlan Thompson, Borneo Project,
E-mail: borneo@earthisland.org

Save the Redwoods League
114 Sansome Street, Room 1200
San Francisco, CA 94104
Phone: (415) 362-2352
Fax: (415) 362-7017
Web: www.savetheredwoods.org
E-mail: redwoods@savetheredwoods.org
Ruskin Hartley, Conservation GIS Program

Save the Elephants
P.O. Box 54667
Nairobi, Kenya
Phone: 2542 891673 / 2542 890596/7
Fax: 2542 890441
Dr. Iain Douglas–Hamilton, director
E-mail: iain@net2000ke.com

Sonoma Ecology Center
205 First Street West
Sonoma, CA 95476
Phone: (707) 996-0712
E-mail: sec@vom.com
Web: www.vom.com/se
Rich Hunter, GIS/GPS project manager

Southern Appalachian Forest Coalition
46 Haywood Street, Suite 323
Asheville, NC 28801
Phone: (828) 252-9223
Fax: (828) 252-9074
Hugh Irwin, conservation planner
E-mail: hugh@safc.org

Southern Rockies Ecosystem Project
Southern Rockies Forest Network
2260 Baseline Road, Suite 205
Boulder, CO 80302
Phone: (303) 258-0433
Fax: (303)546-9922
Web: http://csf.Colorado.EDU/srep/
Bill Martin, GIS manager
E-mail: wwmartin@indra.com

The Surfrider Foundation
Chad Nelson
Mark Rauscher
122 S. El Camino Real, #67
San Clemente, CA 92672
Phone: (949) 492-8170
Fax: (949) 494-8170
E-mail: mrauscher@surfrider.org
Web: http://www.surfrider.org

US Federal Geographic Data Committee,
Bathymetric Subcommittee
Phone: (843) 740-1310
David Stein, executive secretary
E-mail: Dave.Stein@noaa.gov

USGS Alaska Biological Science Center
Philip N. Hooge, Ph.D.
Glacier Bay Field Station
P.O. Box 140, Gustavus, AK
E-mail: philip_hooge@usgs.gov
Web: http://www.absc.usgs.gov/
glba/phooge.htm
Animal Movement Extension To ArcView:
http://www.absc.usgs.gov/glba/gistools/

Vermont Institute of Natural Science and the
Orton Institute
27023 Church Hill Road
Woodstock, VT 05091
Phone: (802) 457-2779, ext. 108
Web: www.vinsWeb.org or
www.communitymap.org
Andrew Toepfer, GIS coordinator,
Community Mapping
E-mail: atoepfer@vinsWeb.org

Western Slope Environmental Resource
Council
Paonia, CO
Phone: (970) 527-5308
Eli Lindsey–Wolcott, mapping and GIS
coordinator
E-mail: Eli@wolcott.to

Wildlife Conservation Society
International Programs / Science Resource
Center
2300 Southern Blvd.
Bronx, NY 10460
Phone: (718) 220-5156
Fax: (718) 220-7114
Web: www.wcs.org
Gillian Woolmer, GIS lab administrator
E-mail: gwoolmer@wcs.org

Wildlife Conservation Society–Bolivia
(ConFauna)
Museo de Historia Natural Noel Kempff
Mercado, Universidad Autonoma Gabriel
Rene Moreno
Av. Irala 565, C.C. 2489, Santa Cruz, Bolivia
Phone: 591-3-366574
Fax: 591-3-366574
E-mail: museo@museo.sczbo.org
Dr. Damian I. Rumiz
E-mail: confauna@scbbs-bo.com
Lic. Lila Sainz
E-mail: aguara@bibosi.scz.entelnet.bo
Alternate: Gillian Woolmer, GIS lab
administrator
E-mail: gwoolmer@wcs.org

Wildlife Conservation Society China Program
East China Normal University,
3663 Zhongshan Road North
Shanghai 200062, China
Phone: 86 (21) 62232361
Fax: 86 (21) 62851359
E-mail: ezhang@guomai.sh.cn
Dr. Yongpei Wu, GIS specialist
Dr. Endi Zhang, associate conservationist
E-mail: ezhang@guomai.sh.cn
Dr. Joshua Ginsberg, director of Asia
Program
E-mail: jginsberg@wcs.org
Alternate: Gillian Woolmer, GIS lab
administrator
E-mail: gwoolmer@wcs.org

Wildlife Conservation Society–Guatemala
Mesoamerican and Caribbean Program
Flores, Peten, Guatemala
c/o Department of Fisheries and Wildlife,
Nash Hall
Oregon State University
Corvallis, OR 97331
Phone: (541) 737-4531
Phone (in Guatemala): 502-926-0569
E-mail: wcspeten@secmas.gua.net
Robin Bjork, research fellow
Alternate: Gillian Woolmer, GIS lab
administrator
E-mail: gwoolmer@wcs.org

Wildlife Conservation Society–Indonesia
Program
Jl. Ceremai # 8
Bogor, W. Java, Indonesia
P.O. Box 311, Bogor 16003, W. Java,
Indonesia
Phone: 62 251 325 664
Fax: 62 251 357 347
E-mail: wcs-ip@indo.net.id
Margaret F. Kinnaird, conservation ecologist
Alternate: Gillian Woolmer, GIS lab
administrator
E-mail: gwoolmer@wcs.org

Wildlife Conservation Society–Laikipia
Project
Mpala Research Centre
P.O. Box 555
Nanyuki, Kenya
Web: www.wcs.org
Dr. Laurence G. Frank
E-mail: lgfrank@uclink.berkeley.edu
Alternate Address: Field Station for
Behavioral Research
3210 Tolman Hall, 1650
University of California
Berkeley, CA 94720-1650
Phone: (510) 643-7821
Alternate: Gillian Woolmer, GIS lab
administrator
E-mail: gwoolmer@wcs.org

Wildlife Conservation Society–Mamirauá
Program Brazil
Avenida Brasil 197
Tefé, Amazonas, 69470-000, Brazil
Phone: 92-743-2736
Fax: 92-743-2736
E-mail: mamiraua@pop-tefe.rnp.br
Web: http://www.pop-tefe.rnp.br
Jose Marcio Ayres, director Instituto de
Desenvolvimento Sustentável Mamiraua and
senior conservation scientist
Joao Paulo Viana, fisheries management and
computing coordinator
Alternate: Gillian Woolmer, GIS lab
administrator
E-mail: gwoolmer@wcs.org

Wildlife Conservation Society–Myanmar
Program
Building C-3, Aye Yeik Mon 1st Street
Hlaing Township
Yangon, MYANMAR
E-mail: wcsmm@mptmail.net.mm
Kyaw Thinn Latt, research assistant, WCS–
Myanmar
Saw Tun Khaing, Director, WCS–Myanmar
Program
Alternate: Gillian Woolmer, GIS lab
administrator
E-mail: gwoolmer@wcs.org

Wildlife Conservation Society–Uganda
Plot 32 Lubowa Estate
P.O. Box 7487
Kampala, Uganda
Phone: 011 256 41 200699
Fax: 718 364 4275
Dr. Andrew J. Plumptre, director of Albertine
Rift Program
E-mail: aplumptre@aol.com
Alternate: Gillian Woolmer, GIS lab
administrator
E-mail: gwoolmer@wcs.org

Woods Hole Oceanographic Institution
R. A. Goldsmith
Dr. A. J. Pluddemann
Phone: (508) 289-2770
E-mail: rgoldsmith.@shoi.edu

World Wildlife Fund, Canada
245 Eglinton Ave. E, Suite 410
Toronto, Ontario M4S 2S6, Canada
Phone: (416) 489-4567
E-mail: panda@wwfcanada.org
Web: http://wwfcanada.org/
Tony Iacobelli, manager, system planning
E-mail: tiacobelli@wwfcanada.org
Kevin Kavanagh, senior manager,
conservation
E-mail: kkavanagh@wwfcanada.org

World Wildlife Fund, Colombia (Ecoregional
Planning Chocó)
E-mail: lfgomez@wwf.org.co
Luis Fernando Gómez–Navia, Conservation
Planning and GIS

World Wildlife Fund, Peru
Av. San Felipe 720, Lima 11 - Peru
P.O. Box 11-0205
Phone: (51-1) 261-5300 / 261-5301
Fax: (51-1) 463-4459
Web: www.wwfcolombia.org
Luis Fernando Gómez–Navia, Conservation
Planning and GIS
E-mail: lfgomez@wwf.org.co
Daniel Arancibia Vega–Centeno,
GIS specialist
E-mail: davc@terra.com.pe

To join the SCGIS, complete the application form below and send it or a photocopy with your membership dues to

The Society for Conservation GIS
P.O. Box 861
Lake Placid, FL 33862
Attn.: Roberta L. Pickert, Treasurer

The SCGIS is a 501(c)3 nonprofit organization; therefore membership dues and donations are tax deductible.

No individual will be turned away for lack of funds. Individuals may elect to sponsor others with little or no budget. For further information please contact Sandra Coveny at sandrac@peak.org or Roberta L. Pickert at rpickert@archbold-station or pickert@strato.net.

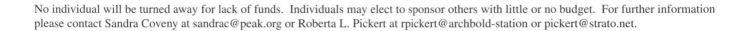

SCGIS Dues (U.S. dollars)

Standard Membership $35 yearly
Sponsoring Membership $50 yearly
Supporting Membership $100 yearly
Sustaining Membership $250 yearly

Conference/ArcView Special Membership $350
Conference/Analyst Special Membership $1,000
Lifetime Membership $2,000 or more one time
Basic (students, seniors, low budget) $20 yearly

Society for Conservation GIS Membership Application Form

Membership Level and Amount Enclosed:_____

Name (First Name, Middle Initial, Family Name):_____

Your Organization's Name:_____

Organization Address (Street, City, State, ZIP/Postcode, Country):_____

_____

Your Work Phone Number:_____ Work Fax Number:_____

Work E-mail Address:_____

Organization Web Site:_____

Because of unreliable e-mail and mail and because people move, we often depend upon alternate address information in order to contact you about membership renewals, the conference, the newsletter. We do not in any way make any member information available to anyone else for any other purpose whatsoever.

Alternate/Home/Permanent Address (Street, City, State, ZIP/Postcode, Country):_____

_____

Alternate/Home/Permanent Phone:_____ Alternate/Home/Permanent Fax:_____

Alternate/Home/Permanent E-mail:_____

Alternate/Home/Permanent Web Site:_____

As an American citizen, you hold a deed to 623 million acres. Yellowstone. The Grand Ganyon. The Arctic National Wildlife Refu Chincoteague. Yosemite. And more.

Unfortunately, many of our greatest places are threatened by oil and gas drilling, logging, subdivision, and mining. Help us fight for their protection so that future generations will inherit America's wilderness, just as we have. The Wilderness Society carries out this mission through public education, advocacy, and scientific and economic analysis. We are proud of our Center for Landscape Analysis, which has earned a reputation as a leader in the application of GIS and remote sensing technologies to land conservation.

Please join us. To become a Wilderness Society member, call 1-800-THE WILD or visit our Web site: www.wilderness.org.

IT'S ALL YOURS.